AF686309

# Research on Indoor Air Environment and Energy Conservation

# Research on Indoor Air Environment and Energy Conservation

Guest Editors

**Kai Zhang**
**Jian Liu**
**Xiaolei Yuan**
**Yizhe Xu**
**Junming Zhou**

MDPI

Basel • Beijing • Wuhan • Barcelona • Belgrade • Novi Sad • Cluj • Manchester

*Guest Editors*

Kai Zhang
College of Urban
Construction
Nanjing Tech University
Nanjing
China

Jian Liu
School of Energy and
Environment
Southeast University
Nanjing
China

Xiaolei Yuan
Department of Mechanical
Engineering
Aalto University
Espoo
Finland

Yizhe Xu
College of Urban
Construction
Nanjing Tech University
Nanjing
China

Junming Zhou
College of Urban
Construction
Nanjing Tech University
Nanjing
China

*Editorial Office*
MDPI AG
Grosspeteranlage 5
4052 Basel, Switzerland

This is a reprint of the Special Issue, published open access by the journal *Buildings* (ISSN 2075-5309), freely accessible at: https://www.mdpi.com/journal/buildings/special_issues/3511OWBJZP.

For citation purposes, cite each article independently as indicated on the article page online and as indicated below:

| Lastname, A.A.; Lastname, B.B. Article Title. *Journal Name* **Year**, *Volume Number*, Page Range. |
|---|

**ISBN 978-3-7258-7020-2 (Hbk)**
**ISBN 978-3-7258-7021-9 (PDF)**
**https://doi.org/10.3390/books978-3-7258-7021-9**

© 2026 by the authors. Articles in this reprint are Open Access and distributed under the Creative Commons Attribution (CC BY) license. The reprint as a whole is distributed by MDPI under the terms and conditions of the Creative Commons Attribution-NonCommercial-NoDerivs (CC BY-NC-ND) license (https://creativecommons.org/licenses/by-nc-nd/4.0/).

# Contents

# About the Editors

**Kai Zhang**

Kai Zhang is an associate professor of Heating, Ventilation, and Air Conditioning in the College of Urban Construction at Nanjing Tech University and was selected for the "Science and Technology Deputy Chief" program in Jiangsu Province. He is dedicated to understanding both the preparation and application of radiative cooling materials, especially the application of radiative cooling in the field of building energy conservation. His primary interests also encompass data center thermal management, chip cooling, building thermal environment, building thermal comfort, indoor air quality, and heat and humidity transfer. He has published approximately 100 peer-reviewed papers and has been granted approximately 30 invention patents. He has presided/participated in projects including the National Natural Science Foundation of China, the National Key Research and Development Project of China, and the Science and Technology Project of Jiangsu Provincial Department of Housing and Urban Rural Construction. He serves as an Editorial Board Member of *Buildings* and *Scientific Reports* and as a Guest Editor of *Building Simulation*. He has also been invited to serve as a scientific committee member of HB2023 Asia and BuildSim Nordic 2024.

**Jian Liu**

Jian Liu is an associate researcher at the School of Energy and Environment, Southeast University, and the technical lead of the university's major scientific facility pre-research project "Polar and Extreme Environment Simulation Experimental Facility". His research focuses on the generation and control of extreme multi-factor coupled environments, as well as industrial high-temperature heat pump technology and applications. He has published approximately 50 high-level papers and been granted 12 invention patents. He has presided over projects including the Jiangsu Province Carbon Peak and Carbon Neutrality Project (total funding 27 million RMB, with 10 million RMB from national allocation), a National Natural Science Foundation Youth Project (Category C), sub-projects under the 14th Five-Year National Key R&D Program, a Jiangsu Provincial Natural Science Foundation Youth Project, two Jiangsu Provincial Key Projects, and five industry collaborative projects. He also serves as a Youth Editorial Board Member of the *Journal of Refrigeration* and a Guest Editor for the "Refrigerant Alternative Technology" column. Under his guidance, students have twice won first prizes in the National Energy Conservation and Emission Reduction Competition.

**Xiaolei Yuan**

Xiaolei Yuan is an independent researcher at Aalto University, Finland. His research focuses on building performance simulation, low-carbon and energy-flexible buildings and communities, and the safe and energy-efficient operation of data centers. He has published over 80 peer-reviewed papers. His work integrates numerical modeling, data-driven approaches, and system-level optimization to support carbon-neutral and resilient building energy systems. He has participated in and led several Finnish national, EU Horizon, and industry-funded research projects and holds 45 patents. He received the Sustainability Young Researcher Award, the Aalto University Dean's Award for Excellent Doctoral Dissertation, and the Chinese Sanhua Award. In terms of academic leadership and service, he served as Scientific Committee Chair of BuildSim Nordic 2024 and was responsible for preparing the official bid proposal for the 18th International IBPSA Conference (Building Simulation 2023), contributing to IBPSA China's successful acquisition of the hosting rights. He has also been invited to serve as Vice Chair of the SCANVAC Cold Climate HVAC Conference 2027 and the REHVA HVAC World Congress CLIMA 2028.

**Yizhe Xu**

Yizhe Xu is currently serving as a lecturer at the College of Urban Construction, Nanjing Tech University. His research centers on the field of building energy, with a particular focus on data center modeling, cloud computing scheduling, the collaborative regulation of computing and cooling resource matching, the design of efficient refrigeration plant rooms, as well as the optimization of their operation. He has carried out a substantial amount of fundamental and applied research work in these related areas. He has published more than 20 academic papers. His SCI papers have been cited over 660 times in total, with an H-index of 13, and the highest citation count for a single paper reaching 160 times. Meanwhile, he has applied for and been granted more than 10 patents (including one U.S. patent). In 2023, he was selected for the "Science and Technology Deputy Chief" program in Jiangsu Province. In 2024, he was included in the "Dual-Appointment Plan for High-Level Talents" in Changzhou City. In 2025, he was honored with the title of "Outstanding Undergraduate Thesis Supervisor" at Nanjing Tech University. Additionally, he received the first prize of the Science and Technology Award from the Jiangsu Society of Refrigeration in 2023 and the second prize in 2024.

**Junming Zhou**

Junming Zhou is an associate professor, Master's supervisor, and Associate Director of the Department in the College of Urban Construction, Nanjing Tech University. His research focuses on thermal and moisture control, advanced dehumidification technology, carbon neutrality, and related energy storage technology. The research direction is focused on thermal and humidity control as well as deep dehumidification technology. Utilizing membrane separation technology to explore efficient air dehumidification schemes holds significant scientific importance for advancing energy conservation and emission reduction in the field of dehumidification air conditioning

# Preface

Indoor air environment and energy conservation have become increasingly important issues due to the rapid development of modern society. With more people spending time indoors, the quality of indoor air has had a direct impact on people's health and well-being. Furthermore, the increasing energy consumption of buildings to satisfy the indoor thermal environment has led to a significant increase in carbon emissions, exacerbating the issue of climate change.

This Reprint focuses on the topic of "Indoor Air Environment and Energy Efficiency," aiming to establish an academic platform that brings together researchers, engineers, and policymakers worldwide. It systematically explores cutting-edge theories, key technologies, and practical applications in (a) advances and reviews of indoor air environment/building energy conversion; (b) passive and positive methods for the energy-efficient provision of clean air; (3) identification and control of indoor pollution sources; (4) novel energy materials for building applications; (5) a renewable/integrated energy system for energy-saving construction; (6) applications of passive cooling/heating technologies in buildings; and (7) modeling and simulation for indoor environment/building energy systems, offering comprehensive solutions to address current challenges in ensuring indoor air quality while advancing building energy conservation.

Through discussions of the aforementioned topics, this Reprint not only highlights innovative achievements in indoor pollution source control, air purification technologies, and thermal environment regulation, providing scientific support for enhancing indoor living health quality, but it also focuses on areas such as renewable energy utilization, application of novel energy-saving materials, and optimization of building energy systems. These efforts aim to promote the transition of the building sector toward low-energy consumption and low-emission models, thereby contributing to the achievement of global climate governance objectives. Furthermore, we anticipate that academic exchanges facilitated by this Reprint will help consolidate industry consensus, offering guidance for policy formulation, engineering design, and technology dissemination, thus accelerating the industrialization and practical implementation of healthy and low-carbon building technologies.

As the Lead Guest Editor, Dr. Kai Zhang would like to thank *Buildings* for its support of this Reprint and acknowledge the Editorial Board Members of this Reprint and Ms. Eveline Cheng for their invaluable assistance in organizing and publishing this issue. We appreciate all the authors for their pioneering research and exploration into frontier issues within this field.

**Kai Zhang, Jian Liu, Xiaolei Yuan, Yizhe Xu, and Junming Zhou**
*Guest Editors*

buildings

MDPI

Review

# Coaxial Pipes Used as Ground Buried Heat Exchangers—A Review of Research in Recent Years

Geng Wang [1], Nai Rong [2,*], Xuefei Li [1,*], Ning Hu [1], Zhi Zhang [1], Yuan Zhang [3] and Yuhan Wang [1]

1 School of Environment and Energy Engineering, Anhui Jianzhu University, Hefei 230601, China
2 Energy Saving Research Institute, Anhui Jianzhu University, Hefei 230002, China
3 College of Architectural Science and Engineering, Yangzhou University, Yangzhou 225127, China
* Correspondence: rongnai@ahjzu.edu.cn (N.R.); lxf@ahjzu.edu.cn (X.L.)

**Abstract:** The efficient utilization of geothermal energy depends heavily on high-performance ground heat exchangers. Coaxial pipe is a high-efficiency heat exchanger composed of two nested tubes of different diameters. In this paper, the structure and thermal exchange characteristics of coaxial pipe geothermal exchangers are introduced, which are superior to single-U and double-U geothermal exchangers in respect of installation, heat transporting, and deep geothermal application. Thermal test research of coaxial pipe geothermal exchangers is investigated. Relevant studies in recent years on the factors affecting the thermal performance of coaxial pipe ground heat exchangers, including exchanger configurations, circulating fluids, subsurface conditions, flow patterns, and operational modes, are reviewed. In addition, research on the impact of coaxial pipe ground heat exchangers on the ground, as well as applications for coaxial pipe ground heat exchangers, is summarized. Recommendations are made for potential future research on coaxial pipe ground heat exchangers. It is believed that the results of these studies will help to raise awareness of coaxial pipe ground heat exchangers and to continue to promote their application.

**Keywords:** coaxial pipes; geothermal exchangers; parametric studies; thermal test; engineering application

## 1. Introduction

Energy and environmental issues are the two major challenges that present-day society has to address [1,2]. The massive consumption of fossil fuels has resulted in large amounts of carbon emissions [3–6] in the face of the dilemma of dwindling reserves. In addition to expanding the area of carbon sinks [7–10] and developing carbon capture and storage technologies [11–14], energy decarbonization has become an inevitable choice. Renewable energy sources with zero carbon emissions in the utilization process, such as air energy [15,16], wind energy [17,18], solar energy [19,20], nuclear energy [21,22], and tidal energy [23,24], have received continuous attention in recent decades.

As a renewable and sustainable energy source, geothermal energy has also been the focus of research [25]. Thermostatic layers are found 10 to 15 m below the earth's surface, and ground temperature increases with depth, which varies with the geographical location [26–28]. Places located at lower latitude usually have higher shallow ground temperatures [29,30]. Hence, coupled with mega mass volume and high thermal inertia, the ground shows great potential as a heating and cooling source and as a seasonal thermal energy storage reservoir. In fact, geothermal energy demonstrates many advantages, e.g., local availability, continuity, smaller temperature fluctuations, meteorological conditions immunity, near-zero $CO_2$ emissions, and low cost, compared to other new energies and fossil fuels, such as coal and petroleum [31–33]. Installed geothermal energy utilization engineering has mostly spread in Europe [34,35], North America [36,37], and Asia [38], followed by Oceania [39] and Africa [40,41].

Practically, the thermal resources contained in the ground can be categorized into three types according to depth: shallow geothermal energy, medium-depth geothermal energy, and deep geothermal energy. Geothermal energy exploitation goes deeper with the development of drilling technology. Obviously, the utilization mode of geothermal energy differs according to the depth [42]. However, the definition of the scope of these three regions is still not uniform. Cai et al. [43] used two values—150 m and 3000 m—as the division index, while Li et al. [44] hold that a depth of 2000 m can be considered as the deep layer. Deng et al. [45] defined medium-depth geothermal energy as referring to the heat embodied in the rocks 2–3 km underground.

One of the approaches to develop geothermal energy is the employment of borehole heat exchangers (BHEs), which are becoming increasingly popular due to their high efficiency, sustainability, and universality [46–48]. Typical BHE designs use a single-U pipe [49,50] or double-U pipes [51–53], as shown in Figure 1a,b. Other common geometries are tiled-tube [54], helical-pipe [55,56] and multi-pipe [57], etc. Typically, the BHEs are buried in holes, and the gaps between the hole wall and the pipe outer surfaces are filled with high thermal conductivity grout to enhance the heat exchange. The fluid enters through the pipe inlet and flows downward, reaches the lowest point, and then flows upward to the outlet. The flow pattern varies in different BHEs. In some types of BHE, the fluid does not come into direct contact with the outside soil/grout but continuously conducts the heat exchange with it through the walls of the pipes during the whole flow process, thus realizing heat exchange goals. There are fewer available studies on the application of horizontal borehole heat exchangers, mainly due to the fact that the construction site often does not provide the required drilling area [58].

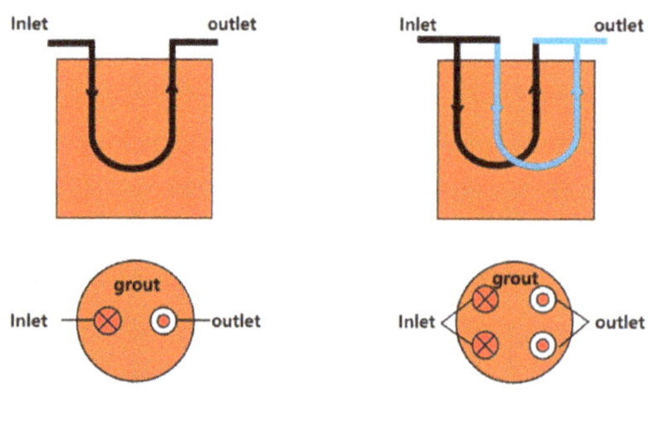


(**a**) (**b**)

**Figure 1.** Diagram of single-U pipe and double-U pipes geothermal heat exchangers. (**a**) Single-U pipe and (**b**) double-U pipes.

A coaxial pipe heat exchanger, also named a pipe-in-pipe heat exchanger [59] or double-tube heat exchanger [60], features high heat-exchanging efficiency. In this type of heat exchanger, a thin center pipe is inserted into a thicker outer pipe so that an annular space is formed between them. In conventional usage, the inner pipe and annulus are, respectively, used as passageways for each of the two streams of fluids, circulating in a parallel-flow or counter-flow pattern. Hot fluid may enter the heat exchanger via the inner pipe and cold fluid flow through the annulus channel and vice versa. Heat from the hot fluid may pass through the wall of the inner pipe and then enter the cold fluid to accomplish heat transportation [61]. Accordingly, the inner pipe should have a high heat conductivity, while the outer pipe usually needs to be thermally insulated to minimize heat exchange with the neighboring environment as much as possible.

A coaxial pipe heat exchanger placed horizontally in a thermostatic water bath was investigated experimentally [62]. The heat exchanger demonstrated a varied flow pattern comparing with the conventional ones. The heat transfer process occurs between the fluid in the annulus and the hot water in the water bath through the outer pipe wall. Similarly, coaxial pipe heat exchangers were also suitable to be used as geothermal exchangers [63]. Furthermore, the use of coaxial pipe geothermal exchangers (CPGEs) could potentially reduce borehole drilling costs. Considering the space that other geothermal heat exchangers have to leave between the "legs" for grout to enhance the thermal process, a CPGE requires a smaller borehole diameter for tubes of the same size. CPGEs are more beneficial for deeper geothermal energy utilization from an installation perspective, because the installation difficulty of keeping legs apart for the above-motioned other kinds of BHEs can be avoided in the case of CPGEs.

The methodology for conducting this study is shown in Figure 2. This study aimed to summarize the thermal performance and application research development of CPGEs in recent years. This would provide a systematic introduction for researchers and designers to component performance and system configuration. This paper is organized as follows: the structure, flow patterns, and heat transfer characteristics of CPGEs are briefly introduced in Section 2; the impact factors on the thermal performance of CPGEs and the influencing zone of CPGEs on the ground are generalized in Section 3; various thermal performance tests of CPGEs are reviewed in Section 4; application sectors of CPGEs are given in Section 5; and a conclusion including future potential research interests is presented in Section 6.

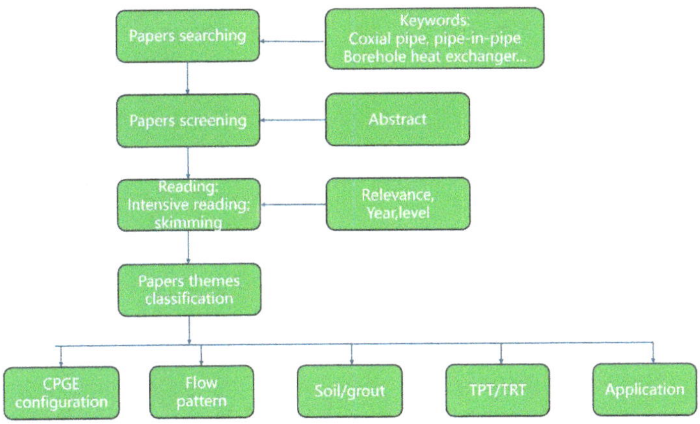


**Figure 2.** Research strategy of this review.

## 2. Coaxial Pipe Used as BHE

In geothermal utilization scenarios, coaxial pipes are usually buried in abandoned petroleum/gas wells or deliberately drilled boreholes. The interior pipe length is slightly shorter, and the lower end of the interior pipe is shallower than that of the exterior pipe to form the turning-around space for circulating fluid [64,65]. The structure of a CPGE is schematically depicted in Figure 3a,b. Concentric pipes dominate the form of the CPGE. In fact, for an equal borehole diameter, a CPGE contains a larger volume of circulating fluid than a single-U or double-U tube heat exchanger [66], which leads to more heat absorption and release with the same temperature difference between the inlet and the outlet of working fluid.

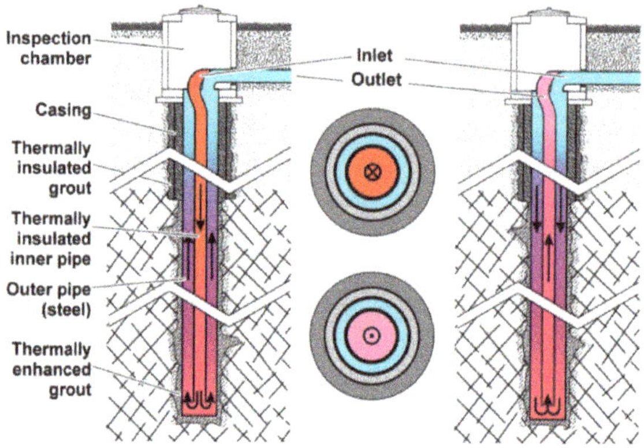


(a) (b)

**Figure 3.** Structure diagram of a CPGE and two flow patterns in a CPGE. (**a**) Mode A, (**b**) Mode B Reprinted with permission from Ref. [67]. 2025, Elsevier.

Unlike conventional coaxial pipe heat exchanger usage, in which two streams of circulating fluid exchange heat mutually, only one kind of working fluid acting as a heat carrier is employed in a CPGE where the heat exchange occurs between fluid and grout/soil. There are two kinds of practicable fluid flow patterns in a CPGE [67]. In one case, called Mode A, the circuiting fluid enters the CPGE via the inlet located at the top of the inner pipe, reverses the flow direction at the lower extremity, travels upwards, and exits from the annular space's upper opening; alternatively, the pattern in which the circuiting fluid enters downwards at the annulus' top and leaves from the inner pipe top after turning around at the bottom is named as Mode B. These two flow patterns are also clearly demonstrated in Figure 3 [67]. These two kinds of working modes employ varied heat exchange processes along the fluid flow path. Due to the ground temperature gradient, the fluid in the annulus faces a variable vertical temperature boundary condition.

The external pipe of the CPGE is circumferentially exposed to the grout/soil. Either in Mode A or in Mode B, or either in a heat extraction or in a heat injection working condition, heat transfer in the ground mainly happens radially inwards or outwards before meeting thermal interference from another CPGE [68].

As shown in Figure 4, the fluid in the annulus space lies in the center of the heat transfer chain in both Mode A and Mode B [69]. For the inner pipe, the temperature difference between the fluid in the central pipe and the fluid in the annulus always exists. Therefore, heat flow crossing the wall of the inner pipe from the hot side to the side with relative lower temperature unavoidably happens, and this is named thermal shunt or thermal shortcut. In this process, the total thermal resistance, including the convective thermal resistance at the inner surface of the inner pipe wall, the inner pipe wall thermal resistance, and the convective thermal resistance at the outer surface of the inner pipe, is overcome. This thermal resistance can be expressed by Equation (1) [70]:

$$R_1 = \frac{1}{2\pi r_{1i} h_{1i}} + \frac{1}{2\pi k_1} \ln \frac{r_{1o}}{r_{1i}} + \frac{1}{2\pi r_{1o} h_{1o}}, \quad (1)$$

The thermal shortcut should be minimized, and it should be ensured that the majority of the heat exchange occurs between the circuiting fluid and the neighboring soil or grout adjacent to the outer surface of the outer pipe wall. According to Equation (1), for a CPGE, properly selected inner pipe materials can weaken the thermal shortcut notably.

To exchange heat with soil, the outer pipe's thermal resistance is overcome, which consists of the convective thermal resistance at the inner surface of the outer pipe wall, and the outer pipe wall's thermal resistance. This thermal resistance can be calculated as Equation (2):

$$R_2 = \frac{1}{2\pi r_{2i} h_{2i}} + \frac{1}{2\pi k_2} \ln \frac{r_{2o}}{r_{2i}}, \tag{2}$$

The thermal resistance network of the CPGE is depicted in Figure 3, which evidently differs from that of a single U-tube [71–73] and double U-tubes [74,75].

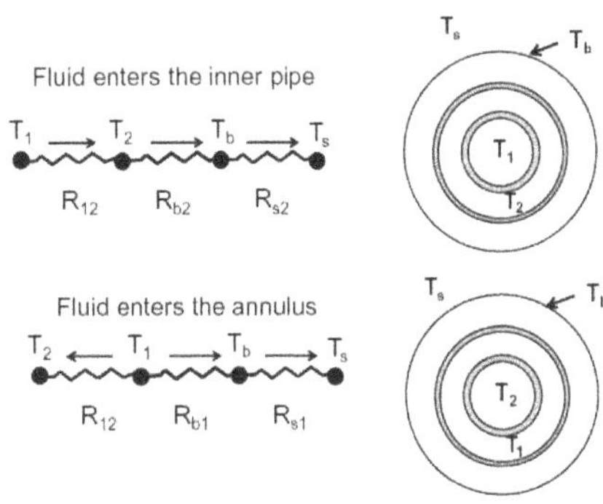


**Figure 4.** Thermal resistance model of a CPGE. Reprinted with permission from Ref. [69], 2025, Elsevier. (T means temperature, R means thermal resistance. Subscripts: 1 means flow path number 1, 2 means flow path number 2, b means borehole, s means ground or soil).

Raymond et al. [76] derived the following analytical equation for a three-dimensional borehole resistance. Equation (3):

$$R_b = R_2 \left( 1 + \frac{H^2}{3(m_w c_w)^2 R_1 R_2} \right), \tag{3}$$

The selection of flow state in the inner pipe and annulus is also of great significance. Keeping the laminar flow in the inner pipe with low *Re* helps to reduce $h_{1i}$ for a minor thermal shunt and decreases fluid flow pressure drop, while rational turbulent flow in the annulus augments $h_{2i}$ to enhance the heat transfer between the working fluid and the soil/grout and confines the flow pressure drop in an acceptable range.

Borehole heat capacity, mainly involving fluid, pipes, grout, and ground soil, had an especially significant influence on ground thermal response and fluid temperature in a short time [77]. During the early operation period, the thermal storage properties of the circulating fluid, the pipes, and the grout/soil play a decisive role for heat transfer. As time passes, the influence of these parameters fades and the ground soil thermal properties dominate the temperature curves.

In heat delivering mode, circuit fluid enters the coaxial pipe with a temperature higher than that of the ground and injects heat to the soil along the path, accompanied with a gradual temperature reduction—the ground acts as a heat sink [30,44]. Conversely, in heat extraction mode, the heat flow direction is reversed and the ground serves as a heat source. It should be noted that, in Mode B, a possible temperature rise may happen at the shallow depth of the CPGE, due to the heat released from high-temperature fluid to the return fluid in the internal pipe.

## 3. Parametric Studies of CPGE Heat Transfer Performance

Heat flows between a BHE and the ground lead to temperature change in the circulating fluid, which disturbs the original ground temperature field nearby the BHE [78]. The heat transfer process consists of the process inside and outside the CPGE. At the same time, the heat transfer process outside the tube includes the heat transfer process in the grout and the heat transfer process outside the borehole, i.e., in the ground. CPGE performance research has centered on the impact factors affecting these two parts [79]. Related studies have centered around four aspects: CPGE configurations, circulating fluid conditions, the borehole surface and subsurface properties, and the flow and operational characteristics.

### 3.1. Impact of CPGE Configurations

CPGE configurations include these factors: (1) material thermophysical properties, mainly the thermal conductivities of the inner and outer pipes; (2) geometric dimensions, i.e., the shape and diameters of the inner and outer pipes; (3) geometry of the CPGE pipe and borehole.

#### 3.1.1. Thermal Conductivity of Inner Pipe

Obviously, low thermal conductivity of the inner pipe results in fewer heat shortcuts between the two sides of the inner pipe, regardless of the heat injection or heat expelling modes. Dong et al. [80] chose a heat-resistant polyethylene (PE-RT) piping system for hot and cold water. Table 1 presents the thermal properties of inner tube materials in several studies. Most of the materials are plastics.

**Table 1.** Materials for inner pipes in research.

| Authors | Materials | Thermal Conductivity [W/(m K)] | Specific Heat Capacity [J/(kg K)] | Density (kg/m$^3$) |
|---|---|---|---|---|
| Wang et al. [31] | polyethylene pipe | 0.18 | 2100 | 930 |
| Cai et al. [43] | high-density polyethylene | 0.45 | 2300 | 950 |
| Gordon et al. [65] | high-density polyethylene | 0.4 | - | - |
| Oh et al. [81] | high-density polyethylene | 0.4 | - | - |
| Śliwa et al. [82] | fibre glass | 0.361 | - | 1900 |
| Śliwa et al. [82] | vacuum tubes filled with insulation | 0.006 to 0.0008 | - | - |
| Chen et al. [83] | 540 polyethylene | 0.42 | 2300 | 950 |

Beier et al. [30] studied the effect of the thermal resistance of the internal pipe wall on the shape of vertical temperature profiles and found the liquid temperature difference between the inlet and outlet decreased significantly for an inner tube with higher thermal resistance.

Śliwa et al. [84] reported a method of enclosing the inner pipe within an insulation layer for heat shortcut reduction. Luo et al. [85] offer a reminder that beyond a certain limit, the effect of further increasing the inner pipe's thermal resistance on preventing thermal shunt tends not to be apparent. Song et al. [86] found that at a depth exceeding a critical value, inner tube insulation became unnecessary, since the temperature of the falling water was already high enough to avoid significantly cooling the warm water going upward within the inner pipe under the heat extraction working condition.

Based on their low thermal conductivity, Śliwa et al. [82] proposed vacuum tubes as inner pipes, as shown in Figure 5. Du et al. [87] used a double-walled gel steel pipe with a thermal conductivity of 0.002 W/(m·K) as the inner pipe. During the manufacturing process, the air between the two pipes was pumped out to keep a desirable inside vacuum level before filling the space with aerogel material. Then the pipes were sealed and lengthened using sleeves. The high initial investment and drilling rig rental costs resulting from the large inner-column weight had to be taken into account in application [31].

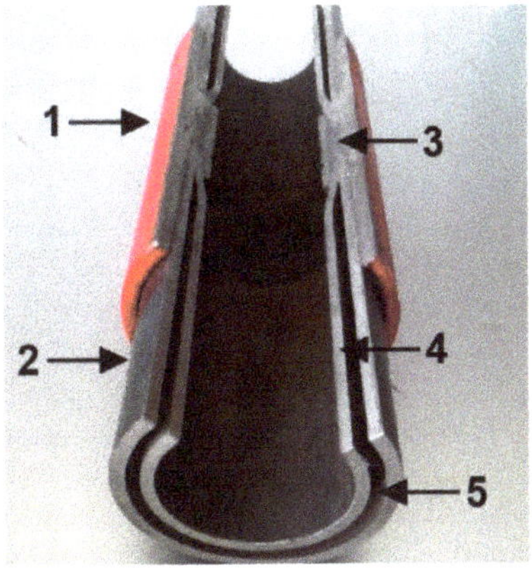


**Figure 5.** Components of vacuum insulating tubing. Reprinted with permission from Ref. [82], 2024, Elsevier. (1—coupling, 2—outer steel pipe (e.g., L80), 3—internal coupling insulator and weld, 4—inner steel pipe (e.g., 13CR85), 5—vacuum with getter material). Based on (Helix Oilfield Services Ltd., 2017, Aberdeen, UK).

#### 3.1.2. Thermal Conductivity of Outer Pipe

In contrast to the inner pipes, high thermal conductivity of the outer pipes is ideal, inducing a more sufficient heat exchange with the ground formation. A high thermal conductivity of the outer pipes leads to an overall thermal resistance reduction of the CPGE and may result in a heat exchange amount superior to that of a single U-tube design [88]. A subsequent benefit is the shortened borehole depth [76] and reduced drilling cost [89].

Material types for the outer pipes are limited. Steel pipes with various additive components have drawn the dominant attention of researchers from different countries due to their popularity. Commonly used pipe steel types are listed in Table 2. Obviously, the thermal conductivity of most typical outer pipe materials is numerically from several dozen up to several thousand times higher than that of the inner pipes' materials. Conversely, the specific heat of outer pipe materials is significantly lower than that of the inner pipes' materials, for time saving of heat transfer.

**Table 2.** Typical materials for outer pipes.

| Authors | Materials | Thermal Conductivity [W/(m K)] | Specific Heat Capacity [J/(kg K)] | Density (kg/m$^3$) |
|---|---|---|---|---|
| Wang et al. [31] | carbon steel | 54 | 470 | 7820 |
| Cai et al. [43] | Seamless steel J55 | 40 | 498 | 7850 |
| Oh et al. [81] | Stainless steel | 15 | - | - |
| Chen et al. [83] | 540Steel | 40 | 498 | 7850 |

Nevertheless, plastic pipes were still used in engineering applications. Raymond et al. [76] adopted thermally enhanced high-density polyethylene with a thermal conductivity of 0.7 W/m·K. Oh et al. [81] compared the heat exchange efficiency of two types of CPGE. The CPGE adopting stainless steel as the external pipe material acquired a greater per meter heat transfer amount than the CPGE employing high-density polyethylene. Acuña and Palm [90] used a thin plastic material as the external pipe. The flexible pipe wall was tightly pressed against the soil by the pressure from the circulating water. The contact thermal resistance between the pipe wall and the borehole wall was minimized and the backfill was omitted.

#### 3.1.3. Diameter Ratio of Inner Pipe to Outer Pipe

Apart from the CPGE pipes' materials, the cross-sectional dimension also has an effect. For single U-tube and double U-tube heat exchangers, each leg is separate and usually has the same diameter [63]. Under this kind of configuration, the passage shape and area for fluid circulation are always fixed. With regard to CPGEs, the outer pipe adopts a larger diameter than the inner pipe. For an outer pipe of fixed dimension, the change of inner pipe diameter unavoidably changes the cross-sectional area of the inner pipe and annulus simultaneously, i.e., the upward and downward passageway areas for the working fluid. Subsequently, the flow field regime and heat transfer inside them are changed. The optimal diameter ratio of the inner pipe to the outer pipe, on the one hand, lowers the convective heat transfer coefficient to reduce heat shortcut and on-way flow loss in the inner pipe, and on the other hand, it guarantees the smooth heat exchange process with the ground while ensuring relative lower pressure drops in the annulus.

Usually, a wider inner tube results in a lower fluid velocity and convective heat transfer coefficient in it. This method helps to reduce thermal shunt. In addition, the fluid velocity increase in the annulus leads to a convective heat transfer coefficient increase on the interior surface of the external pipe. Based on this, the heat transfer rate is strengthened. Iry and Rafee [91] studied the effect of pipe diameter ratio on the characteristic numbers *Re* and *Nu* at the interior surface of the outer pipe. A small ratio was suggested for a high *Nu*. Research by Pan et al. [92] shows that in a heat extraction working status, a small ratio of inner to outer diameter implies lower CPGE bottom fluid temperature. Wang et al. [31] studied the change of outlet temperature and heat removal capacity when the inner pipe diameter was unchanged and the outer pipe diameter was increased. The larger outer tube case demonstrated stronger heat transfer capacity. Yekoladio et al. [93] found that the optimum outer pipe diameter varied according to the ground temperature gradient. A larger gradient needed a smaller external pipe diameter for heat rejection.

Iry and Rafee [91] reported a ratio of 0.65 for minimum pumping power in the case of water circuit fluid, while the figure from Morchio and Fossa [94] was around 0.6. Mokhtari et al. [95] revealed that ratios of 0.675 and 0.353 were the optimum values for pressure drop minimization and thermal efficiency maximization, respectively, in the case of steam return fluid in the inner pipe. Daneshipour and Rafee [96] concluded that a ratio around 0.7 produced a minimized pressure drop for a CuO-water nanofluid heat carrier. Yekoladio et al. [93] found that the optimal ratio varied from 0.683 to 0.653 when the flow regime in both the inner pipe and the annulus changed from laminar flow to turbulence. In summary, a value higher than 0.6 and lower than 0.7 is considered to be appropriate for engineering applications.

#### 3.1.4. Shape of CPGE Pipe

Li et al. [44] compared the performances between corrugated pipes and smooth inner pipes. Minor deviations were observed under the heat extraction mode. Zarrella et al. [97] proposed a type of coaxial heat exchanger, as shown in Figure 6. The central steel pipe was welded around with steel helix fins and the inner pipe was coated using a closed-cell insulation layer. The passage in the annulus is spiral for the heat carrier. It can actually prolong the pathway and thus result in a longer time for heat exchange while minimizing thermal shortcut. Liu et al. [98,99] analyzed the effect of rotational directions and rotational speeds of continuous spiral fins on heat transfer performance, exergy efficiency, and entropy generation. Research related to outer pipes mainly focuses on finding high thermal conductivity materials, and fewer papers concern the topic of the structure optimization of outer pipe walls. The CPGE employed by Rajeh et al. [100] and Li et al. [101] was a kind of multi-external-chamber borehole exchanger, in which the annulus space was evenly divided up by several partitions.

**Figure 6.** Inner pipe welded with steel helix, Reprinted with permission from Ref. [97], 2024, Taylor & Francis Group.

References [102,103], respectively, carried out analyses of the heat extraction performance of a kind of CPGE when the outer pipe diameter was changed to a small value while the inner diameter was maintained along the entire depth.

Jia et al. [59] experimentally explored the impacts of eccentricity on heat transfer and pressure loss of a CPGE. Eccentricity between 0.4 and 0.6 leads to a 20% to 35% increase in heat transfer coefficient and a 33% to 54% decrease in pressure loss.

Here, the eccentricity $e$ is defined as Equation (4):

$$e = \frac{2L_d}{D_h}, \tag{4}$$

Rajeh et al. [104,105] proposed a kind of CPGE having an oval outer pipe and studied the influence of varied inner pipe positions in the outer pipe on heat transfer.

3.1.5. Shape of Borehole

Kurevija and Strpić [106] conducted thermal response tests of a 100-m-long CPGE having a slope angle of 45°. This structure allows the convective heat transfer driven by groundwater movement to be fully taken advantage of. Its superiority is due to a longer length of CPGE placed in an upper permeable layer. DAI et al. [107] experimentally investigated the winter heat extraction performance of a 2070-m-deep well heat exchanger having an angle of 19°, which was not outfitted with a lower end cap. Working fluid cycled between the CPGE and the strata. Cheng et al. [108] pointed out that an open-end CPGE not only suppressed geothermal heat production but also wasted working fluid and posed a pollution risk for the surrounding soil. Yin et al. [109] proposed a 1764 m leaky downhole coaxial open-loop geothermal system and experimentally tested the heat transfer performance with a constant inlet water temperature of 24 °C for 60 h. Wang et al. [58] proposed a kind of well with a horizontal lower part and an open end.

There are published articles involving horizontally buried BHEs of other types [110], while there have still been still few such thermogeological designs of CPGEs reported. This may be explained by the fact that the ground temperature at the same level is almost of equal value, causing a very weak heat transfer driving force along the whole length of the CPGE. Akram et al. [111] experimentally determined the heat exchange per meter of the borehole of a horizontal CPGE in winter and in summer, which was numerically lower than that of vertical ones [45].

### 3.2. *Impact of Circulating Fluid Conditions*

The impact of circulating fluid on the performance of a CPGE mainly results from thermophysical properties, flow condition, and inlet fluid temperature.

#### 3.2.1. Types of Circulating Fluid

Fluids having high heat transfer characteristics and high heat carrying capacity, while having relatively low hydraulic flow pressure loss, are considered to be ideal working carriers for CPGEs. Except for water, other kinds of fluids have featured in research. Wood et al. [89] tested the heat output performance of a heat pump that employed a CPGE as the evaporator and a monopropylene glycol mixture of 76% water as the working fluid for extracting heat from the earth. Alimonti and Soldo [112] compared the heat extraction ability and transporting power consumption of water and diathermic oil. It was found that the heat extraction performance of water was higher than that of diathermic oil under the same flow rate. The power consumption of the two fluids was almost equal under a constant flow rate, while water consumed more energy for transportation under variable flow rates. Thermophysical properties of the used fluids are listed in Table 3.

**Table 3.** Properties of different circulating fluids.

| Authors | Circulating Fluid | Density (kg/m$^3$) | Thermal Conductivity (W/m·K) | Specific Heat (J/kg·K) | Dynamic Viscosity (Pa·s) |
|---|---|---|---|---|---|
| Luo et al. [85] | water (15 °C) | 999 | 0.59 | 4190 | --- |
| Alimonti and Soldo [112] | diathermic oil (60 °C) | 762 | 0.13 | 2500 | 3.3 |
| Ashrae [113] | 30% glycol water solution (30 °C) | 1041.26 | 0.455 | 3674 | 1.69 |

Olson [114] patented the application of nanofluids for ground source heat pumps. Faizal et al. [29] and Liu et al. [115] found that even low concentrations of nanoparticles may lead to substantial increases in fluid thermal properties. Togun [116] confirmed the effect of nanofluids on heat transfer enhancement in the annular passage. Pisarevsky et al. [117] found that when the concentration of aluminum oxide ($Al_2O_3$) nanoparticles was in the range 2% wt to 8% wt, the thermal conductivity of the nanofluid may obtain an increase up to 13%. Daneshipour and Rafee [96] put copper oxide (CuO) nanoparticles into water to produce a nanofluid heat carrier. High nanoparticle concentrations among 0–6% in consideration were found to be contributive to a better heat transfer effect, but a flow pressure drop amplification was simultaneously observed. This means for working fluid selection, a compromise between heat transfer performance and pump work consumption should be reached based on practical need. The thermophysical parameters of $Al_2O_3$ and CuO nanoparticles are shown in Table 4.

**Table 4.** Thermophysical properties of CuO and $Al_2O_3$ nanoparticles.

| Nanoparticle | Density (kg/m$^3$) | Heat Capacity (J/kg·K) | Thermal Conductivity (W/m·K) |
|---|---|---|---|
| CuO [96] | 6320 | 532 | 76 |
| $Al_2O_3$ [96] | 3900 | 880 | 40 |

Zhao et al. [118] used phase-change $CO_2$ as the working fluid to extract heat from the medium-deep geothermal reservoir.

#### 3.2.2. Flow Rates of Circulating Fluid

Deng et al. [45] pointed out that heat carrier flow rate exerts an influence by varying the convective heat transfer coefficient of the outer pipe inner wall, i.e., the thermal resistance of a borehole heat exchanger. Xie et al. [119] defined the heat storage efficiency as the evaluation index of heat injection characterized by the following Equation (5):

$$E_h = \frac{Q_s}{mh_{in}} \tag{5}$$

$E_h$ decreases with the increase in flow rate. This suggests a low flow rate for the heat expelling mode. Templeton et al. [27] observed the fluctuation of fluid temperature at a CPGE outlet against the growing flow rate in the heat removal mode. Wang et al. [31] found that an inlet velocity between 0.3 m/s and 0.7 m/s corresponded to an increased heat output for Mode B. Nian et al. [120] concluded that increased flow rate produces more heat. These are similar to the case of single U-tube BHEs [28]. Meanwhile, for long-duration heat removal operation, the outlet temperature as well as the heat extraction amount of a CPGE declined with time; consequently, a higher flow rate contributed to a heat transfer performance deterioration of the CPGE [121].

#### 3.2.3. Temperature of Inlet Circulating Fluid

Xie et al. [119] gave the effect of inlet circulating fluid temperature on $E_h$ under the heat injection mode. The $E_h$ increases with the increase in inlet fluid temperature. Luo et al. [85] pointed out that for the exothermic mode, the inlet temperature affected the selection of working fluid flow pattern. Adopting the annular pipe as inlet may contribute to better performance, in a case where the input water temperature is higher than the temperature at the bottom ground layer. Song et al. [86] and Niu et al. [122] studied the influence of inlet water temperature on the heat removal performance of a CPGE under Mode B. It was found that as the inlet temperature increased, the outlet temperature also increased. However, the amount of heat yielded showed a downward trend.

### 3.3. *Impact of Surface and Subsurface Properties*

Ground geological and hydrological properties primarily involve the ground surface temperature, thermophysical properties of grout and soil, vertical distribution of ground temperature, and flow condition of underground water.

#### 3.3.1. Ground Surface Properties

Saadi and Gomri [123] investigated thermal interferences between two neighboring boreholes under surface seasonal effects in the heat extraction working condition. The function curves of the evolution of thermal interferences at the bottom and the beginning time of thermal interferences along the depth were given. The obtained results may give a guide for the matrix design of CPGEs. Practically, most studies began from a certain depth, neglecting the temperature fluctuation of the upper variable temperature layer.

3.3.2. Grout

The contact faces between two components in a system usually require special attention [124–126]. To reduce the contact thermal resistance between the pipe and the soil, high thermal conductivity grout is employed. Apart from making use of the soil originally dug out in situ to backfill the borehole, researchers tried other artificial materials as grout or took thermal conductivity enhancement measures for the backfill materials, as presented in Table 5. Sliwa and Rosen [127] and Song et al. [128] concluded that increased thermal conductivity of grout contributed to heat transfer, similar to the cases of single-U [28] and double-U [129] BHEs. However, if amounting to a certain value, the increase of the backfill materials' thermal conductivity was reported not to result in a proportional increase of heat exchange efficiency in the CPGE [81]. Pan et al. [92] examined the impact of grout thermal conductivities on the variation of CPGE outlet and bottom fluid temperatures. In the heat removal working condition, the temperatures of these two positions increased with the increase of borehole grout thermal conductivity for Mode B.

**Table 5.** Summary of varied grouts.

| Authors | Grout | Thermal Conductivity (W/m K) | Thermal Diffusivity ($\times 10^{-6}$ m$^2$/s) | Thermal Capacity (J/kg K) | Density (kg/m$^3$) |
|---|---|---|---|---|---|
| Wang et al. [31] | protoplasm | 4.0 | - | 1172 | 2283 |
| Deng et al. [45] | synthetic mud ball | 2.0 | - | 850 | 2700 |
| Zarrella et al. [130] | --- | 1.4 | 0.70 | | |
| Oh et al. [81] | cement | 1.373 | - | - | - |
| Oh et al. [81] | bentonite | 0.86 | - | - | - |
| Song et al. [128] | conventional cement | 0.7 | - | 2000 | 2140 |
| Xie et al. [119] | cement | 0.73 | - | - | - |
| Song et al. [128] | cement | 0.8 | - | 1900 | 2140 |
| Chen et al. [83] | 540 cement mortar | 0.93 | | 1050 | 1800 |

3.3.3. Soil

Soil characteristics such as density and moisture content affect the thermal properties, which vary from place to place and among strata. Fang et al. [131] pointed out higher thermal resistance of the soil or rock was not favorable for heat removal. The increase of soil thermal conductivity was beneficial to the heat transfer process either toward or away from a CPGE [27]. Lous et al. [132] carried out a sensitivity study of volumetric heat capacity, porosity, and thermal dispersivity of soil on outlet fluid temperature, specific heat extraction rate, and the thermally affected zone. Gordon et al. [65] examined the impact of stratum stratification on the heat transfer performance of a CPGE. Sliwa and Rosen [127] gave the temperature calculation equation of a CPGE neighborhood at a certain time. The equation incorporated the thermal conductivity of the ground.

3.3.4. Ground Temperature Gradient

The drilling depth of a borehole mainly depends on ground temperature and practical purpose. Ground temperature increases with depth due to the heat from the earth's core, having a downward positive direction. The gradient usually falls into the range of 10–50 °C/km, depending on location [93,133]. A high ground temperature means a small temperature difference between the fluid in the annulus and the ground. Therefore, the driving force for heat release is weakened, or even opposite in direction. Luo et al. [85] reported that geothermal gradient exerted a minor negative effect on the heat injection working mode. For CPGEs exclusively for heat extraction applications, it is preferable to choose a depth of more than 1500 m; a depth of between 1789 m [134] and 2000 m [31] is ideal to take full use of the high ground temperature. Fang et al. [131] and Caulk [135] verified that a high gradient of ground temperature produced high energy output. Cai et al. [43] studied the effect of ground temperature gradient on temperature distribution and the performance characteristics of a CPGE under an intermittent heat extraction working condition.

#### 3.3.5. Underground Water

Unlike circulating fluid in single-U and double-U BHEs exchanging heat with ground water along the downward path as well as the upward path, the heat carrier in a CPGE is affected only along one path—either rising or falling—in the annulus. The effect of groundwater flow on CPGE heat performance functions in relative shallower strata [67]. In an aquifer, ground water conducts convective heat exchange with the exterior surface of the outer pipe when passing through [128], enlarging the heat transfer coefficient there. Song et al. [128] and Guo et al. [136] found that groundwater flow brought an increase in both outlet temperature and heat production, and a thick aquifer acted as a positive assistant for the thermal recovery of soil. Beier et al. [30] discovered that axial heat conduction of groundwater-filled boreholes should not be ignored.

Ground water flowing in different directions usually plays varied roles. Lous et al. [132] reported that a high velocity of horizontal groundwater improved the thermal performance of CPGEs. Templeton et al. [27] found that the vertical flow of geofluid exerted a negative effect on the heat transfer between rock and a heat exchanger. Song et al. [128] found that due to groundwater flow facilitating the heat transfer process, the influence radius of a CPGE was doubled from 20 m to 40 m.

### 3.4. *Impact of Flow and Operational Characteristics*

Once established, the performance of a CPGE is mainly affected by circulating fluid flow patterns in the CPGE and by operational modes.

#### 3.4.1. Flow Patterns

As described in Section 2, the flow patterns of a CPGE consist of two modes (Mode A and Mode B, Figure 2). The mode selection depends on the purpose of the geothermal energy utilization. In the injection condition, the high temperature surrounding the soil at the bottom of the CPGE may reheat the heat carrier. However, Mode A can avoid this possibility. Mode B is similar to the flow pattern in the counter-flow type heat exchanger, which helps to extract heat from ground reservoirs. Comparatively speaking, flow pattern selection is unnecessary for single-U or double-U BHEs.

With increasing depth, a CPGE with the annular space as a downtrend channel is superior for heat extraction. Wang et al. [31] reported that a CPGE of 2000 m in Mode B produced 20.5–85.3% higher heat extraction than in Mode A for varied circulating fluid flow rate. The characteristics qualitatively coincided with the results in References [85,135]. Pan et al. [92] compared the outlet temperatures of a 2000-m-depth CPGE under the two flow patterns in the heat extraction condition. It was found that higher outlet temperatures were obtained in Mode B than in Mode A. Also, Holmberg et al. [133] pointed out that for boreholes with a depth of less than 200 m, little performance difference was observed for the two modes.

In terms of working fluid inlet temperature, Luo et al. [85] drew the conclusion that for a heat release objective, Mode A performed better for deeper CPGEs with low inlet temperatures, while Mode B was found to be suitable for shallower ones. Furthermore, if the inlet temperature is higher than the bottom temperature of the tube, Mode B acts better regardless of depth.

Beier et al. [69] compared the influence of Mode A and Mode B on temperature distribution around the CPGE in the heat extraction mode with various fluid flow rates. It was found that the surrounding temperatures in Mode B were always lower than those in Mode A. This also means Mode B showed an advantage for the heat extraction process.

3.4.2. Operational Modes

Operational modes include a continuous mode and intermittent ones. Meanwhile, heat injection and/or heat extraction processes should also be considered.

Saadi and Gomri [123] studied the heat extraction amount and the heat carrier temperature increase under the continuous heat extraction mode. Both parameters tended to decline over time. Cai et al. [43] studied the outlet temperature change under an intermittent mode with various run–stop ratios. A larger proportion of running time produced a lower outlet temperature. Deng et al. [45] obtained a higher per meter heat exchange rate under an intermittent heat exploit mode than a continuous mode. The reason for this was that the rest periods actually served as the recovery stage for the ground temperature. This indicated that intermittent operation had the potential to improve the performance of a CPGE. Also, it provided the possibility of borehole depth reduction. In addition, Xie et al. [119] compared the annual heat extraction amount of a heat removal process and the process with a 50% alternative heat release time. With heat release, the heat extraction gained a noticeable increase.

## 4. CPGE Thermal Performance Tests

Considering the varied structure of ground heat exchangers and various local geological conditions, an on-site heat transfer performance test is the most direct and reliable way to acquire the thermal behavior of CPGEs and the thermal physical properties of the ground (with or without grout) at a specific position. According to the experimental objectives and data processing methods, tests can be divided into two major categories: basic thermal response tests (BTRTs) and thermal exchange performance tests (TPTs). BTRTs were conducted with a constant heat rate to obtain the design parameters (i.e., thermal conductivity and heat capacity of the ground), while TPTs were carried out with a fixed inlet temperature to check the heat exchange performance of BHEs [137,138]. BTRTs can be further subdivided into thermal response tests (TRTs) and distributed thermal response tests (DTRTs) based on the number of temperature measuring points.

In tests, one heat migration direction is to the ground (heat rejection test) and the other is from the ground (heat extraction test). In the former, the circuit fluid is usually heated in a certain way to a temperature higher than that of the ground before entering the CPGE. For the latter, by releasing heat before entering the CPGE, the temperature of the heat carrier fluid is reduced to a value lower than the ground temperature—once entering the CPGE, the heat carrier starts to absorb heat from the ground. These two modes of tests can also be carried out alternately.

Typical test equipment is shown in Figure 7(a) and Figure 7(b), which is applicable to all similar tests. The test system is mainly composed of one CPGE or several parallelly buried CPGEs, connecting pipes, a circling pump, a heat and/or cold source, adjusting valves, temperature measuring meters, flow rate measuring meters, and data recording apparatuses.

Tests may last for several hours or days, in a continuous or artificially intermittent operational mode. Fluid temperatures and flow rates are recorded throughout tests at designated temporal intervals.

In Table 6, a variety of tests on the thermal performance of CPGEs are demonstrated.

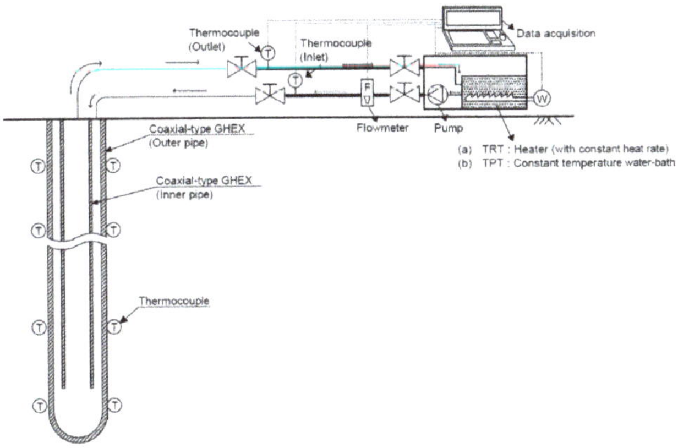


(a)

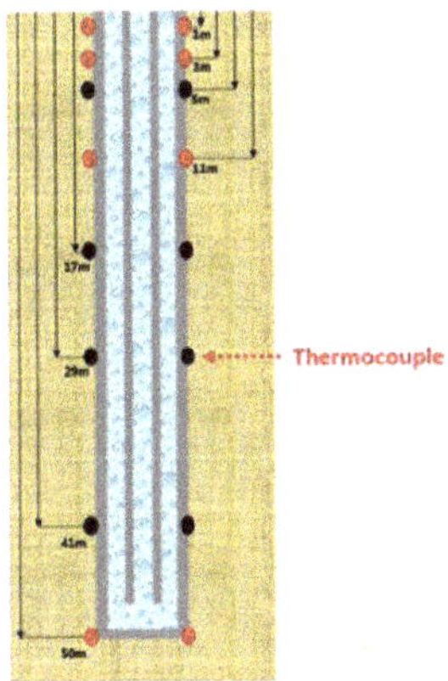


(b)

**Figure 7.** Schematics of CPGE thermal performance tests. Reprinted with permission from Ref. [81], 2024, Elsevier. (**a**) System schematics; (**b**) the unevenly installed thermocouple installation on the outer pipe.

**Table 6.** Summary of thermal performance experiments.

| Authors | Borehole Diameter (mm) | External Diameter (mm) | | Thickness of Pipes (mm) | | Materials of Pipes | | Tube Length (m) | Fluid | Flowrate (L/h) | Flow Pattern | Heat Transfer Mode | Grout or Soil | Time Duration (h) |
|---|---|---|---|---|---|---|---|---|---|---|---|---|---|---|
| | | Inner Pipe | Outer Pipe | Inner Pipe | Outer Pipe | Inner Pipe | Outer Pipe | | | | | | | |
| Wang et al. [31] | 311 | 110 | 177.8 | 10 | 9.19 | high-density polyethylene pipe | J55-special steel | 2000 | water | 5270 | Mode B | extraction | rock-soil (2.5–5.5 W/mK) | 118.5 |
| Deng et al. [45] | 254 | 93 | 159 | 3.0 | 4.5 | polyethylene pipe | carbon steel | 2000/2500 | water | $1.29 \times 10^4$ $-2.88 \times 10^4$ | Mode B | extraction | synthetic mud ball (2.0 W/mK) | 84 |
| Kurevija and Strpi [106] | 110 | 63 | 32/40 | --- | --- | PE | PE | 100 | glycol solution | --- | Mode B | injection | Soil | 50 |
| Gordon et al. [65] | 98.6 | 48. 2 | 89 | 4.6 | 8.6 | high-density polyethylene | high-density polyethylene | 183 | water | 2016 | Mode A | Injection | sand and limestone | 9 |
| Oh et al. [81] | 150 | 40 | 75 | --- | --- | high-density polyethylene | stainless steel | 50 | water | 1788 | Mode A | injection | bentonite (grout 0.86 W/mK) | 48 |
| Beier et al. [30] | 115 | 40 | 114 | 2.4 | 0.4 | medium-density polyethylene | polyethylene | 188 | water | 2088 | Mode A | injection | Soil (3.25 W/mK) | 78 |
| Dai et al. [107] | | 110 | --- | 10 | -- | thermal proof plastic | --- | 1780 | water | $6 \times 10^4$ | Mode B | extraction | soil | 336 |
| Holmberg et al. [133] | 114 | 40 | 114 | 2.4 | 0.4 | polypropylene | polypropylene | 165 | Water | 2088 | Mode B | injection | --- | 78 |

### 4.1. TRTs

TRTs are based on Fourier's law of heat conduction. The purpose of a TRT is to define the thermal conductivity of the ground (with or without grout) and the thermal resistance of the CPGE, or to ascertain the heat exchange rate per meter of tube for engineering design purposes. A typical minimum duration of 10–52 h for a TRT is required, and three phases—an initial pumping phase, a heating phase, and a recovery phase—can be included in the heat injection mode for a TRT [139].

Zarrella et al. [130] reported a type of TRT in which heat is added to the circulating fluid at a constant rate by the testing equipment at the surface. The test's duration time was 72 h and the specific heat injection rate per unit of borehole length was set to be around 60 W/m. Kurevija and Strpi [106] introduced a series of fall-off tests in Mode B with lower heat power. Each step finally reached a steady-state heat flow condition. The used working fluid was glycol solution. Gordon et al. [65] analyzed the short-term fluid temperature variations of a CPGE in a TRT. Li et al. [140] obtained a heat transfer amount per unit meter of 132.18~145.91 W in a 2539-m-deep CPGE in a 72 h experiment. Li et al. [101] carried out an experiment lasting for 24 h, with 3 h being a constant heat flow and the next 21 h being a constant inlet water temperature of 30 °C. The effects of flow direction, inner pipe insulation layer length ratio, and groundwater flow on heat flux per unit length were studied.

As far as data processing is concerned, Gordon et al. [65] pointed out that considering the variable heat flux along the pipe, using the short-term transient average temperature of the CPGE inlet and outlet in a TRT to calculate the thermophysical properties of the borehole and ground was not suitable. Beier et al. [69] found that by adopting the average value, the predicted thermal resistance of a borehole were overestimated to a maximum deviation of three times. So, temperature fluctuation profiles of more locations along the pipe are needed.

### 4.2. DTRTs

Collection of temporal and spatial temperature variations inside and immediately adjacent to the CPGE, including the temperatures of circulating fluid and grout/soil, helps to better understand the thermal process. Considering this, a DTRT was employed. The difference in experimental methods between a DTRT and TRT is that in the former, temperatures at multiple positions along the fluid flow path (not only limited to the inlet and outlet of the CPGE) are monitored [80]. More thermocouples or other temperature sensors are used in tests.

Oh et al. [81] installed thermocouples at various depth intervals close to the outside surface of the outer pipe, as shown in Figure 7(a) and Figure 7(b). The on-way temperature evolutions of the backfill of 50-m-deep CPGEs were collected.

Fiber optic cables can be placed inside the pipe along the whole flow path for the purpose of recording the temperature of multiple points [141]. This optical method was also adopted in CPGE TRTs. Holmberg et al. [102] placed a fiber optic cable inside the central pipe as well as in the annulus to obtain vertical temperature profiles within circulating fluids. Acuña and Palm [90] pointed out that the soil thermal conductivities calculated based on the data during the heating stage and heat recovery stages are different in DTRTs. Also with the aid of distributed fiber optic temperature sensors, Huang et al. [142–144] studied the long-term thermal performance of a deep coaxial borehole heat exchanger in variable-load field testing for building heating purposes.

### 4.3. TPTs

In TPTs, there is no difference in the experimental device and process in comparison with TRTs and DTRTs, but the key concern index is the total heat exchange amount between the working fluid and the soil/rock. The heat exchange amount can be easily obtained via multiplying the temperature difference between the fluid at the inlet and outlet of the CPGE by the specific heat as well as the flow rate of the working fluid.

Pokhrel et al. [145] conducted a 456 h experiment using an inlet temperature of 70 °C in a high-temperature well having an average thermal gradient of 0.38 °C/m. Oh et al. [81] conducted heat exchange amount comparisons of four CPGEs of varied construction and a single-U BHE. All the boreholes were 50 m deep, and the CPGEs in the tests adopted different structures and pipe materials or were backfilled with different grouts. Deng et al. [45] reported that the temperature difference between the inlet and the outlet reached 4~12.3 °C with a depth of 2~3 km and ground temperature of 70~90 °C when extracting heat.

## 5. Applications of CPGEs

As one approach to exploit economical and sustainable geothermal energy, depending on the working fluid temperature at the outlet, CPGEs are mainly used for heating or heat pump systems, heat storage systems, and electricity power generation. In these applications, the borehole depth of CPGEs is an essential factor.

### *5.1. CPGEs Used for Heating or Heat Pump Systems*

CPGEs can be directly used as heat sources. The hot water from the outlet of the CPGE enters the building heating system to release heat and then returns to the CPGE to be reheated, thus forming a heating cycle. The temperature of the working fluid for heat extraction in CPGEs can easily achieve 60 °C with an adequate depth, satisfying the demand of direct heating [146]. Cai et al. [43] employed CPGEs to provide heating for commercial buildings and residential buildings, working according to their schedules. Huchtemann and Müller [102] successfully used the 35~40 °C water from a 2518-m-depth well to heat a university building with a terminal of heating panels (HPs) and concrete core activations (CCAs). Geothermal heat could meet a ratio of 63~69% of total heating energy demand. Li et al. [140] carried out a 72 h field experiment on a CPGE at a depth of 2539 m in Xi'an city, China, and recorded the heat removal amount per meter of pipe. Wang et al. [147] extracted heat using a deep coaxial borehole heat exchanger for clean space heating near Beijing, China. Deng et al. [45] conducted a field test of CPGE heat pump systems having the terminal device of a radiant floor. The user-side water temperature was 45/40 °C and ground-side water temperature was 30/20 °C.

In daily life and production processes, people often need to transfer unwanted heat away [148–154]. Heat pumps can be an efficient means. Similar to single-U BHEs [155,156] and double-U BHEs [51,157], CPGEs can be used as a part of ground source heat pumps for district heating and cooling [158]. The inlet and outlet of a CPGE are respectively linked with the outlet and inlet of the ground-source-side heat exchanger of a ground source heat pump through connecting pipes, forming a closed loop sub-system. The working media circulate in the sub-system and exchange heat with the refrigerant of the heat pump when passing through the heat exchanger. In heating mode, the ground source heat pump extracts heat from underground through CPGEs, while in cooling mode, the ground source heat pump exhausts heat to underground via the CPGEs. Zhou et al. [159] reported a ground source heat pump heating system using mid-deep CPGEs. Wang et al. [31] found such a kind of heat pump unit can obtain an average COP (coefficient of performance) of 6.4. The results from Nian et al. [120] indicated that the CPGE flow rate was the leading influencing factor on the entire COP of CPGE ground source heat pump heating systems. Wood et al. [89] studied the evaporation and discharge pressure of a heat pump under various flow rates in the sub-system, and the CPGEs had a depth of 72 m.

### 5.2. CPGEs Used in Heat Storage Systems

Acting as a thermal storage reservoir, the earth can settle the temporal mismatch between heat supply and demand [160]. Bär et al. [67] suggested CPGEs as devices to expel heat of more than 110 °C from solar panels or thermal power stations to the ground in summer. The ground acts as a thermal storage reservoir, and storage depths lie between 500 m and 1500 m below the surface. The stored thermal energy can be extracted by using CPGEs when needed, thus realizing cross-seasonal thermal energy storage and avoiding heat accumulation.

### 5.3. CPGEs Used in Electricity Generation

BHEs with U-shaped tubes are usually used in shallow depth cases restricted by technical problems of tube dropping [161]. Hence, the formation temperature, and subsequently the temperature of the produced hot water, usually does not reach the target value. Therefore, there have been no applications of them in power generation.

Power generating plants use CPGEs to acquire high temperature water or vapor. Because of this, the depth of boreholes usually amounts to a value larger than 1000 m [93,95,112]. As petroleum wells usually have satisfactory depths, abandoned oil wells can be used again.

There are two methods of utilization. One is direct use. Geothermal energy is directly supplied to the power generating cycle, and the working fluid of the CPGE is also the working fluid of the power generation cycle. Mokhtari et al. [95] introduced a cycle where the circulating fluid continuously circulated in a closed pipe system composed of a CPGE, turbine, condenser, pump, and connecting pipes, as shown in Figure 8a. The circulating fluid was heated to become saturated steam in the CPGE and directly transported to drive a Rankine cycle turbine. After leaving the turbine, the circulating fluid entered the condenser to be cooled down to a liquid state. Finally, the liquid-state working fluid traveled to the CPGE inlet to begin a new cycle. Energy, exergy, and exergo-economic analyses of the system were conducted.

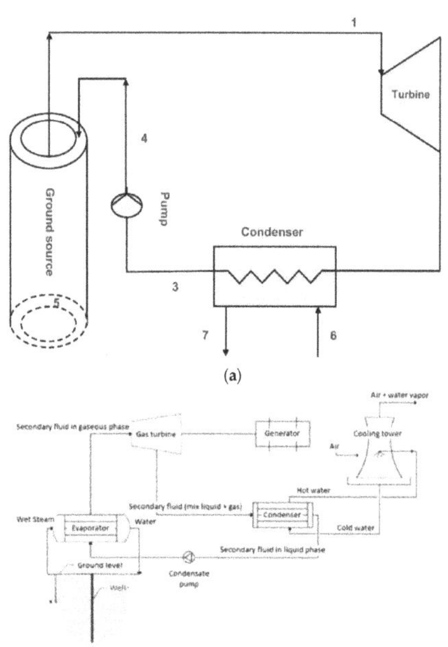


**Figure 8.** *Cont.*

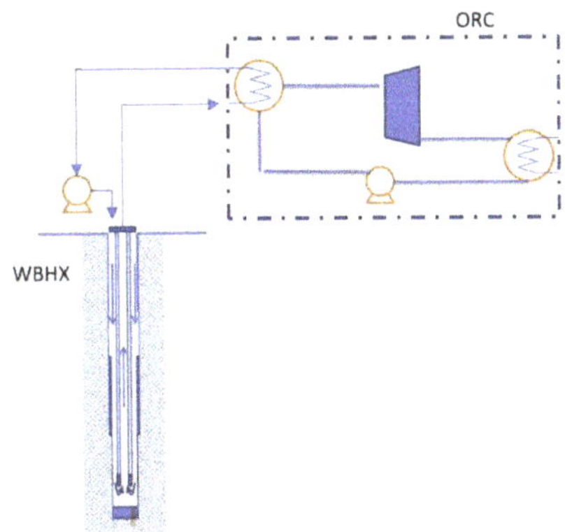


(c)

**Figure 8.** Use of CPGE in electricity power generation. (**a**) Direct use (1–5 are the flow paths of the working fluid, and 6–7 are the flow paths of the cooling medium). Reprinted with permission from Ref. [95], 2024, Elsevier; (**b**) binary cycle power plant, Reprinted with permission from Ref. [93], 2024, Elsevier; (**c**) organic Rankine cycle power plant (W BHX means CPGE), Reprinted with permission from Ref. [112], 2024, Elsevier.

The other is indirect use, in which the heat from the CPGE is used to heat the working fluid of the Rankine cycle. The CPGE, evaporator, pump, and connecting pipes constitute a secondary loop of the power generation system. The working fluid of the CPGE (secondary-loop water) and the working fluid of the power generation cycle (primary-loop water) just exchange heat. Firstly, The CPGE converts the secondary-loop water into wet steam, which is then pumped to the evaporator of the Rankine cycle to heat the primary-loop liquid-phase water from the condenser. The secondary-loop wet steam is reverted to a liquid state in the evaporator and then returns to the CPGE to be reheated. The primary-loop water changes into a gaseous phase in the evaporator and then goes to drive the gas turbine, simultaneously turning into a mix of liquid plus gas, and returns to the condenser to be cooled down to a liquid state by the cold water from the cooling tower. An indirect-use scenario is shown in Figure 8b. Yekoladio et al. [93] utilized a CPGE to exploit geothermal resources from 110 °C to 160 °C for power generation. Energetic and exergetic analyses of the geothermal binary power generation cycle were conducted.

Alimonti and Soldo [112] reported a kind of binary organic Rankine cycle (ORC) adopting a CPGE as the heat source, which is schematically depicted in Figure 8c. The working media in the ORC is R-C318, not water. In fact, refrigerants R123, R134a, R245fa, and R22 can also be employed in an ORC [95].

## 6. Conclusions

The influence of CPGE configurations, features of circulating fluids, subsurface geological and hydrological conditions, flow patterns, and operational modes have been reviewed. In addition, the influence range of CPGEs on the ground and related application sectors have been summarized. Some key conclusions are drawn as follows:

- High inner pipe thermal resistances and low outer pipe thermal resistance values are preferable. Heat transfer efficiency and flow resistance levels were investigated.
- CPGEs of varied geometries were researched.
- Organic aqueous solutions, such as glycol water solution, diathermic oil, and solutions with added nanoparticles, such as CuO and $Al_2O_3$ nanoparticles, were used as circulating fluids by researchers.

- CPGE heat removal and release performance under Mode A and Mode B (as shown in Figure 3) were compared. Mode A is commonly selected for heat removal, and Mode B is used for heat release.
- Various operational modes, i.e., continuous heat removal, intermittent heat removal, and alternative heat removal with heat injection, were reviewed.
- Applications of CPGEs in indoor heating and electricity power generation sectors were studied.

Future studies may mainly refer to new pipe materials and structures, thermoeconomics and thermodynamic evaluation of CPGEs, simple and fast calculation models, unsteady-state working conditions, on-site experimental studies, annual and long-term operational characteristics monitoring studies (including ground endothermic, exothermic, and recovery progresses), CPGE matrices, and new types of heat carriers. These works will contribute to the technical improvement and wider application of CPGEs.

**Author Contributions:** Conceptualization, N.R. and X.L.; resources, N.H. and Z.Z.; writing—original draft preparation, G.W.; writing—review and editing, N.R., Y.Z. and Y.W. All authors have read and agreed to the published version of the manuscript.

**Funding:** This research was funded by the NATIONAL KEY RESEARCH AND DEVELOPMENT PROGRAM OF CHINA, grant number 2023YFC3807700; the UNIVERSITY NATURAL SCIENCE RESEARCH PROJECT OF ANHUI PROVINCE, grant number 2022AH030034; the UNIVERSITY NATURAL SCIENCE RESEARCH PROJECT OF ANHUI PROVINCE, grant number 2022AH010018; and the UNIVERSITY NATURAL SCIENCE RESEARCH PROJECT OF ANHUI PROVINCE, grant number KJ2021JD09.

**Conflicts of Interest:** The authors declare no conflicts of interest.

## Nomenclature

| Variable | | Subscript | |
|---|---|---|---|
| c | specific heat capacity, kJ/(kg·K) | i | inner suface |
| $D_h$ | the hydraulic diameter of the cross-section of the annular space | in | inlet |
| | | o | outer surface |
| e | eccentricity, % | out | outlet |
| $E_h$ | heat storage efficiency, % | s | heat storage |
| h | specific enthalpy, kJ/kg or convective heat transfer, $W/(m^2 \cdot K)$ | w | water |
| | | 1 | inner pipe |
| | | 2 | outer pipe |
| H | length of the CPGE or depth of the borehole, m | Abbreviation | |
| | | BHE | borehole heat exchangers |
| k | thermal conductivity, W/(m·K) | CCA | concrete core activation |
| $L_d$ | center distance of the inner and outer pipe, (m) | COP | coefficient of performance |
| m | mass flow rate, kg/s | CPGE | coaxial pipe geothermal exchanger |
| r | radius, m | | |
| R | thermal resistance, (m·k)/W | DTRT | distributed thermal response test |
| $R_b$ | three-dimensional borehole resistance, (m·k)/W | HP | heating panel |
| | | ORC | organic Rankine cycle |
| $R_e$ | Renold Number | TPT | thermal exchange performance test |
| Q | heat, W | | |
| ΔT | temperature difference, K or °C | TRT | hermal response test |

## References


1. Zhang, X.; Lu, Z.; He, M.; Wang, J. What can Beijing learn from the world megacities on energy and environmental issues? *Energy Rep.* **2022**, *8*, 414–424. [CrossRef]
2. Achkari, O.; El Fadar, A. Latest developments on TES and CSP technologies—Energy and environmental issues, applications and research trends. *Appl. Therm. Eng.* **2020**, *167*, 114806. [CrossRef]
3. Amir Raza, M.; Karim, A.; Aman, M.M.; Ahmad Al-Khasawneh, M.; Faheem, M. Global progress towards the Coal: Tracking coal reserves, coal prices, electricity from coal, carbon emissions and coal phase-out. *Gondwana Res.* **2025**, *139*, 43–72. [CrossRef]
4. Hürpekli, M.; Necati Özsezen, A. Determination of combustion and emission characteristics of liquid Fischer-Tropsch diesel fuel synthesized from coal in a diesel engine. *Energy Conv. Manag.* **2023**, *292*, 117351. [CrossRef]
5. Paraschiv, S.; Paraschiv, L.S. Trends of carbon dioxide ($CO_2$) emissions from fossil fuels combustion (coal, gas and oil) in the EU member states from 1960 to 2018. *Energy Rep.* **2020**, *6*, 237–242. [CrossRef]
6. Avor, E.P.; Supap, T.; Narku-Tetteh, J.; Muchan, P.; Natewong, P.; Appiah, F.A.; Idem, R. Achieving net-zero $CO_2$ emissions from indirect co-combustion of biomass and natural gas with carbon capture using a novel amine blend. *Int. J. Green. Gas Con.* **2023**, *130*, 104005. [CrossRef]
7. Manzini, J.; Hoshika, Y.; Sicard, P.; Anav, A.; De Marco, A.; Sorrentino, B.; Trentanovi, G.; Moura, B.B.; Ferrini, F.; Azzini, L.; et al. Life Cycle Analysis and species-specific net $CO_2$ assimilation model to assess when a new urban forest becomes a carbon sink in a Mediterranean city. *Sci. Total Environ.* **2025**, *959*, 178267. [CrossRef]
8. Zhen, Y.; Zheng, H.; Xiao, Q.; Zhang, C.; Wang, C. Estimation method for karst carbon sinks on the basis of a concentration prediction model. *J. Environ. Manag.* **2025**, *373*, 123845. [CrossRef]
9. Lu, Z.; Liu, Y.; Zhang, J.; Li, Y.; Guo, X.; Li, X. Carbon-oriented energy system planning using forest carbon sink. *Energy* **2024**, *309*, 133010. [CrossRef]
10. Liu, Q.; Wang, S.; Ma, R.; Huang, F.; Li, J.; Ye, S.; Guo, Y. Comparative analysis of forest soil carbon sink and source based on bibliometrics: Development, hotspots, and trends. *J. Clean. Prod.* **2024**, *480*, 144106. [CrossRef]
11. Zayani, I.; Ammari, M.; Ben Allal, L.; Bouhafa, K. Agroforestry olive orchards for soil organic carbon storage: Case of Saiss, Morocco. *Heliyon* **2023**, *9*, e22910. [CrossRef] [PubMed]
12. Tian, X.; Kang, J.-N.; Dai, M.; Li, X.; Ji, Y.-Z.; Liu, L.-C.; Wei, Y.-M. Unleashing tomorrow's potential: A comprehensive exploration of risks in carbon capture and storage. *Renew. Sustain. Energy Rev* **2025**, *210*, 115174. [CrossRef]
13. Rong, N.; Wang, S.; Chu, C.; Guo, Z.; Liu, K.; Han, L.; Ge, L.; Shi, X.; Wang, G.; Wang, Y. Deactivated Ca-based sorbent derived from calcium looping $CO_2$ capture as a partial substitute for cement to obtain low-carbon cementitious building materials. *Constr. Build. Mater.* **2024**, *454*, 139175. [CrossRef]
14. Rong, N.; Wang, J.; Han, L.; Wu, Y.; Mu, Z.; Wan, X.; Wang, G. Effect of steam addition during calcination on $CO_2$ capture performance and strength of bio-templated Ca-based pellets. *J. $CO_2$ Util.* **2022**, *63*, 102127. [CrossRef]
15. Li, J.; Li, M.; Dang, C.; Wang, Q.; Dong, L.; Liu, X. Energy and environment analysis of R1234yf/R245fa cascade air source heat pump system with double ejectors. *Energy Conv. Manag.* **2025**, *325*, 119404. [CrossRef]
16. Hu, Y.; Shen, B. Development and field demonstration of residential air source integrated heat pump using a three-stage compressor. *Energy Build.* **2025**, *328*, 115202. [CrossRef]
17. Hielscher, S.; Wittmayer, J.M.; Progscha, S.; Wientjes, A.; Sharp, H. Speeding-up wind energy developments: Exploring notions of acceleration and justice in regions within Germany and the Netherlands. *Energy Res. Soc. Sci.* **2025**, *119*, 103909. [CrossRef]
18. Brouwer, B.; van Bergem, R.; Renes, S.; Kamp, L.M.; Hoppe, T. Does local ownership matter? A comparative analysis of fourteen wind energy projects in the Netherlands. *Energy Res. Soc. Sci.* **2025**, *120*, 103891. [CrossRef]
19. Muwanga, R.; Namugenyi, I.; Wabukala, B.M.; Tibesigwa, W.; Katutsi, P.V. Examining social-cultural norms affecting the adoption of solar energy technologies at the household level. *Clean. Energy Syst.* **2024**, *9*, 100164. [CrossRef]
20. Khademi, M.M.; Koohshoori, M.S.; Kasaeian, A. Development of a Renewable Energy System Utilizing Solar Dish Collector, Multi Effect Desalination and Supercritical $CO_2$ Brayton Cycle to Produce Fresh Water and Electricity. *Energy* **2024**, *314*, 134285. [CrossRef]
21. Bashir, M.F.; Ma, B.; Sharif, A.; Ao, T.; Koca, K. Nuclear energy consumption, energy access and energy poverty: Policy implications for the COP27 and environmental sustainability. *Technol. Soc.* **2023**, *75*, 102385. [CrossRef]
22. Vurim, A.; Mukhamedova, N.; Baklanova, Y.; Syssaletin, A.; Akayev, A. Information and analytical system as a promising database used to justify the safety of nuclear energy. *Nucl. Eng. Des.* **2023**, *415*, 112704. [CrossRef]
23. Xu, T.; Haas, K.A.; Gunawan, B. Estimating annual energy production from short tidal current records. *Renew. Energy* **2023**, *207*, 105–115. [CrossRef]
24. Xu, S.; Yu, F.; Zhang, X.; Wei, D.; Diao, Y.; Li, G.; Huang, H. Investigation of temporal and spatial distribution of tidal energy in Liuheng waterway via coastal acoustic tomography. *Renew. Energy* **2025**, *240*, 122180. [CrossRef]
25. Anya, B.; Mohammadpourfard, M.; Akkurt, G.G.; Mohammadi-Ivatloo, B. Exploring geothermal energy based systems: Review from basics to smart systems. *Renew. Sustain. Energy Rev.* **2025**, *210*, 115185. [CrossRef]

26. Shahsavar, A.; Goodarzi, A.; Mohammed, H.I.; Shirneshan, A.; Talebizadehsardari, P. Thermal performance evaluation of non-uniform fin array in a finned double-pipe latent heat storage system. *Energy* **2020**, *193*, 116800. [CrossRef]
27. Templeton, J.D.; Ghoreishi-Madiseh, S.A.; Hassani, F.; Al-Khawaja, M.J. Abandoned petroleum wells as sustainable sources of geothermal energy. *Energy* **2014**, *70*, 366–373. [CrossRef]
28. Kerme, E.D.; Fung, A.S. Heat transfer simulation, analysis and performance study of single U-tube borehole heat exchanger. *Renew. Energy* **2020**, *145*, 1430–1448. [CrossRef]
29. Faizal, M.; Bouazza, A.; Singh, R.M. Heat transfer enhancement of geothermal energy piles. *Renew. Sustain. Energy Rev.* **2016**, *57*, 16–33. [CrossRef]
30. Beier, R.A.; Acuña, J.; Mogensen, P.; Palm, B. Transient heat transfer in a coaxial borehole heat exchanger. *Geothermics* **2014**, *51*, 470–482. [CrossRef]
31. Wang, Z.; Wang, F.; Liu, J.; Ma, Z.; Han, E.; Song, M. Field test and numerical investigation on the heat transfer characteristics and optimal design of the heat exchangers of a deep borehole ground source heat pump system. *Energy Convers. Manag.* **2017**, *153*, 603–615. [CrossRef]
32. Ikeda, S.; Choi, W.; Ooka, R. Optimization method for multiple heat source operation including ground source heat pump considering dynamic variation in ground temperature. *Appl. Energy* **2017**, *193*, 466–478. [CrossRef]
33. Chen, X.; Rahaman, M.A.; Murshed, M.; Mahmood, H.; Hossain, M.A. Causality analysis of the impacts of petroleum use, economic growth, and technological innovation on carbon emissions in Bangladesh. *Energy* **2023**, *267*, 126565. [CrossRef]
34. Radioti, G.; Cerfontaine, B.; Charlier, R.; Nguyen, F. Experimental and numerical investigation of a long-duration Thermal Response Test: Borehole Heat Exchanger behaviour and thermal plume in the heterogeneous rock mass. *Geothermics* **2018**, *71*, 245–258. [CrossRef]
35. Alexander, K.; Phillip, S.; Dirk, M. Development of a Long-Term Operational Optimization Model for a Building Energy System Supplied by a Geothermal Field. *J. Therm. Sci.* **2022**, *31*, 1293–1301.
36. Anand, J.; Liu, X.; Anees, F.; Li, Y.; Eckman, B.; Malhotra, M. Assessing the impacts of air-sealing on the sizing, operation, and economic feasibility of ground-source heat pumps for electrifying single-family houses in the US. *J. Build. Eng.* **2024**, *98*, 111149. [CrossRef]
37. Abdel-Salam, M.R.H.; Zaidi, A.; Cable, M. Field study of heating performance of three ground-source heat pumps in Canadian single-family houses. *Energ. Build.* **2021**, *247*, 110959. [CrossRef]
38. Kumar, S.; Murugesan, K. Experimental investigation of thermal performance of ground source heat pump system for summer and monsoon seasons of Himalayan region of India: A case study. *Renew. Energy* **2024**, *237*, 121842. [CrossRef]
39. Weeratunge, H.; Aditya, G.R.; Dunstall, S.; de Hoog, J.; Narsilio, G.; Halgamuge, S. Feasibility and performance analysis of hybrid ground source heat pump systems in fourteen cities. *Energy* **2021**, *234*, 121254. [CrossRef]
40. Benti, N.E.; Woldegiyorgis, T.A.; Geffe, C.A.; Gurmesa, G.S.; Chaka, M.D.; Mekonnen, Y.S. Overview of geothermal resources utilization in Ethiopia: Potentials, opportunities, and challenges. *Sci. Afr.* **2023**, *19*, e01562. [CrossRef]
41. Venomhata, H.D.V.; Oketch, P.O.; Gathitu, B.B.; Chisale, P. Working fluid selection for the geothermal-solar hybrid cycle at Olkaria II power plant in Kenya. *Heliyon* **2023**, *9*, e12762. [CrossRef] [PubMed]
42. Agemar, T.; Weber, J.; Schulz, R. Deep Geothermal Energy Production in Germany. *Energies* **2014**, *7*, 4397–4416. [CrossRef]
43. Cai, W.; Wang, F.; Liu, J.; Wang, Z.; Ma, Z. Experimental and numerical investigation of heat transfer performance and sustainability of deep borehole heat exchangers coupled with ground source heat pump systems. *Appl. Therm. Eng.* **2019**, *149*, 975–986. [CrossRef]
44. Li, C.; Guan, Y.; Yang, R.; Lu, X.; Xiong, W.; Long, A. Effect of inner pipe type on the heat transfer performance of deep-buried coaxial double-pipe heat exchangers. *Renew. Energy* **2020**, *145*, 1049–1060. [CrossRef]
45. Deng, J.; Wei, Q.; Liang, M.; He, S.; Zhang, H. Field test on energy performance of medium-depth geothermal heat pump systems (MD-GHPs). *Energy Build.* **2019**, *184*, 289–299. [CrossRef]
46. Brettschneider, A.L.; Perković, L. Theoretical analysis of using multiple borehole heat exchangers for production of heating and cooling energy in shallow geothermal reservoirs with underground water flow. *Appl. Therm. Eng.* **2024**, *254*, 123914. [CrossRef]
47. Hassan Al-Kbodi, B.; Rajeh, T.; Li, Y.; Zhao, J.; Zhao, T.; Zayed, M.E. Heat extraction analyses and energy consumption characteristics of novel designs of geothermal borehole heat exchangers with elliptic and oval double U-tube structures. *Appl. Therm. Eng.* **2023**, *235*, 121418. [CrossRef]
48. Huang, S.; Li, J.; Bai, Z.; Dong, J. Assessment of the effect of heat storage on the production of clean geothermal energy using the medium and deep U-type borehole heat exchanger system. *J. Clean. Prod.* **2024**, *447*, 141471. [CrossRef]
49. Biglarian, H.; Abbaspour, M.; Saidi, M.H. Evaluation of a transient borehole heat exchanger model in dynamic simulation of a ground source heat pump system. *Energy* **2018**, *147*, 81–93. [CrossRef]
50. Song, X.; Shi, Y.; Li, G.; Yang, R.; Xu, Z.; Zheng, R.; Wang, G.; Lyu, Z. Heat extraction performance simulation for various configurations of a downhole heat exchanger geothermal system. *Energy* **2017**, *141*, 1489–1503. [CrossRef]

51. Al-Khoury, R.; Focaccia, S. A spectral model for transient heat flow in a double U-tube geothermal heat pump system. *Renew. Energy* **2016**, *85*, 195–205. [CrossRef]
52. Pärisch, P.; Mercker, O.; Oberdorfer, P.; Bertram, E.; Tepe, R.; Rockendorf, G. Short-term experiments with borehole heat exchangers and model validation in TRNSYS. *Renew. Energy* **2015**, *74*, 471–477. [CrossRef]
53. Shah, S.K.; Aye, L.; Rismanchi, B. Validations of a double U-tube borehole model and a seasonal solar thermal energy storage system model. *Renew. Energy* **2022**, *201*, 462–485. [CrossRef]
54. Shi, Y.; Cui, Q.; Song, X.; Xu, F.; Song, G. Study on thermal performances of a horizontal ground heat exchanger geothermal system with different configurations and arrangements. *Renew. Energy* **2022**, *193*, 448–463. [CrossRef]
55. Hasan, N.; Ali, M.H.; Pratik, N.A.; Lubaba, N.; Miyara, A. Improving the thermal performance of vertical ground heat exchanger by modifying spiral tube geometry: A numerical study. *Heliyon* **2024**, *10*, e35718. [CrossRef]
56. Agbossou, A.; Souyri, B.; Stutz, B. Modelling of helical coil heat exchangers for heat pump applications: Analysis of operating modes and distance between heat exchangers. *Appl. Therm. Eng.* **2018**, *129*, 1068–1078. [CrossRef]
57. Choi, H.K.; Yoo, G.J.; Pak, J.H.; Lee, C.H. Numerical study on heat transfer characteristics in branch tube type ground heat exchanger. *Renew. Energy* **2018**, *115*, 585–599. [CrossRef]
58. Wang, G.; Song, X.; Shi, Y.; Zheng, R.; Li, J.; Li, Z. Production performance of a novel open loop geothermal system in a horizontal well. *Energy Convers. Manag.* **2020**, *206*, 112478. [CrossRef]
59. Jia, G.; Wang, J.; Ma, C.; Tao, Z.; Zhang, Z.; Ma, Z.; Jin, L. Influence of eccentricity on the thermal performance of pipe-in-pipe heat exchanger utilized for geothermal heating, an experimental study. *Appl. Therm. Eng.* **2024**, *252*, 123723. [CrossRef]
60. Jalaluddin; Miyara, A. Thermal performance investigation of several types of vertical ground heat exchangers with different operation mode. *Appl. Therm. Eng.* **2012**, *33–34*, 167–174. [CrossRef]
61. Fu, Q.; Ding, J.; Lao, J.; Xie, W.; Wang, W.; Lu, J. Numerical Simulation of Heat Transfer Performance between Molten Salt and Supercritical $CO_2$ in Double-pipe Heat Exchanger. In Proceedings of the 10th International Conference on Applied Energy, Hong Kong, China, 22–25 August 2018; pp. 5741–5746.
62. Gordon, D.; Bolisetti, T.; Ting, D.S.K.; Reitsma, S. Experimental and analytical investigation on pipe sizes for a coaxial borehole heat exchanger. *Renew. Energy* **2018**, *115*, 946–953. [CrossRef]
63. Cui, Y.; Zhu, J.; Twaha, S.; Riffat, S. A comprehensive review on 2D and 3D models of vertical ground heat exchangers. *Renew. Sustain. Energy Rev.* **2018**, *94*, 84–114. [CrossRef]
64. Bu, X.; Ran, Y.; Zhang, D. Experimental and simulation studies of geothermal single well for building heating. *Renew. Energy* **2019**, *143*, 1902–1909. [CrossRef]
65. Gordon, D.; Bolisetti, T.; Ting, D.S.K.; Reitsma, S. Short-term fluid temperature variations in either a coaxial or U-tube borehole heat exchanger. *Geothermics* **2017**, *67*, 29–39. [CrossRef]
66. Brown, C.S.; Kolo, I.; Banks, D.; Falcone, G. Comparison of the thermal and hydraulic performance of single U-tube, double U-tube and coaxial medium-to-deep borehole heat exchangers. *Geothermics* **2024**, *117*, 102888. [CrossRef]
67. Bär, K.; Rühaak, W.; Welsch, B.; Schulte, D.; Homuth, S.; Sass, I. Seasonal High Temperature Heat Storage with Medium Deep Borehole Heat Exchangers. *Energy Procedia* **2015**, *76*, 351–360. [CrossRef]
68. Fang, L.; Diao, N.; Shao, Z.; Wang, Z.; Fang, Z. Study on Thermal Resistance of Coaxial Tube Boreholes in Ground-Coupled Heat Pump Systems. *Procedia. Eng.* **2017**, *205*, 3735–3742. [CrossRef]
69. Beier, R.A.; Acuña, J.; Mogensen, P.; Palm, B. Borehole resistance and vertical temperature profiles in coaxial borehole heat exchangers. *Appl. Energy* **2013**, *102*, 665–675. [CrossRef]
70. Cengel, Y.A.; Ghajar, A.J. *Heat and Mass Transfer: Fundamentals and Applications*, 6th ed.; McGraw-Hill Education: New York, NY, USA, 2020.
71. Jingjing, W.; Juntao, D.; Jianguo, Z.; Yongtang, Y.; Jianmin, Z.; Xin, H. A segmented analytical solution heat transfer model of U-tube ground heat exchanger based on finite solid cylindrical heat source method. *Geothermics* **2025**, *125*, 103170. [CrossRef]
72. Luo, Y.; Yan, T.; Yu, J. Integrated analytical modeling of transient heat transfer inside and outside U-tube ground heat exchanger: A new angle from composite-medium method. *Int. commun. Heat. Mass. Transf.* **2020**, *162*, 120373. [CrossRef]
73. Soltani, M.; Farzanehkhameneh, P.; Moradi Kashkooli, F.; Al-Haq, A.; Nathwani, J. Optimization and energy assessment of geothermal heat exchangers for different circulating fluids. *Energy Conv. Manag.* **2021**, *228*, 113733. [CrossRef]
74. Kerme, E.D.; Fung, A.S. Comprehensive simulation based thermal performance comparison between single and double U-tube borehole heat exchanger and sensitivity analysis. *Energ. Build.* **2021**, *241*, 110876. [CrossRef]
75. Salilih, E.M.; Abu-Hamdeh, N.H.; Oztop, H.F. Analysis of double U-tube ground heat exchanger for renewable energy applications with two-region simulation model by combining analytical and numerical techniques. *Int. Commun. Heat Mass.* **2021**, *123*, 105144. [CrossRef]
76. Raymond, J.; Mercier, S.; Nguyen, L. Designing coaxial ground heat exchangers with a thermally enhanced outer pipe. *Geotherm. Energy* **2015**, *3*, 7. [CrossRef]

77. Nian, Y.-L.; Cheng, W.-L. Analytical g-function for vertical geothermal boreholes with effect of borehole heat capacity. *Appl. Therm. Eng.* **2018**, *140*, 733–744. [CrossRef]
78. Ma, Z.D.; Zhang, Y.P.; Saw, L.H.; Cui, X.; Jia, G.S.; Jin, L.W. Investigation on local geothermal energy attenuation after long-term operation of ground heat exchanger with considering aquifer effect. *Geothermics* **2023**, *107*, 102608. [CrossRef]
79. Lee, S.; Park, S.; Kang, M.; Oh, K.; Choi, H. Effect of tube-in-tube configuration on thermal performance of coaxial-type ground heat exchanger. *Renew. Energy* **2022**, *197*, 518–527. [CrossRef]
80. Dong, S.; Yu, Y.; Li, B.; Ni, L. Field test and geologic-thermal-economic analysis of medium-depth borehole heat exchanger. *J. Clean. Prod.* **2024**, *447*, 141381. [CrossRef]
81. Oh, K.; Lee, S.; Park, S.; Han, S.-I.; Choi, H. Field experiment on heat exchange performance of various coaxial-type ground heat exchangers considering construction conditions. *Renew. Energy* **2019**, *144*, 84–96. [CrossRef]
82. Śliwa, T.; Kruszewski, M.; Zare, A.; Assadi, M.; Sapińska-Śliwa, A. Potential application of vacuum insulated tubing for deep borehole heat exchangers. *Geothermics* **2018**, *75*, 58–67. [CrossRef]
83. Chen, H.; Liu, H.; Yang, F.; Tan, H.; Wang, B. Field measurements and numerical investigation on heat transfer characteristics and long-term performance of deep borehole heat exchangers. *Renew. Energy* **2023**, *205*, 1125–1136. [CrossRef]
84. Sapinska-Sliwa, A.; Rosen, M.A.; Gonet, A.; Sliwa, T. Deep Borehole Heat Exchangers—A Conceptual and Comparative Review. *Int. J. Air-Cond.* **2016**, *24*, 1630001. [CrossRef]
85. Luo, Y.; Guo, H.; Meggers, F.; Zhang, L. Deep coaxial borehole heat exchanger: Analytical modeling and thermal analysis. *Energy* **2019**, *185*, 1298–1313. [CrossRef]
86. Song, X.; Wang, G.; Shi, Y.; Li, R.; Xu, Z.; Zheng, R.; Wang, Y.; Li, J. Numerical analysis of heat extraction performance of a deep coaxial borehole heat exchanger geothermal system. *Energy* **2018**, *164*, 1298–1310. [CrossRef]
87. Du, D.; Li, Y.; Wang, K.; Zhao, Y.; Hu, Z.; Zhang, W.; Wang, Q. Experimental and numerical simulation research on heat transfer performance of coaxial casing heat exchanger in 3500m-deep geothermal well in Weihe Basin. *Geothermics* **2023**, *109*, 102658. [CrossRef]
88. Harris, B.E.; Lightstone, M.F.; Reitsma, S.; Cotton, J.S. Analysis of the transient performance of coaxial and u-tube borehole heat exchangers. *Geothermics* **2022**, *101*, 102319. [CrossRef]
89. Wood, C.J.; Liu, H.B.; Riffat, S.B. Comparative performance of 'U tube' and 'coaxial' loop designs for use with a ground source heat pump. *Appl. Therm. Eng.* **2012**, *37*, 190–195. [CrossRef]
90. Acuña, J.; Palm, B. Distributed thermal response tests on pipe-in-pipe borehole heat exchangers. *Appl. Energy* **2013**, *109*, 312–320. [CrossRef]
91. Iry, S.; Rafee, R. Transient numerical simulation of the coaxial borehole heat exchanger with the different diameters ratio. *Geothermics* **2019**, *77*, 158–165. [CrossRef]
92. Pan, A.; Lu, L.; Cui, P.; Jia, L. A new analytical heat transfer model for deep borehole heat exchangers with coaxial tubes. *Int. commun. Heat. Mass. Transf.* **2019**, *141*, 1056–1065. [CrossRef]
93. Yekoladio, P.J.; Bello-Ochende, T.; Meyer, J.P. Design and optimization of a downhole coaxial heat exchanger for an enhanced geothermal system (EGS). *Renew. Energy* **2013**, *55*, 128–137. [CrossRef]
94. Morchio, S.; Fossa, M. Thermal modeling of deep borehole heat exchangers for geothermal applications in densely populated urban areas. *Therm. Sci. Eng. Prog.* **2019**, *13*, 100363. [CrossRef]
95. Mokhtari, H.; Hadiannasab, H.; Mostafavi, M.; Ahmadibeni, A.; Shahriari, B. Determination of optimum geothermal Rankine cycle parameters utilizing coaxial heat exchanger. *Energy* **2016**, *102*, 260–275. [CrossRef]
96. Daneshipour, M.; Rafee, R. Nanofluids as the circuit fluids of the geothermal borehole heat exchangers. *Int. Commun. Heat. Mass. Transf.* **2017**, *81*, 34–41. [CrossRef]
97. Zarrella, A.; Scarpa, M.; Carli, M.D. Short-time step performances of coaxial and double U-tube heat exchangers: Modeling and measurements. *HVAC&R Res.* **2011**, *17*, 959–976.
98. Liu, Q.; Zhang, Y.; Zhang, X.; Luo, J.; Zheng, J.; Liu, Y.; Cheng, Y.; Lou, J. Numerical study on the heat transfer performance evaluation, flow characteristics, exergy efficiency, and entropy generation analysis of a novel coaxial geothermal heat exchanger. *J. Build. Eng.* **2024**, *84*, 108555. [CrossRef]
99. Liu, Q.; Zhang, Y.; Luo, J.; Zheng, J.; Cheng, Y. Numerical study on heat transfer and flow characteristics of coaxial geothermal heat exchangers with helical finned inner tubes. *J. Build. Eng.* **2023**, *65*, 105752. [CrossRef]
100. Rajeh, T.; Al-Kbodi, B.H.; Li, Y.; Zhao, J.; Zhang, Y. Modeling and techno-economic comparison of two types of coaxial with double U-tube ground heat exchangers. *Appl. Therm. Eng.* **2023**, *225*, 120221. [CrossRef]
101. Li, Y.; Ma, L.; Xu, W.; Zhu, Q.; Li, W.; Zhao, J.; Zhu, J. Multi-external-chamber coaxial borehole heat exchanger: Dynamic heat transfer and energy consumption analysis. *Energy Convers. Manag.* **2020**, *207*, 112519. [CrossRef]
102. Huchtemann, K.; Müller, D. Combined simulation of a deep ground source heat exchanger and an office building. *Build. Environ.* **2014**, *73*, 97–105. [CrossRef]

103. Mottaghy, D.; Dijkshoorn, L. Implementing an effective finite difference formulation for borehole heat exchangers into a heat and mass transport code. *Renew. Energy* **2012**, *45*, 59–71. [CrossRef]
104. Rajeh, T.; Hassan Al-Kbodi, B.; Li, Y.; Zhao, J.; Zayed, M.E.; Rehman, S. Comparative numerical modeling complemented with multi-objective optimization and dynamic life cycle assessment of coaxial ground heat exchangers with oval-shaped and typical circular-shaped configurations. *Appl. Therm. Eng.* **2024**, *244*, 122673. [CrossRef]
105. Rajeh, T.; Al-Kbodi, B.H.; Zayed, M.E.; Li, Y.; Zhao, J.; Rehman, S. Local entropy generation optimization and thermodynamic irreversibility analysis of oval-shaped coaxial ground heat exchangers: A detailed numerical investigation. *Int. Commun. Heat. Mass. Transf.* **2024**, *228*, 125650. [CrossRef]
106. Kurevija, T.; Strpić, K. Hydraulic and thermogeological design differences between two-loop vertical and inclined coaxial borehole heat exchangers. *Renew. Energy* **2018**, *117*, 314–323. [CrossRef]
107. Dai, C.; Li, J.; Shi, Y.; Zeng, L.; Lei, H. An experiment on heat extraction from a deep geothermal well using a downhole coaxial open loop design. *Appl. Energy* **2019**, *252*, 113447. [CrossRef]
108. Cheng, W.-L.; Liu, J.; Nian, Y.-L.; Wang, C.-L. Enhancing geothermal power generation from abandoned oil wells with thermal reservoirs. *Energy* **2016**, *109*, 537–545. [CrossRef]
109. Yin, H.; Song, C.; Ma, L.; Gao, L.; Yang, X.; Li, W.; Zhao, J. Analysis of flow and thermal breakthrough in leaky downhole coaxial open loop geothermal system. *Appl. Therm. Eng.* **2021**, *194*, 117098. [CrossRef]
110. Mohammadzadeh Bina, S.; Fujii, H.; Tsuya, S.; Kosukegawa, H.; Naganawa, S.; Harada, R. Evaluation of utilizing horizontal directional drilling technology for ground source heat pumps. *Geothermics* **2020**, *85*, 101769. [CrossRef]
111. Akram, M.W.; Chen, Q.; Nortz, G.; Nortz, P. Experimental investigation and numerical modeling of an innovative horizontal coaxial ground heat exchanger (HCGHE) for geothermal heat pump applications. *Appl. Therm. Eng.* **2024**, *257*, 124492. [CrossRef]
112. Alimonti, C.; Soldo, E. Study of geothermal power generation from a very deep oil well with a wellbore heat exchanger. *Renew. Energy* **2016**, *86*, 292–301. [CrossRef]
113. ASHRAE. *2017 Handbook—Fundamentals*, SI ed.; ASHRAE: Atlanta, GA, USA, 2017.
114. Olson, J.M. Nanofluids and a Method of Making Nano Fluids for Ground Source Heat Pumps and Other Applications. U.S. Patent 8,580,138, 12 November 2013.
115. Liu, Q.; Tao, Y.; Shi, L.; Zhou, T.; Huang, Y.; Peng, Y.; Wang, Y.; Tu, J. Experimental investigation on the use of CuO/water nanofluid in horizontal spiral-coil ground heat exchanger. *Int. J. Refrig.* **2023**, *149*, 204–223. [CrossRef]
116. Togun, H.; Abdulrazzaq, T.; Kazi, S.N.; Badarudin, A.; Kadhum, A.A.H.; Sadeghinezhad, E. A review of studies on forced, natural and mixed heat transfer to fluid and nanofluid flow in an annular passage. *Renew. Sustain. Energy Rev.* **2014**, *39*, 835–856. [CrossRef]
117. Pisarevsky, M.I.; Struchalin, P.G.; Balakin, B.V.; Kutsenko, K.V.; Maslov, Y.A. Experimental study of nanofluid heat transfer for geothermal applications. *Renew. Energy* **2024**, *221*, 119631. [CrossRef]
118. Zhao, G.; Wang, L.; Liang, Z.; Liu, Q.; Jiang, F. Thermal response analysis of a medium-deep coaxial borehole heat exchanger by circulating $CO_2$. *Geothermics* **2023**, *112*, 102746. [CrossRef]
119. Xie, K.; Nian, Y.-L.; Cheng, W.-L. Analysis and optimization of underground thermal energy storage using depleted oil wells. *Energy* **2018**, *163*, 1006–1016. [CrossRef]
120. Nian, Y.-L.; Cheng, W.-L.; Yang, X.-Y.; Xie, K. Simulation of a novel deep ground source heat pump system using abandoned oil wells with coaxial BHE. *Int. commun. Heat. Mass. Transf.* **2019**, *137*, 400–412. [CrossRef]
121. Li, M.; Shi, Y.; Chen, H.; Liu, C.; Li, H. Study on the heat transfer performance of coaxial casing heat exchanger for medium and deep geothermal energy in cold regions. *Renew. Energy* **2024**, *237*, 121666. [CrossRef]
122. Niu, Q.; Ma, K.; Wang, W.; Pan, J.; Wang, Q.; Du, Z.; Wang, Z.; Yuan, W.; Zheng, Y.; Shangguan, S.; et al. Multifactor analysis of heat extraction performance of coaxial heat exchanger applied to hot dry rock resources exploration: A case study in matouying uplift, Tangshan, China. *Energy* **2023**, *282*, 128277. [CrossRef]
123. Saadi, M.S.; Gomri, R. Investigation of dynamic heat transfer process through coaxial heat exchangers in the ground. *Int. J. Hydrogen. Energy* **2017**, *42*, 18014–18027. [CrossRef]
124. Liu, Z.; Liang, J.; He, Z.; Liu, X.; Liu, H.; Shao, Z. A developed fatigue analysis approach for composite wind turbine blade adhesive joints using finite-element submodeling technique. *Eng. Fail. Anal.* **2024**, *164*, 108701. [CrossRef]
125. Liu, Z.; Liang, J.; He, Z.; Liu, X.; Liu, H.; Shao, Z. Finite element submodeling technique-based fatigue analysis and reliability modeling of wind turbine blade trailing edge. *Compo. Struct.* **2025**, *352*, 118699. [CrossRef]
126. Liu, Z.; Liu, X.; Zhu, S.-P.; Zhu, P.; Liu, W.; Correia, J.A.F.O.; De Jesus, A.M.P. Reliability assessment of measurement accuracy for FBG sensors used in structural tests of the wind turbine blades based on strain transfer laws. *Eng. Fail. Anal.* **2020**, *112*, 104506. [CrossRef]
127. Sliwa, T.; Rosen, M.A. Efficiency analysis of borehole heat exchangers as grout varies via thermal response test simulations. *Geothermics* **2017**, *69*, 132–138. [CrossRef]

128. Song, X.; Zheng, R.; Li, G.; Shi, Y.; Wang, G.; Li, J. Heat extraction performance of a downhole coaxial heat exchanger geothermal system by considering fluid flow in the reservoir. *Geothermics* **2018**, *76*, 190–200. [CrossRef]
129. Casasso, A.; Sethi, R. Sensitivity Analysis on the Performance of a Ground Source Heat Pump Equipped with a Double U-pipe Borehole Heat Exchanger. *Energy Procedia* **2014**, *59*, 301–308. [CrossRef]
130. Zarrella, A.; Emmi, G.; Graci, S.; Carli, M.D.; Cultrera, M.; Santa, G.D.; Galgaro, A.; Bertermann, D.; Müller, J.; Pockelé, L.; et al. Thermal Response Testing Results of Different Types of Borehole Heat Exchangers: An Analysis and Comparison of Interpretation Methods. *Energies* **2017**, *10*, 801. [CrossRef]
131. Fang, L.; Diao, N.; Shao, Z.; Zhu, K.; Fang, Z. A computationally efficient numerical model for heat transfer simulation of deep borehole heat exchangers. *Energ. Build.* **2018**, *167*, 79–88. [CrossRef]
132. Le Lous, M.; Larroque, F.; Dupuy, A.; Moignard, A. Thermal performance of a deep borehole heat exchanger: Insights from a synthetic coupled heat and flow model. *Geothermics* **2015**, *57*, 157–172. [CrossRef]
133. Holmberg, H.; Acuña, J.; Næss, E.; Sønju, O.K. Thermal evaluation of coaxial deep borehole heat exchangers. *Renew. Energy* **2016**, *97*, 65–76. [CrossRef]
134. Dai, C.; Shi, Y.; Zeng, L.; Li, J.; Lei, H. Heat Extraction Performance of a Deep Downhole Heat Exchanger. *Energy Procedia* **2019**, *158*, 5602–5607. [CrossRef]
135. Caulk, R.A.; Tomac, I. Reuse of abandoned oil and gas wells for geothermal energy production. *Renew. Energy* **2017**, *112*, 388–397. [CrossRef]
136. Guo, L.; Zhang, J.; Li, Y.; McLennan, J.; Zhang, Y.; Jiang, H. Experimental and numerical investigation of the influence of groundwater flow on the borehole heat exchanger performance: A case study from Tangshan, China. *Energ. Build.* **2021**, *248*, 111199. [CrossRef]
137. Choi, W.; Kikumoto, H.; Ooka, R. Critical comparison between thermal performance test (TPT) and thermal response test (TRT): Differences in heat transfer process and extractable information. *Energy Conv. Manag.* **2019**, *199*, 111967. [CrossRef]
138. Choi, W.; Kikumoto, H.; Ooka, R. Two thermal performance test (TPT) datasets of a single U-tube borehole heat exchanger with inlet setpoint temperatures of 30 °C and 40 °C. *Data Brief* **2018**, *20*, 1769–1774. [CrossRef]
139. Liu, Y.D.; Beier, R.A. Required duration for borehole test validated by field data. *ASHRAE Trans.* **2009**, *115*, 782–792.
140. Li, C.; Jiang, C.; Guan, Y.; Chen, H.; Yang, R.; Wan, R.; Shen, L. Comparison of the experimental and numerical results of coaxial-type and U-type deep-buried pipes' heat transfer performances. *Renew. Energy* **2023**, *210*, 95–106. [CrossRef]
141. Amer, R.; Xue, Z.; Hashimoto, T.; Nagata, T. Distributed fiber optic temperature and strain sensing in cementing and water injection: Insights to well integrity monitoring and multisensing optical fiber cable design. *Gas Sci. Eng.* **2024**, *130*, 205430. [CrossRef]
142. Huang, Y.; Zhang, Y.; Xie, Y.; Zhang, Y.; Gao, X. Thermal performance analysis on the composition attributes of deep coaxial borehole heat exchanger for building heating. *Energ. Build.* **2020**, *221*, 110019. [CrossRef]
143. Huang, Y.; Zhang, Y.; Xie, Y.; Zhang, Y.; Gao, X.; Ma, J. Field test and numerical investigation on deep coaxial borehole heat exchanger based on distributed optical fiber temperature sensor. *Energy* **2020**, *210*, 118643. [CrossRef]
144. Huang, Y.; Zhang, Y.; Xie, Y.; Zhang, Y.; Gao, X.; Ma, J. Long-term thermal performance analysis of deep coaxial borehole heat exchanger based on field test. *J. Clean. Prod.* **2021**, *278*, 123396. [CrossRef]
145. Pokhrel, S.; Sasmito, A.P.; Sainoki, A.; Tosha, T.; Tanaka, T.; Nagai, C.; Ghoreishi-Madiseh, S.A. Field-scale experimental and numerical analysis of a downhole coaxial heat exchanger for geothermal energy production. *Renew. Energy* **2022**, *182*, 521–535. [CrossRef]
146. Bu, X.; Ma, W.; Li, H. Geothermal energy production utilizing abandoned oil and gas wells. *Renew. Energy* **2012**, *41*, 80–85. [CrossRef]
147. Wang, H.; Xu, Y.; Sun, Y.; Zhao, S. Heat extraction by deep coaxial borehole heat exchanger for clean space heating near Beijing, China: Field test, model comparison and operation pattern evaluation. *Renew. Energy* **2022**, *199*, 803–815. [CrossRef]
148. Luo, F.; Ma, C.; Liu, J.; Yang, L.; Zhou, W. Effect of gas–liquid phase change of axial rotating heat pipe on fluid-thermal-solid behaviors of high-speed spindle. *Appl. Therm. Eng.* **2023**, *232*, 121117. [CrossRef]
149. Zhou, W.; Ma, C.; Yang, L.; Luo, F.; Liu, J. Regulation of thermo-fluid-solid coupling characteristics in high-speed spindle-bearing system for boring machine tool based on sintered-core heat pipes. *Int. commun. Heat. Mass. Transf.* **2024**, *157*, 107717. [CrossRef]
150. Zeng, S.; Ma, C.; Liu, J.; Li, M.; Gui, H. Sequence-to-sequence based LSTM network modeling and its application in thermal error control framework. *Appl. Soft Comput.* **2023**, *138*, 110221. [CrossRef]
151. Li, M.; Zeng, S.; Hu, J.; Ma, C. Free-mounted cooling plate multi-objective topology optimization method towards precision machine tool heat dissipation: An experimental and numerical study. *Int. J. Heat Mass. Tran.* **2023**, *214*, 124394. [CrossRef]
152. Zeng, S.; Liu, J.; Ma, C. Topology optimization in cooling moving heat sources for enhanced precision of machine tool feed drive systems. *Int. J. Therm. Sci.* **2024**, *202*, 109065. [CrossRef]
153. Li, M.; Ma, C.; Liu, J. Topology optimization design of cooling water jacket structure for highspeed spindle-bearing system. *J. Manuf. Proc.* **2023**, *102*, 1–22. [CrossRef]

154. Li, M.; Ma, C.; Liu, J.; Gui, H.; Zeng, S.; Luo, F. Thermal error prediction of precision boring machine tools based on extreme gradient boosting algorithm-improved sailed fish optimizer-bi-directional ordered neurons-long short-term memory neural network model and physical-edge-cloud system. *Eng. Appl. Artif. Intell.* **2024**, *127*, 107278. [CrossRef]
155. Serageldin, A.A.; Sakata, Y.; Katsura, T.; Nagano, K. Performance enhancement of borehole ground source heat pump using single U-tube heat exchanger with a novel oval cross-section (SUO) and a novel spacer. *Sustain. Energy Technol. Assess.* **2020**, *42*, 100805. [CrossRef]
156. Kimiaei, S.; Kazemi-Ranjbar, S.; Jalali, A.; Ahmadi, P. A novel three-dimensional numerical model to simulate heat transfer inside a double U-tube borehole with two independent circuits. *Int. J. Heat Mass. Tran.* **2022**, *184*, 122243. [CrossRef]
157. Tarrad, A.H. An Analytical Model for the Thermal Assessment of a Vertical Double U-Tube Ground-Coupled Heat Pump System in Steady-State Conditions. *Fluid Dyn. Mater. Proc.* **2022**, *18*, 1111–1127. [CrossRef]
158. Quirosa, G.; Torres, M.; Becerra, J.A.; Jiménez-Espadafor, F.J.; Chacartegui, R. Energy analysis of an ultra-low temperature district heating and cooling system with coaxial borehole heat exchangers. *Energy* **2023**, *278*, 127885. [CrossRef]
159. Zhou, W.; Li, R.; Chen, Y.; Zhu, S. Numerical Simulation of Mid-Deep Buried Casing Heat Exchanger and its Heating System Application. *J. Therm. Sci.* **2023**, *32*, 1445–1454. [CrossRef]
160. Cetin, A.; Kadioglu, Y.K.; Paksoy, H. Underground thermal heat storage and ground source heat pump activities in Turkey. *Sol. Energy* **2020**, *200*, 22–28. [CrossRef]
161. Shah, S.K.; Aye, L.; Rismanchi, B. Seasonal thermal energy storage system for cold climate zones: A review of recent developments. *Renew. Sustain. Energy Rev.* **2018**, *97*, 38–49. [CrossRef]

**Disclaimer/Publisher's Note:** The statements, opinions and data contained in all publications are solely those of the individual author(s) and contributor(s) and not of MDPI and/or the editor(s). MDPI and/or the editor(s) disclaim responsibility for any injury to people or property resulting from any ideas, methods, instructions or products referred to in the content.

*buildings* MDPI

*Article*

# A Multi-Objective Optimization Method for Enhancing Outdoor Environmental Quality in University Courtyards in Hot Arid Climates

**Amr Sayed Hassan Abdallah [1,*], Randa Mohamed Ahmed Mahmoud [2], Ayman Ragab [3] and Mohammed M. Gomaa [3,4]**

[1] Department of Architectural Engineering, Imam Mohammad Ibn Saud Islamic University (IMSIU), Riyadh 11432, Saudi Arabia
[2] Department of Architecture, Faculty of Engineering, Assiut University, Assiut 71516, Egypt; randa.m@aun.edu.eg
[3] Department of Architectural Engineering, Faculty of Engineering, Aswan University, Aswan 81542, Egypt; ayman.ragab@aswu.edu.eg (A.R.); mgomaa@dah.edu.sa (M.M.G.)
[4] Department of Architecture, School of Engineering, Computing& Design, Dar Al-Hekma University, Jeddah 22246, Saudi Arabia
* Correspondence: asabdullah@imamu.edu.sa

**Abstract**

Enhancing urban air quality and thermal comfort involves addressing multifaceted environmental and design challenges. Investigating the effects of urban morphological and building geometrical parameters on enhancing air quality and thermal comfort is a multifaceted problem, influenced by different parameters. This study aims to develop optimized design solutions for university buildings and courtyards to enhance outdoor thermal comfort and reduce $CO_2$ concentration levels as an indicator of air quality. Consequently, the methodology involved a combination of field monitoring at two university faculties in Egypt and a computational parametric methodology using Rhino 3D+Grasshopper(V8) for enhancing thermal comfort, reducing $CO_2$ concentration levels, and improving wind velocity. The in situ measurements revealed significantly high $CO_2$ levels (780 ppm) and wind speed (3.8 m/s). The parametric methodology's findings revealed a substantial reduction in the Universal Thermal Climate Index (UTCI) by 2.04 to 10.3 °C, a decrease in $CO_2$ concentration by 57 to 197 ppm, and an increase in wind speed by 0.4 to 4.07 m/s. The most suitable vegetation ratio for trees within narrow courtyard designs was found to be 30%. This ratio effectively enhances thermal comfort (UTCI) and reduces $CO_2$ concentrations, while also maintaining adequate airflow and avoiding excessive obstruction of natural ventilation within the courtyard. These findings provide valuable guidance for optimizing courtyard designs in hot arid climates.

**Keywords:** parametric urban design; wind velocity; $CO_2$ concentration; UTCI; Rhino+Grasshopper

## 1. Introduction

In recent decades, outdoor air quality has become a significant global issue due to rapid urbanization. Improving outdoor air quality offers numerous benefits, including enhanced occupant health and productivity, reduced environmental harm, and lower energy consumption costs. In hot arid climates, such as those found in Egypt, courtyards have historically played an essential role in architectural design. Their enclosed yet open form allows for better shading, passive cooling, and improved ventilation, making them highly effective in mitigating harsh outdoor conditions. Courtyards promote air circulation, reduce heat buildup, and create comfortable microclimates that enhance outdoor thermal comfort and air quality. These features make courtyards a practical and necessary design element in educational buildings within such regions [1,2]. In Egyptian universities, integrating courtyards into building designs is crucial due to their effectiveness in enhancing wind velocity and air temperature, as well as the positive impact of courtyard orientation on airflow rates [1–5]. Several studies have investigated the role of courtyard configurations (trees, shading, water fountains, etc.) in improving the outdoor thermal comfort and air quality in universities' courtyards, thereby improving student health and performance [5–11]. Hence, previous studies have indicated that the inner courtyard is the most suitable design for university buildings in hot, dry regions. Additionally, courtyards can lower $CO_2$ concentrations in adjacent indoor spaces, thereby contributing to a healthier and more comfortable environment for occupants [12,13]. Accordingly, investigating the synergistic effects of urban morphology and building geometry (e.g., courtyard design, vegetation, building height) collectively on the outdoor air quality in universities' courtyards constitutes a significant knowledge gap. Consequently, the relevant literature is categorized into three main areas: (1) studies on improving outdoor thermal comfort and air quality, (2) studies on enhancing indoor thermal comfort and air quality, and (3) vegetation strategies for optimizing indoor and outdoor thermal performance and air quality.

### 1.1. *Background and Literature*

University buildings require a high level of outdoor and indoor environmental quality [14]. The environmental quality depends on several factors: air quality, thermal comfort, acoustic comfort, and visual comfort. The three main areas of the literature review are detailed below.

#### 1.1.1. Role of Geometry Parameters in Enhancing Outdoor Environmental Quality

Firstly, there is a set of studies that have illustrated strategies for improving outdoor thermal comfort in the urban spaces of educational buildings. Abdallah et al. [15] studied the thermal performance of courtyards in university buildings at Sohag University during the hot period. They found that courtyards with a height-to-width (H/W) ratio of 1.2 provided better thermal comfort and performance compared to those with a ratio of 0.7. Also, Eid et al. [16] conducted simulations of eight urban canyon configurations with varying aspect ratios to assess their impact on outdoor thermal comfort and $CO_2$ concentrations. The study revealed that these configurations led to a reduction in $CO_2$ concentration ranging from 0.34 to 12.33 ppm. Additionally, Li et al. [17] have studied the thermal comfort and air quality in five frontal area densities according to the $CO_2$ concentration and physiologically equivalent temperature (PET). It was found that when the frontal area density was increased, the PET value fell below 38 °C, and the $CO_2$ concentration was lower than 30,000 $\mu g/m^3$. Moreover, Su et al. [18] developed a multi-objective methodology using Rhino and Grasshopper to assess the impact of dynamic urban variables, such as building orientation, height, spacing width, and layout, on the life-cycle carbon emissions of outdoor spaces between buildings. Their results indicated a reduction in carbon emissions by 15%. Similarly, Bedra et al. [19] introduced a parametric-simulation framework that explores the impact of three urban form variables—street aspect ratio, building density, and street orientation—on UTCI decreasing. The results indicate that each 10% increase in building density lowers the UTCI by about 1.02 °C. Finally, Abdallah et al. [20] have developed a multi-objective urban optimization framework to optimize the outdoor thermal comfort in outdoor recreational areas. The results indicated that the hybrid solutions between shadings and trees (with diameters between 10 and 15 m) could reduce UTCI by 11.6 °C. Several studies have illustrated strategies for enhancing indoor thermal comfort, air quality, y, and $CO_2$ concentration inside the classrooms of educational buildings, such as [21–23]. Also, van der Walt et al. [21] have conducted field measurements of air temperature and $CO_2$ concentration in three types of classrooms—brick, container, and prefabricated—in Stellenbosch, South Africa. It was found that the brick classroom has the best ventilation and thermal performance in contrast with the container classroom. In conclusion, while previous studies have effectively demonstrated the influence of outdoor improvement strategies to enhance outdoor thermal comfort and air quality, their application across diverse urban and building parameters in university buildings remains insufficiently explored and needs deep investigation.

Secondly, there is a set of studies that have illustrated strategies for improving air quality and wind effects in educational buildings. For example, Wu et al. [24] investigated the impact of varying air temperature and wind speed on air quality, and the findings showed that the most influential variables on air quality in suburban and urban areas were wind speed and air temperature, respectively. Also, Mahyuddin and Essah [22] have used Computational Fluid Dynamics (CFD) in investigating the internal and external airflow in two different classrooms. The results indicated that the external wind direction, speed, and pressure played the main role in improving internal ventilation, as well as the upward convergence between $CO_2$ concentration inside the classrooms and thermal performance. Further, Maiques et al. [25] have investigated the impact of four different ventilation strategies in four classrooms inside two building orientations. The results indicated that natural ventilation assisted in improving indoor thermal comfort and air quality by 5% and 2%, respectively. Li et al. [26] have integrated Green Design Studio with Rhino+Grasshopper (V8) to establish a multi-objective modular platform to analyze building and urban parameters to improve indoor environment quality, indoor energy efficiency, and outdoor air quality. In addition, Moghadam et al. [27] have assessed the impact of a novel integration ventilation system (transpired solar collectors and mechanical ventilation heat recovery) on indoor wind flow rate, air temperature, and $CO_2$ concentration in a university classroom in Ireland. The results proved the efficiency of the novel ventilation system in reducing air temperature and $CO_2$ concentration, in addition to reducing heating energy demand by 20%. Eventually, Yao et al. [28] proposed a multi-objective optimization and low-carbon framework to improve indoor air quality, thermal comfort, and daylight illuminance based on changing five design parameters: layout dimension, building orientation, building envelope, WWR, and openable window area ratio. The results revealed that the $CO_2$ concentration decreased by 14%. In conclusion, while numerous studies have examined the impact of indoor strategies on enhancing thermal comfort and air quality, there is a notable gap in research focusing on the effects of outdoor strategies on the indoor environment. This area warrants further investigation to better understand how outdoor environmental factors influence indoor conditions.

#### 1.1.2. Role of Vegetation in Enhancing Outdoor Environmental Quality

Thirdly, a set of previous studies have explained vegetation strategies for optimizing indoor and outdoor thermal performance and air quality. For instance, Gebreyesus et al. [29] used the i-Tree Eco model to investigate the impact of green infrastructure on enhancing air quality. The results indicated that the annual removal reached 274.2 t of pollutants, a percentage of 37.4%. Furthermore, Abdeen and Rafaat [30] conducted a study at the Canadian University in Egypt, utilizing DesignBuilder (V.5.0.3.007) software to assess the impact of vertical green walls on $CO_2$ emissions, heat gain, and energy consumption. Their findings indicated that the $CO_2$ emissions in the inner courtyard decreased by 13%

to 28.43%. Additionally, Gustafsson et al. [31] have integrated two specific models, called PALM and VIDA, to study the impact of different tree types on reducing $NO_2$ and particulate matter PM concentration. It was found that the tree species of sparse crowns with sparse tree arrangements achieved the highest results in providing natural ventilation and reducing $NO_2$ concentration and PM levels. Also, i-Tree Eco was used to assess the impact of different tree diversity and spatial distribution on reducing PM10 to improve air quality in Geneva, Switzerland [32]. It was found that the tree density was playing the main role in reducing PM10 from 45 to 19%. Kaveh et al. [33] have investigated the influence of building height and diversity of greenery positions as a novel ventilation pattern on improving air quality (PM2.5, PM10, $NO_2$) by the simulation process. It was found that stepped buildings with green walls caused the highest air temperature of 18.0 °C and high air quality, but two rows of deciduous trees caused the lowest air temperature of 17.5 °C and low air quality. An assessment of the impact of green space percentage on enhancing air quality was conducted by [34]. The results indicated that a vegetation percentage of 27% has a significant influence in reducing air pollutants (PM2.5, PM10), and that impact would increase by increasing the green space percentage. In conclusion, while existing studies have separately examined the roles of vegetation and green infrastructure in reducing air temperature and air pollution, there is a notable gap in research investigating how varying vegetation ratios simultaneously influence reduction in UTCI and $CO_2$ concentrations, besides increasing wind velocity.

### *1.2. Research Objectives*

Recently, the improvement process for air quality has become a dilemma that needs creative solutions from urban and building designers. The ratio of air pollution is rising daily because of rapid urbanization, the population is increasing, and there is an increase in polluted resources such as industrial areas [24,30,35]. While recent studies emphasize innovative strategies to mitigate air pollution in urban public spaces [15,25,27], there is limited research that focuses on outdoor spaces of educational buildings in hot arid climates [21,30], which represents a significant research gap. Furthermore, research on improving the outdoor spaces of educational buildings to mitigate air pollution in hot arid climates remains limited [21,30]. Therefore, the problem is the complexity of modeling and evaluating a diverse number of urban morphological and building geometrical parameters together under varying meteorological conditions and investigating their effect on improving air quality in the outer spaces of educational buildings [18,26,28]. Hence, the main aim of this study is to develop optimized design solutions for university buildings and courtyards to enhance outdoor thermal comfort and reduce $CO_2$ concentration levels as an indicator of air quality. Consequently, the methodology involved a combination of field monitoring at two university faculties in Egypt and a computational parametric methodology using Rhino 3D+Grasshopper(V8), which contributes to the following aims:

1. Investigating the dynamic influence of a set of urban morphological and building geometrical parameters on optimizing air quality in university buildings.
2. Improving configurations for university buildings with outdoor courtyards to enhance the students' experience in the outdoor spaces.
3. Improving the students' progress and scientific performance, besides boosting social activities.

Nevertheless, the key novelty lies in evaluating the combined influence of urban morphology and building geometry on enhancing air quality in university buildings within hot arid climates. Ultimately, this study equips designers with evidence-based guidelines to optimize university buildings in hot arid climates, particularly in Upper Egypt, by identifying key parameters (e.g., vegetation density, building height, courtyard openings) that collectively enhance thermal comfort, air quality, and wind velocity.

## 2. Research Methodology

### 2.1. Case Study and Weather Climate Description

The case study selected for monitoring and calibration is two educational buildings in Sohag University, Sohag, Egypt. The location coordinates of Sohag are 26.56° N and 31.69° E. The city is located in the southern part of Upper Egypt, according to the Egyptian Code of Improving Energy Use [36]. In winter, outdoor temperatures were between 7.3 °C and 22 °C. In the summer, the average was 39.6 °C during the day and 23.1 °C at night. Relative humidity was 30% in May and 57% in January [37]. Winter wind direction is north and northwest, and northwest/north in summer [38]. Due to a clear sky all year long, Sohag has strong, direct solar radiation [38]. Over an area of 750 acres, the New Sohag University is located in New Sohag City. The educational buildings selected contained passive strategies. They were selected during the post-occupancy stage. Others were excluded as they were not yet in full operational capacity, and did not have passive features. As a result, the Faculty of Agriculture and the Faculty of Education were selected. Figure 1 shows the Faculty of Education, near the main entrance of New Sohag University. A 4711 $m^2$ ground floor and four upper floors make up the building. It has two courtyards totaling 1531 $m^2$. Some openings overlook the courtyards. Figure 1 also shows the Faculty of Agriculture, near the main entrance of the university. Five upper floors and a ground floor make up the building, with a total area of 7492 $m^2$. The building has three courtyards, totaling an area of 3147 $m^2$. Some openings in classes, offices, and labs overlook the courtyards.

**Figure 1.** Satellite image and outside view of the two faculties in New Sohag University.

During the hot period (March–May 2018), $CO_2$ and wind speed were monitored. This period was selected due to its position at the beginning of the hot period and before the end of the second semester. Measurements were taken over six days per month, from 9 a.m. to 3 p.m., due to the schedule of the sessions and lectures and the presence of students. This study is part of detailed evaluations of the two buildings in past research for a one-year evaluation and monitoring [15]. The two buildings were selected because their design and included passive strategies (cross ventilation, inner courtyards, and vegetation therein). Several parameters were measured with a variety of datalogger devices depicted in Table 1. Calibration was carried out by the author in the Environmental Laboratory at Assiut University prior to the commencement of measurements, using reference devices to ensure accuracy. Measurement devices were installed in both classrooms and courtyards at heights of 1.7 m (to represent standing occupants) and 1.1 m (for seated occupants), in accordance with ASHRAE Standard [39]. All devices were mounted on tripods to achieve the required heights and maintain stability. To minimize the influence of human presence—particularly for wind-speed sensors—devices were positioned away from students, ensuring that occupancy conditions did not affect the measurements. Monitoring was conducted in several locations in three courtyards with an aspect ratio (H/W) of 0.7 in the Faculty of Agriculture and 1.2 in the Faculty of Education. The measurements were conducted to evaluate the current situation and develop a parametric method based on measurement results to achieve indoor air quality. Secondly, when investigating the impact of dynamic variable parameters on enhancing the air quality in university buildings, a parametric method should be relied on, using the simulation Rhinoceros software Version 8, the Grasshopper plug-in, and the Ladybug plug-in. Calibration of the simulation software was conducted based on real measurements of $CO_2$ and wind speed in the Faculty of Agriculture.

**Table 1.** Devices used for measurements.

| Parameters | Instruments | Accuracy | Range | Measurement Interval |
|---|---|---|---|---|
| Wind speed in courtyards | EA3000 Standard Handheld Anemometer | ±5%, +0.1 m/s | 0.2 m/s to 30 m/s | Every 1 min |
| Concentration of $CO_2$ in spaces and courtyards | TR-76Ui | 1%, ±1 °C | (0 to 45 °C) (10% to 90% RH) | Every 30 s |

*2.2. The Proposed Parametric Methodology*

In light of the above considerations, this paper assumed a parametric methodology based on a multi-objective algorithm to achieve the high fitness of objectives. The proposed parametric methodology consisted of 3 main stages as shown in Figure 2. For implementing the methodology computationally, Rhinoceros + Grasshopper as a parametric platform was chosen due to its flexibility and efficiency in establishing parametric models [26,28,40]. The first stage was generating a prototypical model of the base case and using the weather data file of the location. In this stage, Rhinoceros software Version 8 and the Grasshopper plug-in were used with the Ladybug plug-in for inserting the weather data file. Then, a set of 32 design solutions varying between urban morphological and building geometrical parameters was established using Grasshopper to act as optimization parameters in the next stage. In addition, the three optimization objectives were defined as improving thermal performance, reducing the $CO_2$ concentration, and improving wind velocity using the Ladybug plug-in, Honeybee plug-in, and Butterfly plug-in, respectively. In the second stage, a multi-objective optimization process took place based on the optimization parameters to achieve the optimization objectives using the Wallacei plug-in to obtain the optimal solutions and the Colibri plug-in to collect all the results. Thus, the third stage was the visualization of the results of the optimal solutions, in addition to exporting the results of all design solutions in an Excel sheet by the TT Toolbox plug-in for analyzing the correlation determination. As a result, the workflow of the proposed parametric methodology was established and elaborated in Figure 3 to achieve the desired objectives:

1. Improving the thermal comfort in the courtyard by reducing the UTCI.
2. Improving the air quality in the courtyard by reducing $CO_2$ concentration levels. The concentration of $CO_2$ was selected as a key indicator due to its relevance to ventilation effectiveness and its frequent use as a proxy for air stagnation in urban outdoor environments.
3. Improving the air quality in the courtyard by increasing wind velocity.

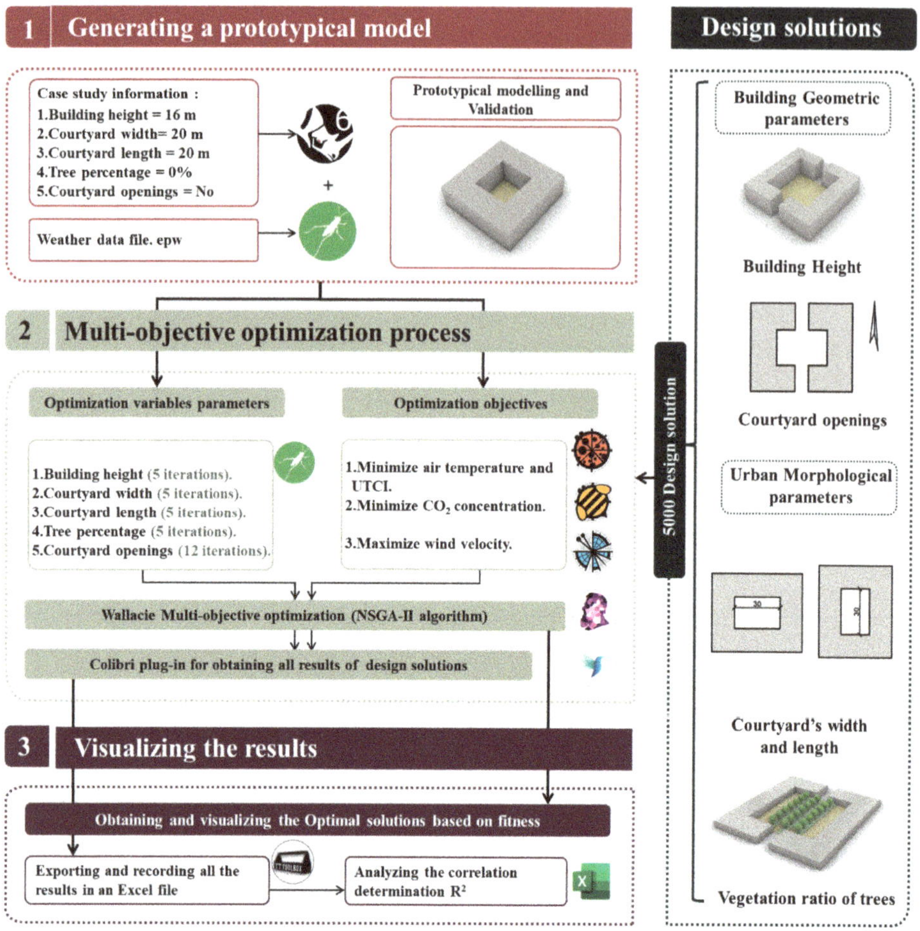


**Figure 2.** The proposed parametric methodology for optimizing air quality in a university building.

### 2.2.1. Generating a Prototypical Model: Stage 1

In the first stage, the previous base cases of the courtyards in the university buildings were utilized for two purposes. The first purpose was to computationally generate the prototypical model of the university's building and its courtyard as shown in Figure 4. Hence, Rhino+Grasshopper(V8) as a parametric platform and the Ladybug plug-in were used to define the location of the model in Sohag City, Egypt, as well as inserting the weather data file. Ladybug is an analysis tool that is used to allow the extraction of EnergyPlus Weather data from a wide range of locations and produce analysis about microclimate conditions in outdoor spaces, such as UTCI and PET thermal comfort indexes. The validation Ladybug plug-in was calculated and found to be 0.83 [20], despite its limitations in the simulation of indoor environments. In addition, the hottest week was determined as the analysis period of the simulation: from 10 July to 16 July, especially in the scheduled hours from 8:00 to 17:00. The second purpose was to validate the accuracy of the prototypical model by comparing the in situ measurements of the wind speed and $CO_2$ concentration level in the Faculty of Agriculture with the simulation results with the same parameters of the real faculty and on the same day. Thus, the Honeybee plug-in was used to determine the $CO_2$ concentration level based on the EnergyPlus engine. Honeybee is a Grasshopper plug-in that provides data regarding $CO_2$ emission, daylight, and energy performance inside spaces. As shown in Figure 5, the Coefficient of Determination ($R^2$) of $CO_2$ concentration was 0.78. In addition, Butterfly, which is a Grasshopper/Dynamo plug-in and object-oriented Python (V7) library that creates and runs computational fluid dynamics (CFD) simulations using OpenFOAM, was used to measure the wind velocity in the courtyard. As shown in Figure 5, the Coefficient of Determination ($R^2$) of wind speed was 0.89. Furthermore, the Root Mean Squared Error (RMSE) values for $CO_2$ concentration and wind speed were 0.986 and 0.18, respectively, indicating a relatively low level of error. This suggests that the model provides accurate predictions, thereby validating the numerical prototypical model.

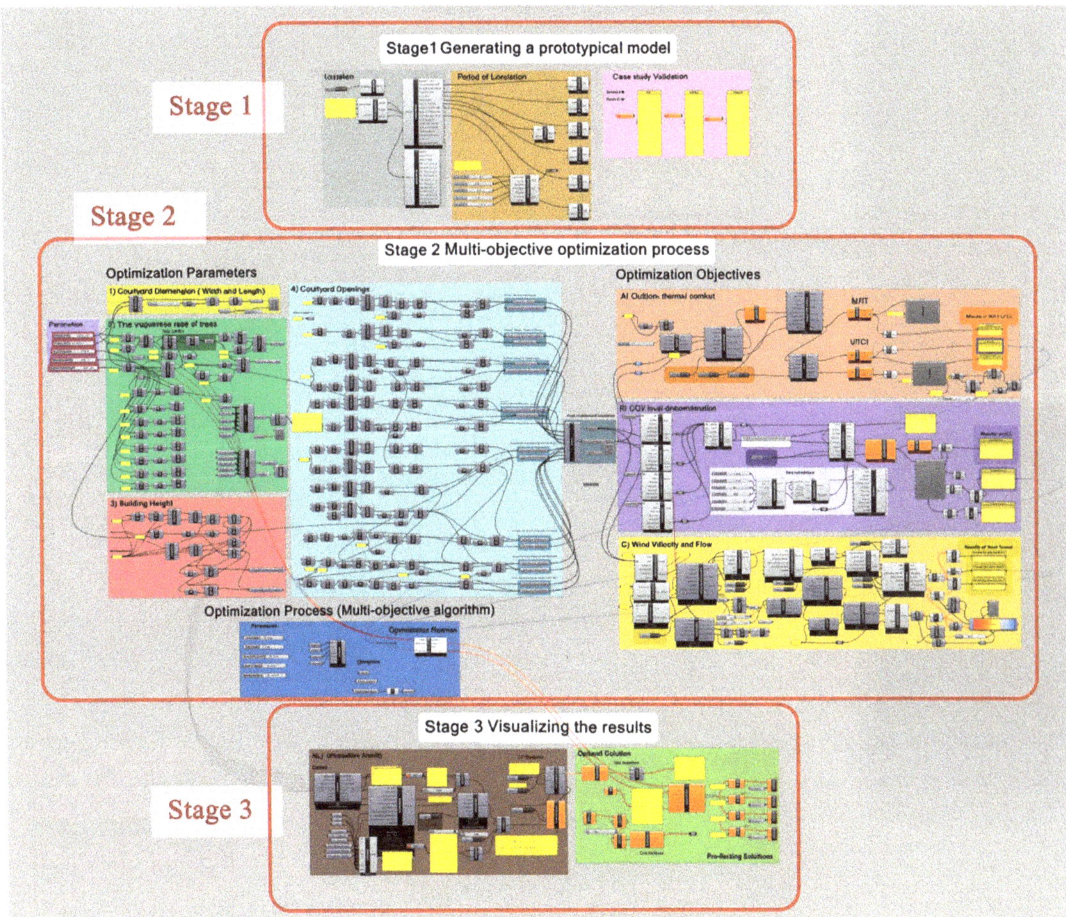


**Figure 3.** The workflow of the parametric methodology for optimizing air quality in the Grasshopper environment.

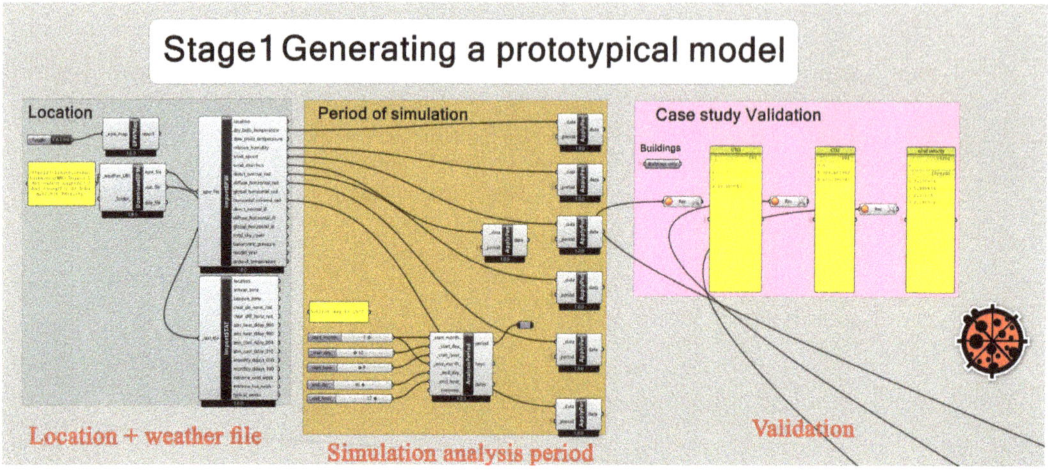


**Figure 4.** Workflow of stage 1: Generating a prototypical model in the Grasshopper environment.

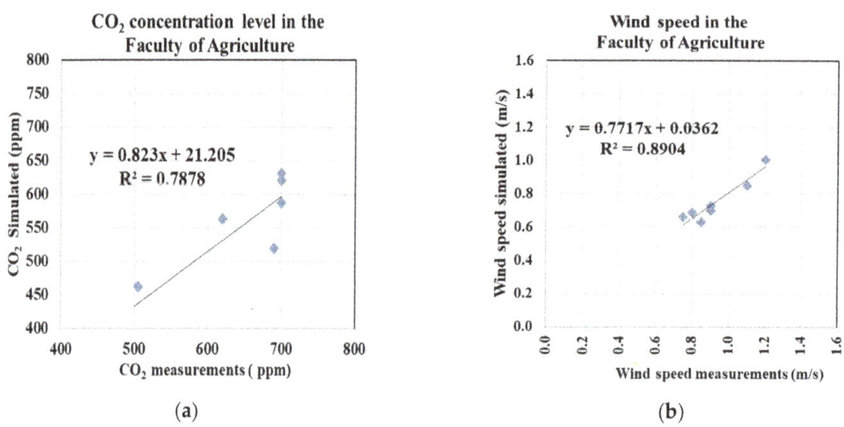


**Figure 5.** The validation of Grasshopper simulation tools compared with field measurements in the Faculty of Agriculture; (**a**) $CO_2$ concentration level and (**b**) wind speed.

### 2.2.2. Multi-Objective Optimization Process: Stage 2

The initial simulation of the Faculties of Agriculture and Education highlighted critical issues concerning air quality and thermal performance within the courtyards. The average UTCI, $CO_2$ concentration level, and wind speed in the courtyard of the Faculty of Agriculture during the hottest week were 42.7 °C, 600 ppm, and 0.9 m/s, respectively. The average UTCI, $CO_2$ concentration level, and wind speed in the courtyard of the Faculty of Education during the hottest week were 44.04 °C, 680 ppm, and 1.5 m/s, respectively. Consequently, the issue was substantiated, and the necessity for optimized design solutions was thoroughly discussed. Hence, a set of design solutions was proposed with diverse building geometric parameters and urban morphological parameters, as illustrated in Figure 6. Firstly, we have the design solutions with different building geometric parameters, which are the following: building height and courtyard opening in the building mass. The building height varied between 4 m and 20 m based on the common height in the university buildings (from 1 to 5 floors). Additionally, eight courtyard opening cases were proposed, each incorporating two opposing or perpendicular openings to harness the cross-ventilation effect, thereby enhancing air quality. Additionally, the building's mass featured an 8 m wide opening, representing 20% to 40% of the inner courtyard-facing façade. Secondly, we proposed design solutions with different urban morphological parameters, which are as follows: the courtyard's width, the courtyard's length, and the vegetation tree ratio. Through observation of the Egyptian universities' courtyards, it was found that the range of the courtyards' widths and lengths was between 20 m and 40 m. Consequently, the proposed courtyard's width and length were from 20 m to 40 m. To improve the air quality in the courtyard, a vegetation area should be implemented inside the courtyard. Thus, a vegetation area with a tree ratio ranging between 10% and 50% was implemented to purify the air and provide shade for the users, in addition to providing spaces for seating and social activities. Certain parameters for tree vegetation ratio were derived from prior studies [20,33,34] which demonstrated their effectiveness in enhancing outdoor thermal comfort and air quality. Meanwhile, building height and courtyard dimensions (expressed as height-to-width, H/W ratios) were adopted from earlier research [15,18], with values adjusted to reflect typical heights in Egyptian university buildings. In contrast, courtyard opening parameters were entirely proposed by the authors.

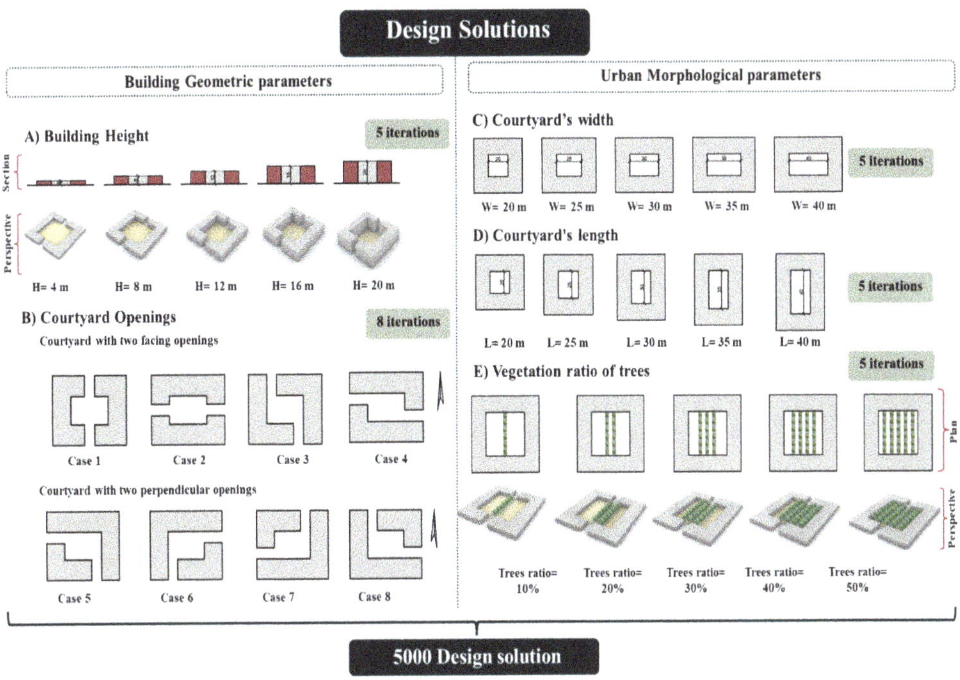


**Figure 6.** The proposed design solutions.

Moreover, the three main objectives were to minimize the UTCI, minimize the $CO_2$ concentration level, and maximize the wind velocity. Hence, the Ladybug plug-in was used to calculate the UTCI, the Honeybee plug-in was used to determine the $CO_2$ concentration level, and the Butterfly plug-in was used to measure the wind velocity (Figure 7). Furthermore, the optimization process relied on the NSGA-II algorithm, which was presented through the Wallacei plug-in as a multi-objective optimization algorithm. Therefore, the Wallacei plug-in sought to optimize the parameters of all the design solutions to achieve the best results across the three objectives. While the simulations conducted using Rhino+Grasshopper (V8) provide valuable results, several factors may introduce uncertainties into the results. For example, the precision of sensors used to collect input data of wind speed or air temperature and the fluctuations in environmental conditions, including changes in wind direction and speed, can affect the consistency of the simulation results over time. However, this approach ensures transparency regarding the limitations of the simulation and underscores the importance of considering various factors that can influence the accuracy of analyses in Rhino and Grasshopper. So, the settings of the optimization process are illustrated in Table 2.

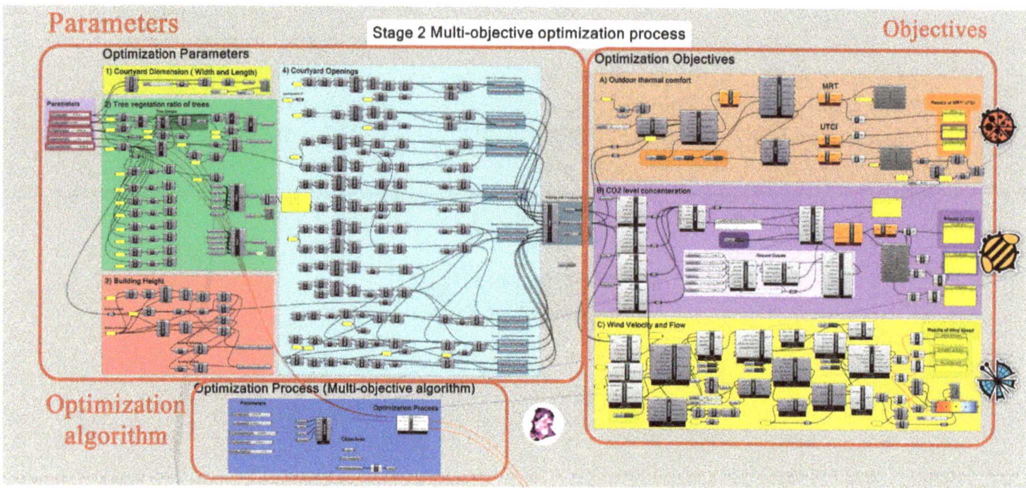


**Figure 7.** Workflow for stage 2 of the multi-objective optimization process in the Grasshopper environment.

**Table 2.** The settings of the optimization process.

| Inputs Data | Value |
|---|---|
| Location | Sohag City, Egypt (26°32′59″ N, 31°42′0.003″ E) |
| Weather file | Sohag.AP SJ EGY 623980 TMYx |
| Simulation period | The hottest week from 10 July to 16 July in the scheduled hours from 9:00 to 17:00 |
| North angle | 0° |
| Simulation grid | 1 m × 1 m |
| Simulation height | Pedestrian level 1.8 m |
| Algorithm | Non-Dominated Sorting Genetic Algorithm II (NSGA-II) |
| Generation number | 50 |
| Population size | 10 |
| Random seed | 1 |

### 2.2.3. Visualization of the Results: Stage 3

To improve the thermal comfort and air quality inside the courtyard of university buildings, three measurable objectives were optimized: decreasing the average UTCI, decreasing the average $CO_2$ concentration level, and increasing the average wind velocity. In the third stage, the results of the optimization process were collected and visualized. Meanwhile, the Wallacei plug-in was used to obtain the optimal solutions based on the fitness value of the three objectives, as shown in Figure 8. The Colibri plug-in collected all the results of the 5000 design solutions and exported them to an Excel file using the TT Toolbox plug-in to analyze the coefficient of correlation determination ($R^2$). Accordingly, the results of the parametric methodology will be explained in the next section.

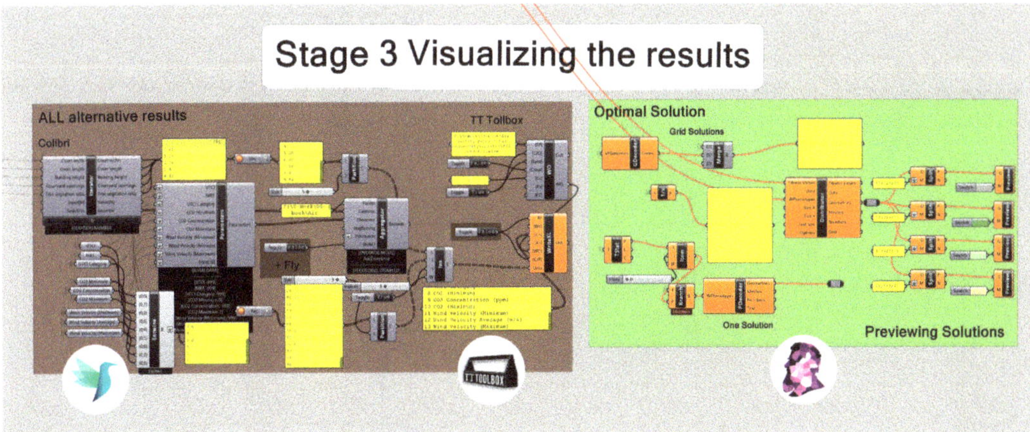


**Figure 8.** Workflow for stage 3 of visualizing the results in the Grasshopper environment.

## 3. Results and Discussion

### *3.1. Monitoring and Evaluation of Wind Speed and $CO_2$ Concentration Levels of the Real Case Study*

The following data is represented for a full understanding of the actual indoor situation and air quality in classrooms overlooking the courtyards. Figure 9 shows the Faculty of Agriculture, with wind speed monitored in the middle courtyard. In this courtyard wind speed is higher than in the other two. This is explained by the effect of the outer street environment, which drives air from high-pressure zones to low-pressure ones. Wind speed increased by an average of 1.1, 2.6, and 1.5 m/s for 0.7 (H/W) aspect ratio in courtyards 1, 2, and 3, respectively. Meanwhile, in the Faculty of Education, courtyards had an aspect ratio (H/W) of 1.2. During the hot period, the highest mean wind speed was 3.8 m/s at point (2), whereas the lowest wind speed recorded was 0.2 m/s at point (3), as shown in Figure 10. The effect of building geometry and street environment on ventilation strategies/courtyard ventilation is discussed by [41,42].

Opening the entrance doors of courtyards in the Faculty of Education for cross ventilation causes the wind speed to be 3.0 m/s. This raises $CO_2$ concentration levels to 780 ppm in the afternoon in different courtyards with an aspect ratio of 1.2. This can be attributed to the high student density in the courtyards, with different activities including resting. In the remainder of the day, $CO_2$ concentration fell to 500 ppm as a result of high wind speed in the two courtyards. The result of opening doors was increased wind speed inside courtyards, $CO_2$ concentration reduction, and better air quality for students in outer environments to practice different activities. This agrees with previous studies by [43,44]. Figure 11 illustrates different $CO_2$ concentrations in spaces overlooking courtyards within both faculties. In a class overlooking "courtyard 2" (H/W = 0.7), $CO_2$ concentration rose to 960 ppm. The reason behind this was the closing windows and students' occupancy. Further, lab activities led to the same result in the laboratory. Further, indoor $CO_2$ concentration increased as doors between courtyards were closed. This caused an increase in outdoor concentration, leading to a corresponding indoor concentration. In the Faculty of Education, in courtyards with a 1.2 aspect ratio, outdoor $CO_2$ concentration exerted a strong effect on indoor $CO_2$ concentration in spaces overlooking the courtyards. The maximum value (1200 ppm) fell within the acceptable range of the ASHRAE standard [39]. The result is reduced fresh air supply to such spaces, thus impacting students' productivity in the classroom, in agreement with previous studies by [44].

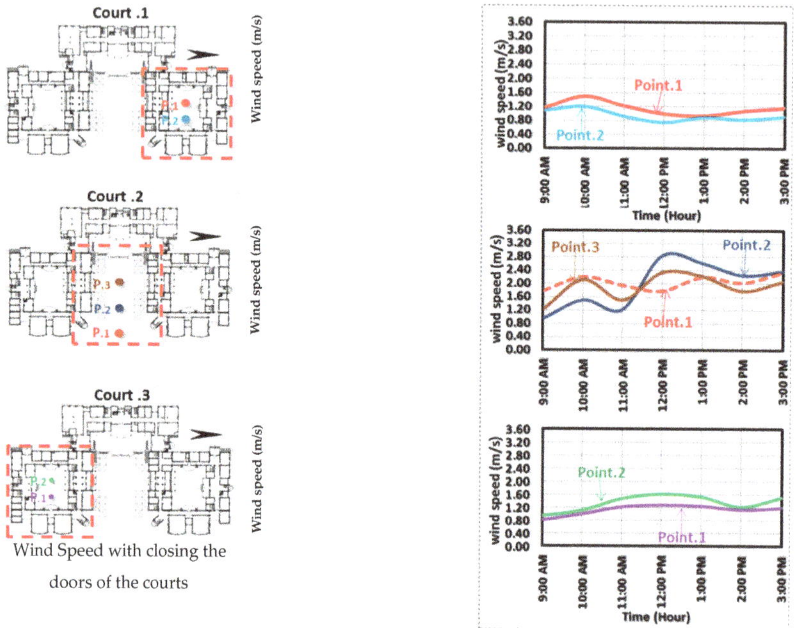


**Figure 9.** Wind speed monitoring in different locations inside the courtyards of the Faculty of Agriculture.

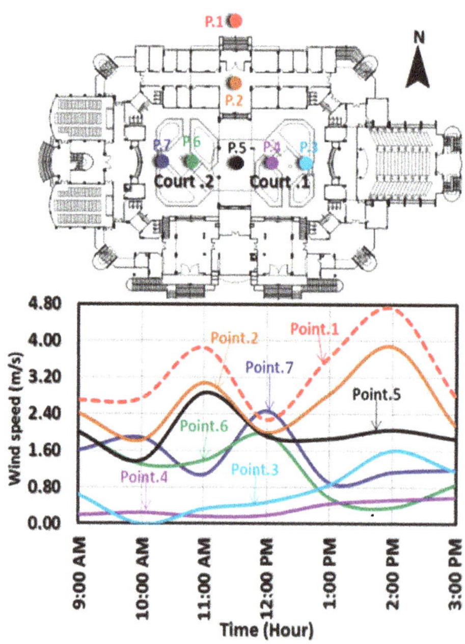


**Figure 10.** Wind speed patterns in different locations inside the courtyards of the Faculty of Education.

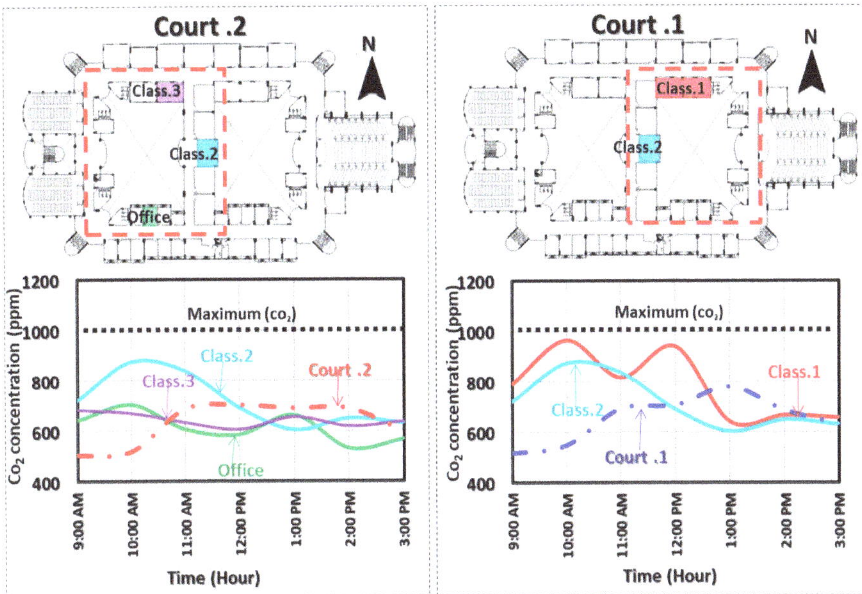


**Figure 11.** $CO_2$ concentration in the spaces overlooking the courtyards in the Faculty of Education building.

Cross ventilation and opening doors between courtyards decrease $CO_2$ concentration inside courtyards, resulting in better air quality therein, leading to a corresponding improvement in spaces overlooking courtyards. The indoor environment of classrooms has to be carefully considered. Comfortable classrooms with good air quality boost learning, and vice versa. Further, urban morphological and building geometrical parameters need to be optimized, particularly inside courtyards, to improve air quality and foster students' experience in the outer spaces of such university buildings.

### *3.2. Analysis of the Simulation Results*

One of the main results of the optimization process by the Wallacei plug-in is the Pareto front of the solutions, as shown in Figure 12. Moreover, the 3D Pareto front could visualize the results of the proposed parametric methodology (the 5000 design solutions) based on the three objectives (Figure 12a). Consequently, the optimal solution can be determined by achieving a balance between the rank of the the objectives at the same time. Also, the top 100 design solutions were presented in the 3D Pareto front and the 2D Pareto front. Therefore, the results of all feasible solutions ranged from 33.6 °C to 40.6 °C for the UTCI, from 482.7 ppm to 542.5 ppm for the $CO_2$ concentration level, and from 1.30 to 5.63 m/s for the wind velocity. The 2D Pareto front indicated that the findings of the top 100 design solutions ranged from 33.6 °C to 36.1 °C for the UTCI, from 483.2 ppm to 492.9 ppm for the $CO_2$ concentration level, and from 2.70 to 3.76 m/s for the wind velocity. Hence, the results of the parametric methodology are very satisfactory, as it would reduce the UTCI and $CO_2$ concentration level, besides increasing the wind velocity, as will be detailed in the following section. The parallel coordinates graph of the feasible solutions is shown in Figure 13. The parallel coordinates present all the combinations of feasible design solutions to achieve the desired objectives and improve the air quality in the university buildings' courtyards. The decrease in UTCI and $CO_2$ concentration level in the last optimized Generation (Gen 49) is noted in contrast to the increase in wind velocity, which is needed for reaching a balance between the three objectives.

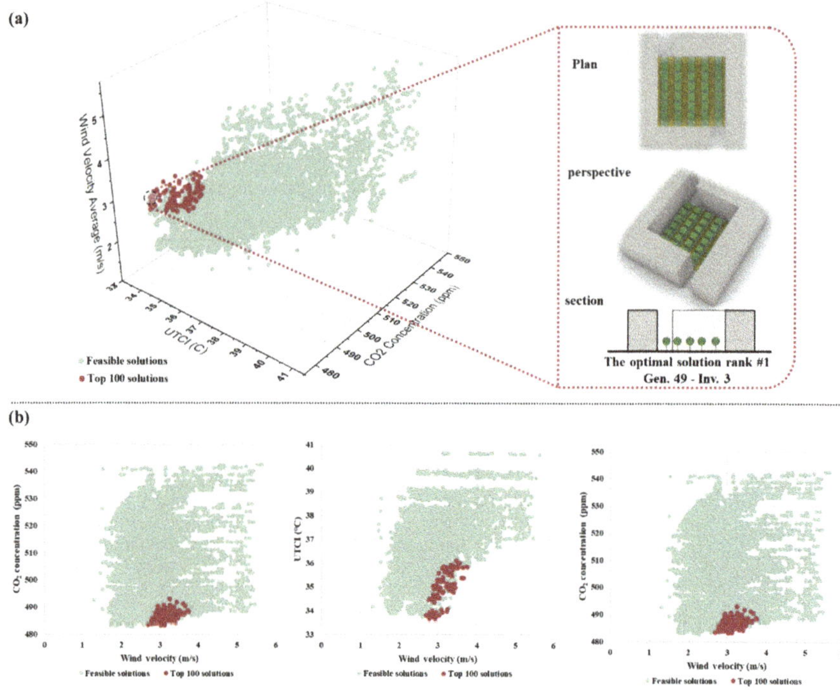


**Figure 12.** Pareto front solutions; (**a**) 3D Pareto, and (**b**) 2D Paretos.

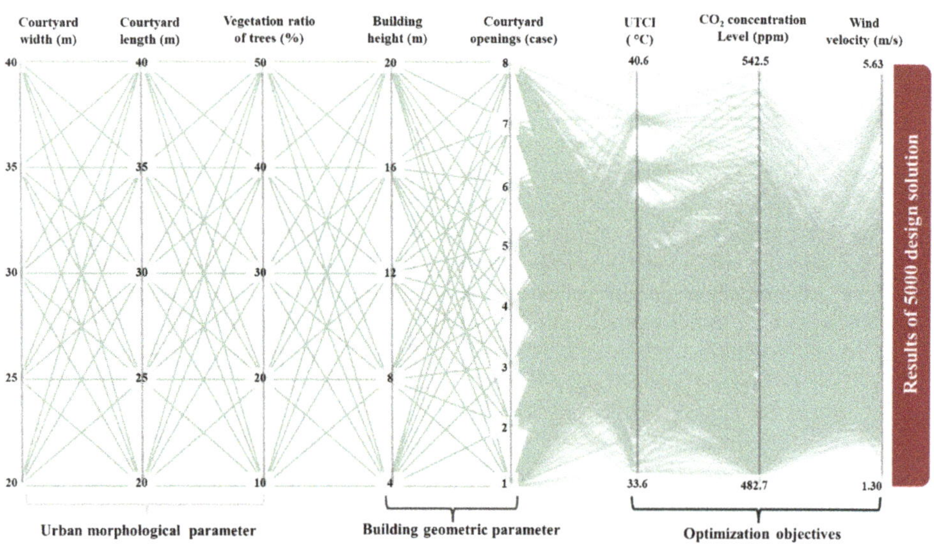


**Figure 13.** Parallel coordinates graph of a feasible solution of the parametric methodology.

The optimal solution for a university building's courtyard, as the outcome of the optimization parametric methodology, is illustrated in Figure 14. The ranking of the design solutions was calculated based on achieving the best fitness values of the three objectives together. Hence, the optimal solution has achieved a fitness value of 34.8 °C for the objective of reducing the UTCI, 485 ppm for the objective of reducing the $CO_2$ concentration level, and 3.35 m/s for the objective of increasing the wind velocity. Despite the fitness ranks of each objective for the optimal solution being 512 for UTCI, 101 for $CO_2$ concentration level, and 1215 for wind velocity, the total rank of the optimal solution is the first. Also, the urban morphological parameters of the optimal solution are 35 m for courtyard width and length, and a 50% vegetation ratio, in addition to the building geometric parameters: 20 m for building height and case 3 of courtyard openings with two facing openings in the north–south direction. Consequently, the optimal solution could reduce UTCI and $CO_2$ concentration levels by 9.24 °C and 195 ppm, respectively, and increase the wind velocity by 1.5 m/s.

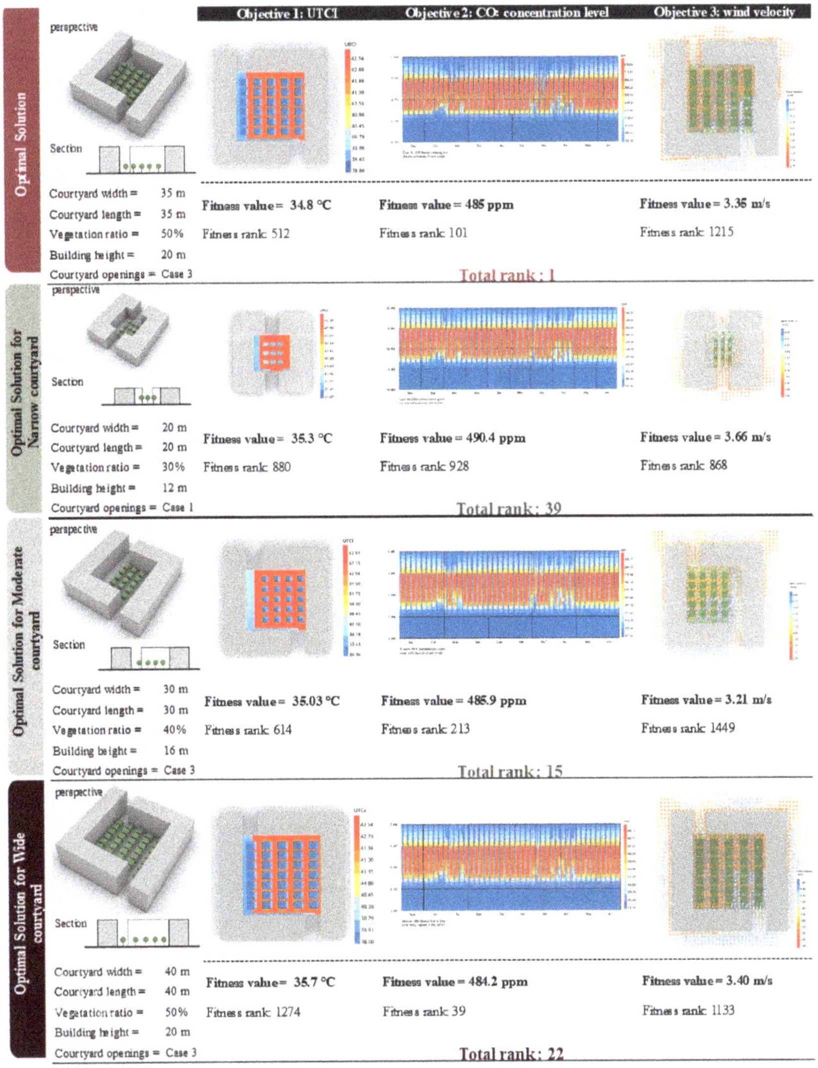


**Figure 14.** The optimal solutions of the proposed parametric methodology.

Moreover, Figure 14 elaborates the optimal solution for each courtyard size as follows: (a) a narrow courtyard with 20 m width and 20 m length, (b) a medium courtyard with 30 m width and 30 m length, and (c) a wide courtyard with 40 m width and 40 m length. For the narrow courtyard (20 × 20 m), the optimal solution with a rank of 39 achieved fitness values of 35.3 °C, 490.4 ppm, and 3.66 m/s for UTCI, $CO_2$ concentration levels, and wind velocity, respectively. At the same time, its parameters were a 30% vegetation ratio, 12 m for building height, and case 1 of courtyard openings. For the medium courtyard (30 × 30 m), the optimal solution with a rank of 15 achieved fitness values of 35.03 °C, 485.9 ppm, and 3.21 m/s for UTCI, $CO_2$ concentration levels, and wind velocity, respectively. The parameters were a 40% vegetation ratio, 16 m for building height, and case 3 of courtyard openings. For the wide courtyard (40 × 40 m), the optimal solution with a rank of 22 achieved fitness values of 35.7 °C, 484.2 ppm, and 3.4 m/s for UTCI, $CO_2$ concentration levels, and wind velocity, respectively. The parameters were a 40% vegetation ratio, 16 m for building height, and case 3 of courtyard openings. It can be concluded that low vegetation ratio and building height are appropriate for narrow courtyards, and when increasing the courtyard size to medium or wide, high vegetation ratio and high-rise buildings can effectively improve the air quality. Finally, the typical and last solutions for each objective will be discussed separately in the following section.

#### 3.2.1. Optimization Objective 1: Improving Outdoor Thermal Comfort

Improving outdoor thermal comfort is an essential aspect of improving the air quality in outdoor urban spaces, as it links environmental variables with potential implications. Thus, reducing the thermal stress is the first objective in the proposed parametric methodology, because of its direct influence on the students' social activities. Accordingly, the UTCI, as an index of outdoor thermal comfort, was simulated for each design solution, and its average was calculated during the hottest week (from 10 July to 16 July). As shown in Figure 15, the average UTCI of the proposed design solutions ranged from 33.66 °C to 40.65 °C. The decrease in UTCI is highly commendable when compared to the base case prior to implementing the design solutions (42.7 °C and 44.04 °C). The reduction in UTCI ranged from 2.04 °C to 10.3 °C, as a result of increasing the vegetation ratio of trees and raising buildings' heights. In addition, the stress classification in almost all design solutions was reduced to strong heat stress (from 32 °C to 38 °C) instead of the classification of very strong heat stress (from 38 °C to 46 °C) in the base case [45]. Figure 15c shows the typical design solution in UTCI, whose morphological parameters are 20 m, 20 m, and 50% for courtyard width, courtyard length, and tree ratio, respectively. In addition, the geometric parameters are 20 m and case 3 (courtyard with two facing openings facing north) for building height and courtyard openings, respectively. Hence, the average UTCI of the optimal solution during the hottest week was 33.66 °C, which predicts its efficiency during the rest of the year. These results are compatible with the results of [3], which indicated that the outdoor thermal comfort has been enhanced by 11.2 °C after applying a vegetation scenario in the school courtyard. Although the results are consistent with the results of [17,34], they are inconsistent with the results of 1.02 °C UTCI reduction [19].

The top 100 design solutions, based on UTCI values, were selected to investigate the impact of the common parameters (Figure 16). The range of UTCI was between 33.6 °C and 34.04 °C. It can be observed that the narrow courtyards, with widths such as 20 m and 25 m, are the most efficient, because they provide the smallest sunny area. Conversely, the variety of courtyard lengths is not considered an efficient parameter. It is obvious that a high ratio of vegetation and trees led to a significant reduction in UTCI, particularly the ratio of 50% of trees, due to the wide shade area provided by the high tree ratio. Similarly, a building height of 20 m assisted in shading almost the entire courtyard area, thus reducing the UTCI. On the other hand, in courtyards with two facing openings, case 1 and case 3, where the openings face the north, the wind velocity inside the courtyard was improved and thus the UTCI was improved. These results are compatible with the results of [20], which indicated that trees with diameters between 10 and 15 m could reduce UTCI by 11.26 °C. In conclusion, among the urban morphological building geometric parameters, courtyard width, vegetation ratio of trees, and building height are the most efficient for improving UTCI.

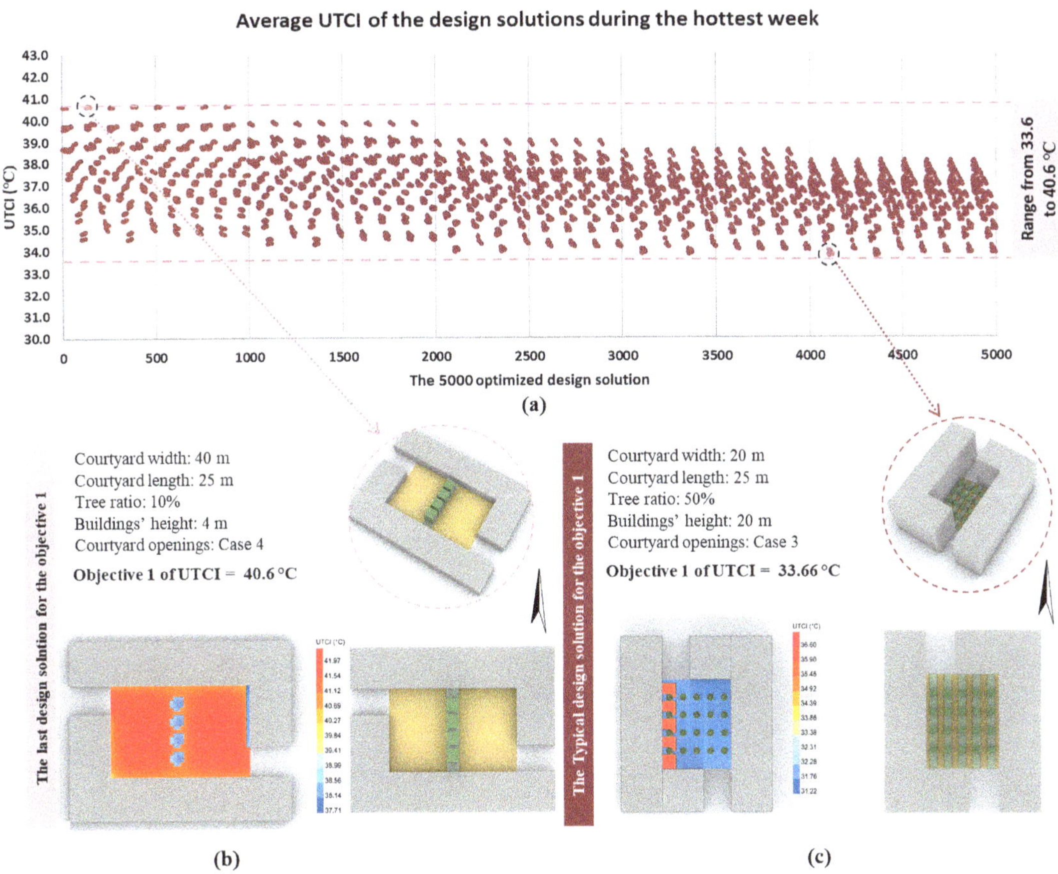


**Figure 15.** The average UTCI of (**a**) the 5000-design solution, (**b**) the last design solution, and (**c**) the typical design solution.

### 3.2.2. Optimization Objective 2: $CO_2$ Concentration Level

The purity of air and lowering the $CO_2$ concentration are critical aspects in improving the air quality in outdoor spaces. Hence, reducing the $CO_2$ concentration level is the second objective in the parametric methodology to provide low-pollutant air for students during their use of the courtyard, which will assist in improving the air quality in indoor spaces, too. Based on this, the $CO_2$ concentration level was simulated for each design solution and its average was calculated during the hottest week (from 10 July to 16 July). Figure 17 illustrates the $CO_2$ concentration level for the 5000 design solutions. It can be observed that the range of $CO_2$ concentration levels is between 482.7 ppm and 542.5 ppm. The observed decrease in $CO_2$ concentration levels is notably adequate when compared to the base case before applying the design solutions (600 ppm and 680 ppm). Because of adding vegetation areas and various tree ratios, the reduction of the $CO_2$ concentration level was between 57.4 ppm and 197.2 ppm. Figure 17c shows the typical design solution with the lowest $CO_2$ concentration level, whose morphological parameters are 40 m, 40 m, and 50% for courtyard width, courtyard length, and tree ratio, respectively. Besides, the geometric parameters are 20 m and case 2 (courtyard with two facing openings facing east) for building height and courtyard openings, respectively. Thus, the reduction in $CO_2$ concentration levels reached 9.5–29%, closely aligning with the 28.43% decrease reported by [30] and the 15% decrease reported by [18], and the 14% decrease reported by [28]. However, these findings are inconsistent with those of [16], which reported a modest 3.07% reduction in $CO_2$ concentration. Hence, the average $CO_2$ concentration level of the optimal solution during the hottest week was 482.7 ppm, which predicts its efficiency during the rest of the year. This shows the influence of built form on air quality.

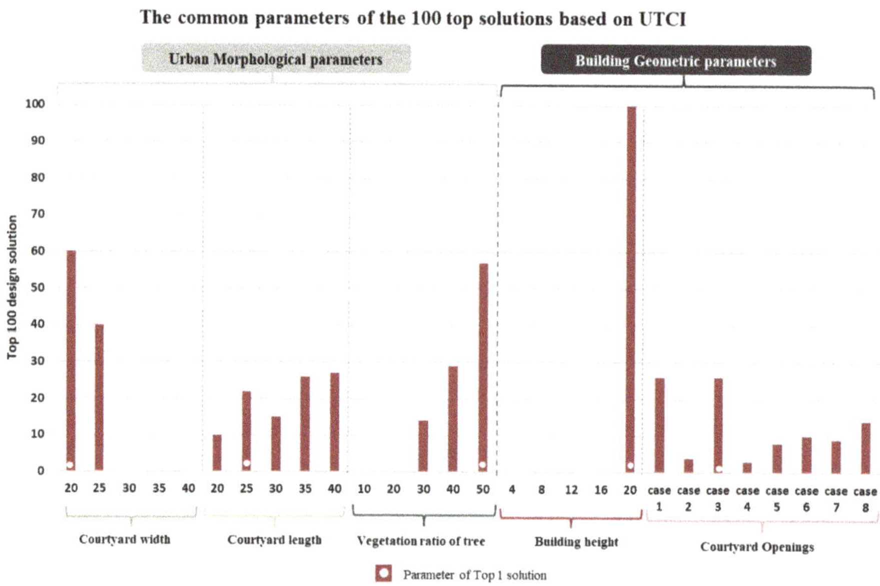


**Figure 16.** The common parameters of the 100 top solutions based on UTCI results.

Figure 18 shows the top 100 design solutions based on $CO_2$ concentration levels that were selected to investigate the impact of the common parameters. The range of $CO_2$ concentration levels was between 482.7 ppm and 484.9 ppm. It can be observed that the width and length of the courtyard are not extremely efficient on $CO_2$ concentration, despite the wide courtyard dimensions (40 m width and length) achieved in reducing the $CO_2$ concentration level. On the other hand, the parameter of vegetation and tree ratio was the most efficient in reducing the $CO_2$ concentration level due to its $CO_2$-absorbing phenomenon. Nevertheless, the building heights of 16 m and 20 m assisted in providing a shaded area in the courtyard and reducing the wind speed, which led to a lowering of $CO_2$ concentration levels. However, the various courtyard opening cases were sufficiently effective in reducing $CO_2$ concentration levels. In conclusion, the urban morphological parameters include the vegetation ratio of trees. In addition, the building geometric parameters include the building height parameter, which is the most efficient for improving the $CO_2$ concentration level.

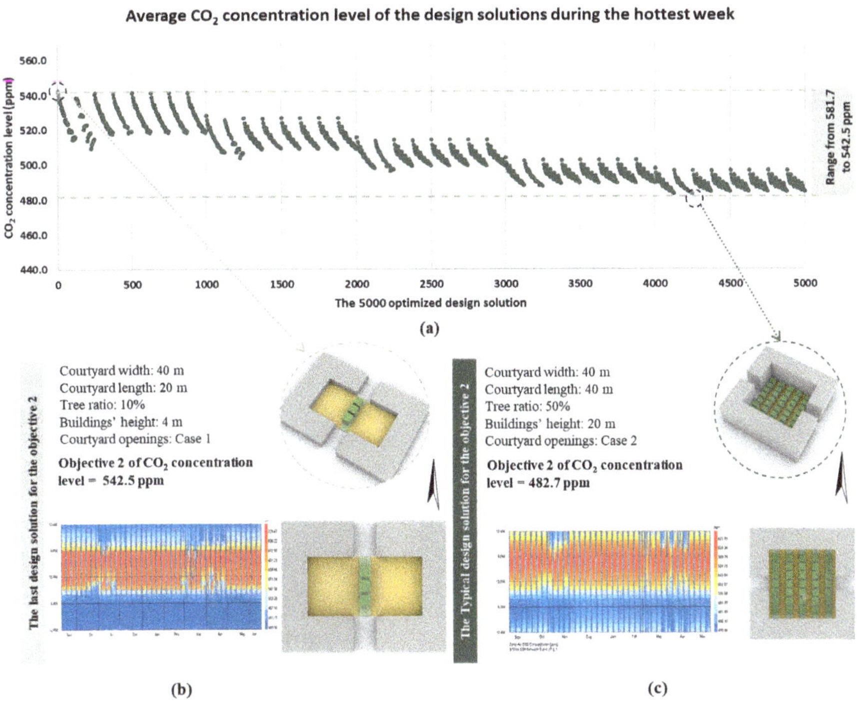


**Figure 17.** The average $CO_2$ concentration level of (**a**) the 5000-design solution, (**b**) the last design solution, and (**c**) the typical design solution.

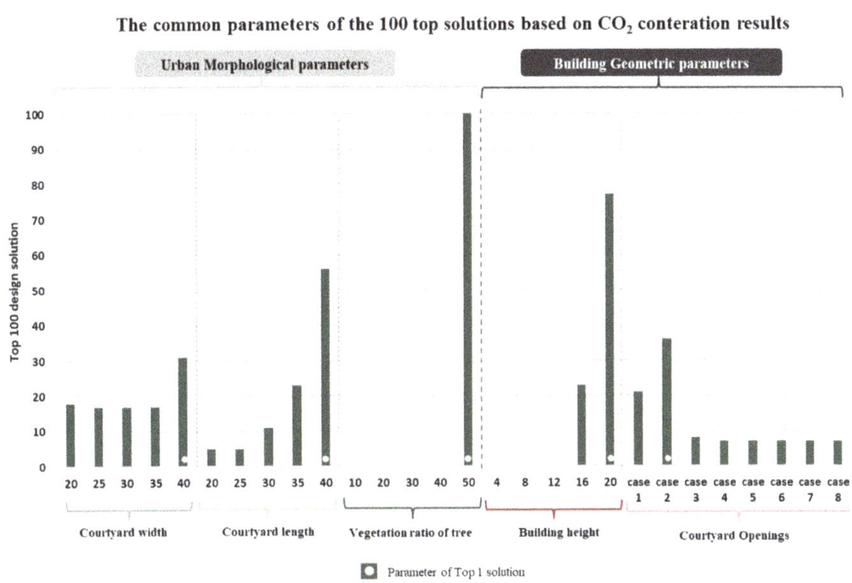


**Figure 18.** The common parameters of the 100 top solutions based on $CO_2$ concentration levels.

### 3.2.3. Optimization Objective 3: Wind Velocity

Improving wind speed to the appropriate limits for sitting for a short time (from 2.5 m/s to 4 m/s), or for walking and strolling (from 4 to 6 m/s), is more critical for enhancing the air quality [46]. Thus, increasing the wind velocity is the third objective of the parametric methodology to boost the students' experience inside the courtyards of the university buildings. Accordingly, the wind velocity was simulated for each design solution and its average was calculated during the hottest week (from 10 July to 16 July). In Figure 19, the average wind velocity of the proposed design solutions ranged from 1.3 m/s to 5.6 m/s. That increase in wind velocity is highly adequate when compared to the wind velocity values in the base cases, which were 0.9 m/s and 1.5 m/s inside the courtyard, although the wind speed out of the courtyard (layout campus) was 6.2 m/s in the hottest week. The increase in wind velocity ranged between 0.4 m/s and 4.7 m/s as a result of applying courtyard openings facing the prevailing wind direction (north and northwest). In addition, lowering the building height led to an increase in the airflow and the wind velocity inside the courtyard. Therefore, the typical design solution is shown in Figure 19b; its morphological parameters are 40 m, 20 m, and 10% for courtyard width, courtyard length, and tree ratio, respectively. In addition, the geometric parameters are 4 m and case 1 (courtyard with two facing openings facing north) for the building height and courtyard openings, respectively. Consequently, the proposed design solutions could enhance wind velocity throughout the year, and these results align well with those reported by [24].

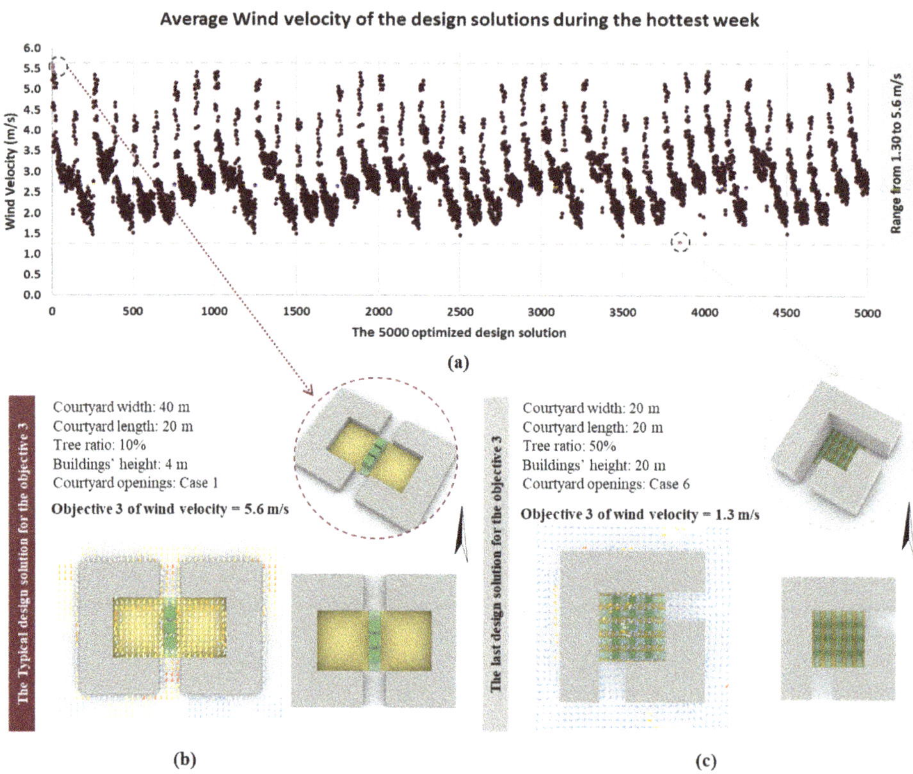


**Figure 19.** The average wind velocity of (**a**) the 5000-design solution, (**b**) the typical design solution, and (**c**) the last design solution.

Figure 20 shows the top 100 design solutions based on wind velocity that were selected to investigate the impact of the common parameters. The range of wind velocity was between 5.6 m/s and 5.1 m/s. Obviously, wide courtyard widths, such as 35 m and 40 m, are more efficient in increasing wind velocity than narrow widths. Increasing the courtyard width will lead to the area being oriented toward the prevailing wind, and consequently the airflow and wind velocity will increase. Conversely, courtyard length is not an efficient parameter because the increase in length is in the direction of the wind and not perpendicular to it. Also, the vegetation ratio of trees parameter is an uninfluential parameter, especially in solutions where the building height is low. The high ratios of vegetation and trees negatively affect wind velocity in solutions where the buildings are high and the courtyards are narrow, because they restrict the movement of air and airflow. On the other hand, the low building height allows wind to flow into the courtyard, thus improving the wind velocity. But at the same time, a building height of 4 m is not recommended in university buildings. Thus, the proposed parameter of courtyard openings toward the prevailing wind (north and northwest) is efficient and contributes to enhancing wind velocity. In conclusion, the urban morphological parameters include courtyard width; the building geometric parameters include courtyard openings, which are the most efficient for improving wind velocity.

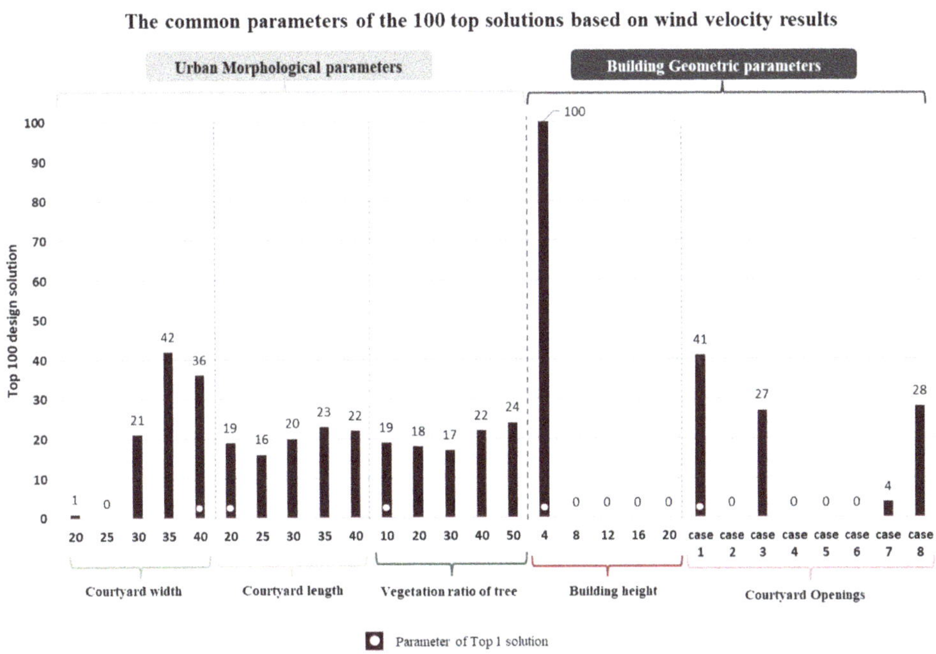


**Figure 20.** The common parameters of the 100 top solutions based on wind velocity.

### 3.3. *Analyzing the Correlation Determination*

In this section, the correlation determination between the three main objectives and between the optimization parameters and objectives will be analyzed. To begin with, the correlation analysis between the three objectives—reducing UTCI and $CO_2$ concentration and increasing wind velocity—is demonstrated in Figure 21. It can be observed that a moderate positive relationship exists between UTCI values and $CO_2$ concentration levels because the Coefficient of Determination ($R^2$) was 0.49 [47]. Otherwise, there are weak positive relationships between wind speed values and UTCI values, with an $R^2$ of 0.27, in addition to a negligible relationship between wind speed values and $CO_2$ concentration levels with an $R^2$ of 0.17. The results indicate that by improving UTCI values, the $CO_2$ concentration levels may be improved, but wind speed values may not improve. Therefore, it was necessary to analyze the correlation between the optimization parameters and objectives to determine the correlation characteristics among them.

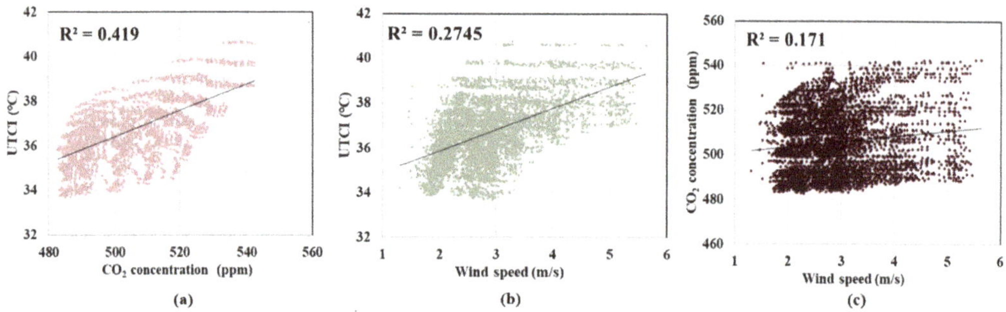


**Figure 21.** Correlation analysis between the optimization objectives: (**a**) UTCI and $CO_2$ concentration level, (**b**) UTCI and wind speed, and (**c**) $CO_2$ concentration level and wind speed.

Furthermore, Figure 22a presents a Sensitivity Map based on Standardized Regression Coefficients that assesses how different urban and building parameters affect the three objectives. In general, the most influential and sensitive parameters for reducing UTCI and $CO_2$ concentration and increasing wind velocity are building height, vegetation ratio of trees, and courtyard openings, respectively. For the objective of decreasing the UTCI, the building height parameter exhibits a very strong negative relationship, with an $R^2$ value of 0.743 (Figure 22b). The higher the building, the wider the shaded area, which leads to a reduction in air temperature and UTCI accordingly. The same applies to the increase in vegetation and ratio of trees, particularly with a high ratio of trees, such as 50%. Also, the courtyard width parameter exhibits a moderate positive relationship with reduced UTCI values, with an $R^2$ value of 0.345, because the narrow width of a courtyard provides the smallest sunny area. For reducing $CO_2$ concentration levels, the vegetation ratio of trees parameter exhibits a very strong negative relationship, with an $R^2$ value of 0.919 (Figure 22c). This was evident because trees are expected to absorb huge amounts of carbon dioxide, unlike the other parameters, which have weak and negligible relationships with reducing $CO_2$ concentration levels. Likewise, the parameter of courtyard openings exhibits a strong negative relationship with increasing wind velocity inside the courtyard, with an $R^2$ value of 0.612 (Figure 22d). Due to the courtyard openings toward the prevailing wind (north and northwest) in almost all of the proposed cases, the airflow and wind velocity could be successfully increased as a result of applying the cross-ventilation phenomenon. In addition, the high-rise-building parameter is negatively correlated with increasing wind velocity, as it blocks airflow within the courtyard and restricts air movement. It can be concluded that improving both urban morphological parameters and building geometrical parameters together was extremely influential in optimizing the university buildings' courtyards.

Finally, the parameters with strong relationships with each objective will be accurately illustrated in this section. Figure 23a shows the minimum, maximum, and average UTCI values based on various building heights. It can be observed that, at a building height of 4 m, the UTCI values ranged between 36.46 °C and 40.65 °C, which could be considered relatively high. In contrast, a building height of 20 m caused a reduction in the UTCI to between 33.67 °C and 37.67 °C. Figure 23b shows the thermal map of the best solutions for UTCI for each value of building height at 13:00, which is the main break time for students. Despite the efficiency of increased building height in reducing UTCI, it conflicts with improving other objectives, such as wind velocity. Consequently, the findings indicated that a more suitable building height is between 12 m and 16 m, and 20 m in the design solution of a wide courtyard with openings oriented towards the prevailing wind.

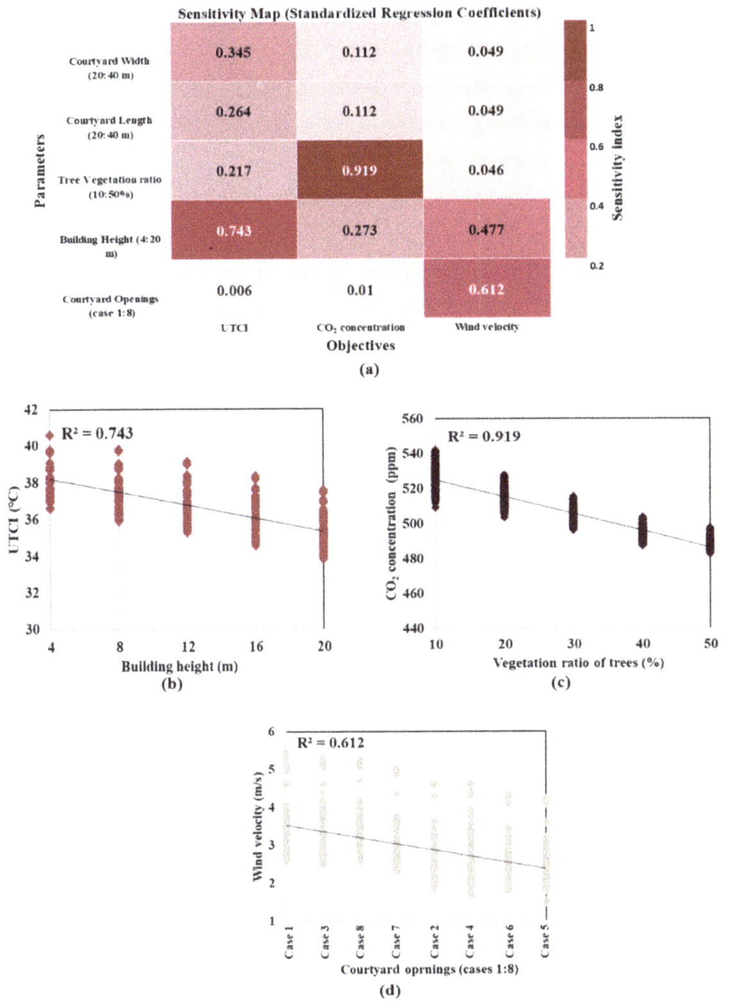


**Figure 22.** The Coefficient of Determination values and interpretation: (**a**) sensitivity analysis between the optimization parameters and objectives, (**b**) UTCI and building height, (**c**) $CO_2$ concentration and vegetation ratio, and (**d**) wind velocity and courtyard openings.

Figure 24a shows the minimum, maximum, and average $CO_2$ concentration levels based on various vegetation ratios of trees. Thus, with a vegetation ratio of 10%, the $CO_2$ concentration levels ranged between 509.24 ppm and 542.5 ppm. In contrast, a vegetation ratio of 10% caused a reduction in the $CO_2$ concentration levels to between 482.7 ppm and 496.5 ppm. Hence, Figure 24b shows the $CO_2$ concentration map of the best solutions of $CO_2$ levels for each ratio of vegetation of trees during the study months from September to June. However, the critical role of vegetation ratios of trees in reducing $CO_2$ concentration levels, besides reducing UTCI, in a few solutions, caused restriction of the movement of air and airflow inside the courtyard. Consequently, the findings indicated that the most suitable vegetation ratios of trees are 30%, 40%, and 50% in the design solutions of narrow, medium, and wide courtyards, respectively.

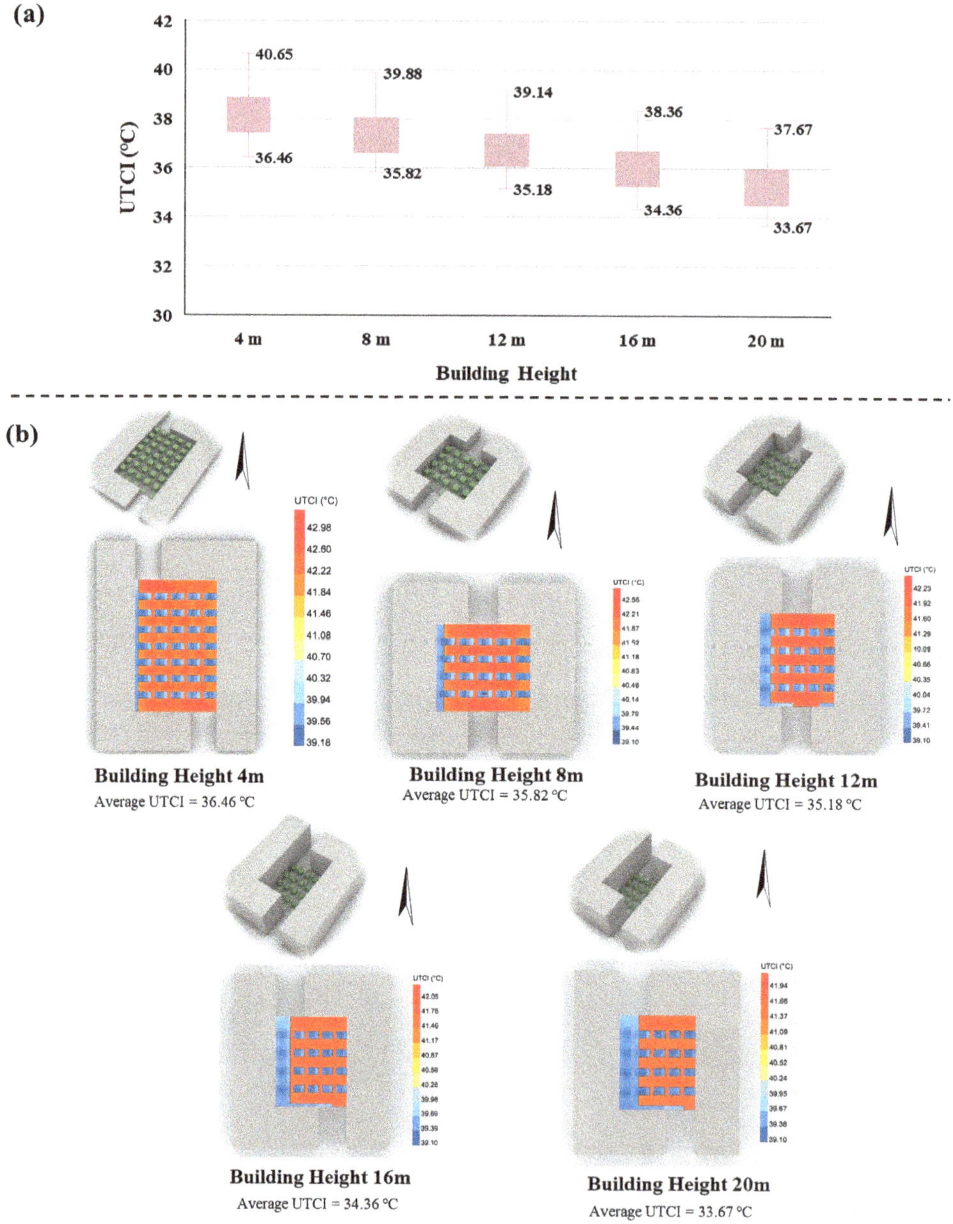


**Figure 23.** (**a**) The range and average values of (**a**) UTCI based on various building heights and (**b**) the thermal map of the best solutions of UTCI for each value of building height.

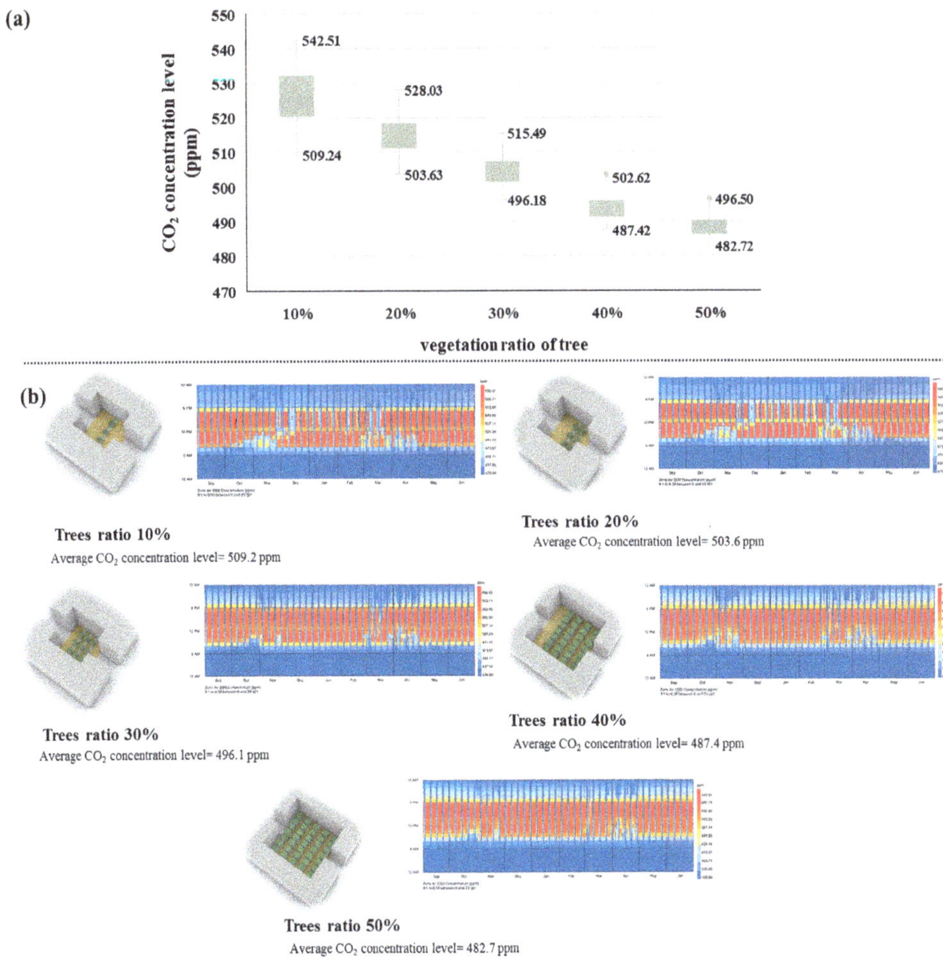


**Figure 24.** (**a**) The range and average values of $CO_2$ concentration levels based on various vegetation ratios of trees, and (**b**) the thermal map of the best solutions for $CO_2$ concentration for each value of vegetation ratio of trees.

Figure 25a shows the minimum, maximum, and average wind velocity levels based on various courtyard openings. So, a courtyard with two facing openings in the north–south direction, such as in case 1 and case 3, contributed to an increase in the wind velocity between 2.53 m/s and 5.64 m/s. But a courtyard with two facing openings in the east–west direction, such as in case 2 and case 4, contributed to an increase in the wind velocity between 1.49 m/s and 4.75 m/s. On the other hand, a courtyard with two perpendicular openings, one of which was toward the north, such as in case 7 and case 8, caused an increase in the wind velocity by a range between 1.31 m/s and 5.43 m/s. In contrast, case 5 and case 6 slightly improved the wind velocity by a range between 1.4 m/s and 3.95 m/s because the openings were not oriented towards the prevailing wind. Moreover, Figure 25b shows the airflow map of the best solutions of wind velocity for each case of the courtyard openings during the hottest week. Although the courtyard openings were not directly influential parameters for improving UTCI and $CO_2$ concentration levels, they play a critical role in improving wind velocity and air quality accordingly. Therefore, the findings indicated that the more efficient courtyard openings are the ones oriented towards the prevailing wind, whether facing or perpendicular openings.

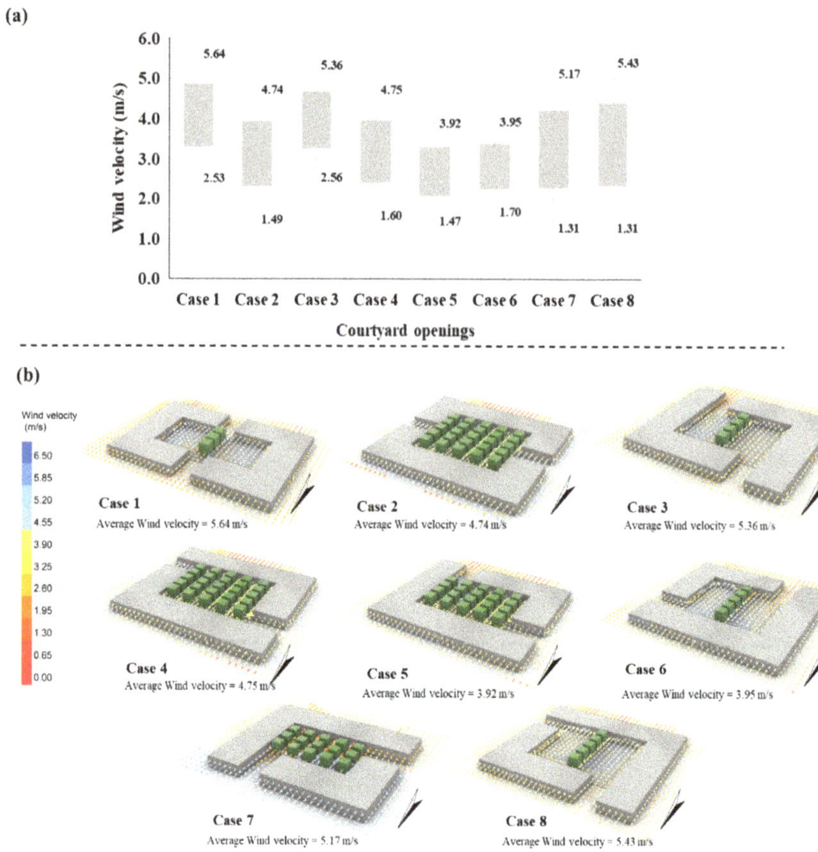


**Figure 25.** (**a**) The range and average values of wind velocity based on various courtyard openings, and (**b**) the airflow map of the best solutions for wind velocity for each case of courtyard openings.

Based on the main aim, the courtyard with narrow width and integration of 30% tree density achieves reduction of $CO_2$, improved student thermal comfort, and increased wind speed, and these results overcome the results of the monitoring in the real situation.

### 3.4. Potentials and Limitations of the Parametric Methodology

The proposed parametric methodology has been discussed, so its potentials are categorized into the following:

(a) Practical potentials include providing guidelines for future design of university buildings in hot arid climates, based on typical courtyard and building dimensions found in Egyptian educational institutions, thereby enabling more efficient courtyard openings instead of closed designs.

(b) Technical potentials include providing numerous design solutions based on various urban and building parameters for bioclimatic design principles, enhancing air quality through metrics such as UTCI, $CO_2$ concentration, and wind velocity, identifying the most effective parameter for each objective through correlation analysis, and determining the optimal placement of courtyard openings.

(c) Conceptual potentials include presenting optimal solutions for either all objectives simultaneously or individual objectives. Consequently, the flexibility of the proposed methodology allows for the expansion of its parameters and objectives by integrating them into the simulation workflow, enabling the steps of the methodology to be carried out smoothly and effectively.

On the other hand, the limitations of the proposed parametric methodology are as follows:

(a) Practical limitations include the reliance on a linear tree distribution, which may restrict the arrangement of seating areas.
(b) Technical limitations include restricting courtyard dimensions and building heights, as well as excluding other urban and building parameters (e.g., shading, façade materials). Due to computational simulation constraints by Rhino and Grasshopper, additional pollutant gases (e.g., CO, $NO_2$, $SO_2$, PM levels) were not considered, and occupant behavior was not accurately addressed. For fast and feasible CFD simulations, the geometry of trees was simplified, which led to low accuracy.
(c) Conceptual limitations include the initial framing of the methodology within the context of Egyptian university buildings to ensure relevance and accuracy based on local data. Additionally, the methodology relies on linear courtyard forms, excluding non-linear configurations such as curved courtyards. Also, the cost, space availability, or architectural regulations will be studied in future studies.

## 4. Conclusions and Recommendations

Improving air quality in university buildings is crucial for safeguarding students' health, enhancing cognitive performance, and fostering a learning environment. Thus, this study aims to develop optimized design solutions for university buildings and courtyards to enhance outdoor thermal comfort and reduce $CO_2$ concentration levels as an indicator of air quality. Consequently, the study developed a parametric methodology by using Rhino 3D+Grasshopper (V8) software to optimize urban and building parameters to achieve three objectives: improving thermal comfort, reducing the $CO_2$ concentration level, and improving wind velocity. The results can be summarized as follows:

- The aspect ratio (H/W) significantly influences airflow patterns, with lower ratios (0.7) promoting better ventilation when doors are open (average airspeed increase: 1.1–2.6 m/s), while higher ratios (1.2) create uneven airflow distribution (0.2–3.8 m/s) based on measurements.
- The most efficient courtyard dimensions are 20 × 20 m, providing the smallest sunny area, while courtyard openings facing north, the prevailing wind direction, significantly contributed to the following: (a) student thermal comfort, with a reduction in the Universal Thermal Climate Index (UTCI) ranging between 2.04 and 10.3 °C, (b) air quality, with a $CO_2$ concentration reduction between 57 and 197 ppm, and (c) ventilation, with a wind speed increasing by 0.4–4.07 m/s.
- A building height of 20 m could significantly reduce UTCI, because of maximizing the shaded area, but it limits airflow and reduces wind speed, which led to lowered $CO_2$ concentration levels. The most suitable vegetation ratio of trees integrated inside the courtyards is 30% in the design solution of a narrow courtyard for improving UTCI and $CO_2$ concentration, besides avoiding restricting the movement of air and airflow inside the narrow courtyard.
- Courtyard openings facing north, the prevailing wind direction, contributed to increasing the wind velocity between 1.03 m/s and 4.7 m/s in addition to indirectly reducing $CO_2$ by 195 ppm with appropriate vegetation ratios, and so enhancing thermal comfort.

In conclusion, it is recommended to use narrow courtyards (20 × 20 m) with a maximum building height equal to 20 m in new university building designs, and to implement a vegetation density ratio equal to 30% inside the courtyards of current and new buildings. This research helps enhance current and future courtyard design only with total building height in university buildings in a hot arid climate, particularly in Upper Egypt, with the help of the university administration, for outdoor air quality and thermal comfort. Nevertheless, future work could expand this methodology's applicability to other climates and regions to ensure broader generalizability by adjusting input parameters and weather data files. Additionally, it could explore diverse courtyard configurations and mitigation strategies, such as kinetic shading systems, to enhance outdoor air quality and thermal comfort. The findings of this study can potentially be extended to other climate zones or varying courtyard dimensions and shapes without requiring further simulations.

**Author Contributions:** Conceptualization, A.S.H.A. and R.M.A.M.; methodology, A.S.H.A. and R.M.A.M.; software, R.M.A.M.; validation, R.M.A.M. formal analysis, A.S.H.A. and R.M.A.M.; investigation, A.S.H.A. and R.M.A.M.; resources, A.S.H.A. and R.M.A.M.; data curation, A.S.H.A. and R.M.A.M.; writing—original draft preparation, A.S.H.A. and R.M.A.M.; writing—review and editing, A.S.H.A., R.M.A.M., A.R. and M.M.G.; visualization, R.M.A.M.; supervision, A.S.H.A. & A.R. and M.M.G.; project administration, A.S.H.A., A.R. and M.M.G.; funding acquisition, A.R. and M.M.G. All authors have read and agreed to the published version of the manuscript.

**Funding:** This work was supported and funded by the Deanship of Scientific Research at Imam Mohammad Ibn Saud Islamic University (IMSIU) (grant number IMSIU-DDRSP2502).

**Data Availability Statement:** Data is contained within the article: The original contributions presented in this study are included in the article. Further inquiries can be directed to the corresponding author.

**Conflicts of Interest:** There are no conflicts of interest. The authors declare that they have no known competing financial interests or personal relationships that could have appeared to influence the work reported in this paper.

## References

1. Salameh, M. Courtyards as Passive Design Solution for School Buildings in Hot Areas: UAE as a Case Study. Doctoral Dissertation, The British University in Dubai (BUiD), Dubai, United Arab Emirates, 2018.
2. Martinelli, L.; Matzarakis, A. Influence of height/width proportions on the thermal comfort of courtyard typology for Italian climate zones. *Sustain. Cities Soc.* **2017**, *29*, 97–106. [CrossRef]
3. Mahmoud, R.M.A.; Abdallah, A.S.H. Assessment of outdoor shading strategies to improve outdoor thermal comfort in school courtyards in hot and arid climates. *Sustain. Cities Soc.* **2022**, *86*, 104147. [CrossRef]
4. Abdallah, A. Passive design strategies to improve student thermal comfort in Assiut University: A field study in the Faculty of Physical Education in hot season. *Sustain. Cities Soc.* **2022**, *86*, 104110. [CrossRef]
5. Soflaei, F.; Shokouhian, M.; Abraveshdar, H.; Alipou, A. The impact of courtyard design variants on shading performance in hot-arid climates of Iran. *Energy Build.* **2017**, *143*, 71–83. [CrossRef]
6. Zamani, Z.; Heidari, S.; Hanachi, P. Reviewing the thermal and microclimatic function of courtyards. *Renew. Sustain. Energy Rev.* **2018**, *93*, 580–595. [CrossRef]
7. Salameh, M.; Taleb, H. Courtyard as Passive Design Solution for School Buildings in Hot Area. In Proceedings of the 2nd World Congress on Civil, Structural, and Environmental Engineering, Barcelona, Spain, 2–4 April 2017.
8. Ghaffarianhoseini, A.; Berardi, U.; Ghaffarianhos, A. Thermal performance characteristics of unshaded courtyards in hot and humid climates. *Build. Environ.* **2015**, *87*, 154–168. [CrossRef]
9. Taleghani, M.; Tenpierik, M.; van den Dobbelsteen, A. Heat in courtyards: A validated and calibrated parametric study of heat mitigation strategies for urban courtyards in the Netherlands. *Sol. Energy* **2014**, *103*, 108–124. [CrossRef]
10. Zango, M.; Modi, S. The Effect of Vegetation in Enhancing the Performance of Courtyard in Buildings of Tropical Climate. *J. Appl. Sci. Environ. Sustain.* **2017**, *3*, 34–42.
11. Soflaei, F.; Shokouhian, M.; Shemirani, S. Investigation of Iranian traditional courtyard as passive cooling strategy (a field study on BS climate). *Int. J. Sustain. Built Environ.* **2016**, *5*, 99–113. [CrossRef]

12. Kapalo, P.; Mečiarová, Ľ.; Vilčeková, S.; Krídlová Burdová, E.; Domnita, F.; Bacotiu, C.; Péterfi, K.E. Investigation of $CO_2$ production depending on physical activity of students. *Int. J. Environ. Health Res.* **2019**, *29*, 31–44. [CrossRef]
13. Forouzandeh, A. Numerical modelling validation for the microclimate thermal condition of semi-closed courtyard spaces between buildings. *Sustain. Cities Soc.* **2018**, *36*, 327–345. [CrossRef]
14. Shaari, N.A.; Zaki, S.A.; Ali, M.; Sukri, M. Investigation of the PMV and TSV Models of Thermal Comfort in Air-Conditioned University Classrooms in Malaysia. *Mech. Mater.* **2016**, *819*, 207–211. [CrossRef]
15. Abdallah, A.S.H.; Mohamed, D.; Mohamed, A. The influence of different courtyard ratios in university buildings on their thermal performance during the hot period: (Faculties of Agriculture and Education, New Sohag University, Egypt as a case study). *IOP Conf. Ser. Earth Environ. Sci.* **2019**, *397*, 012019. [CrossRef]
16. Eid, M.A.; Mahmoud, R.M.A.; Abdallah, A.S.H. A New Planning Proposal for Achieving Residents' Thermal Comfort in Hot Arid Climate-based on Simulation Model. *Mansoura Eng. J.* **2023**, *48*, 1. [CrossRef]
17. Li, Z.; Zhang, H.; Wen, C.Y.; Yang, A.S.; Juan, Y.H. Effects of frontal area density on outdoor thermal comfort and air quality. *Build. Environ.* **2020**, *180*, 107028. [CrossRef]
18. Su, S.; Feng, J.; Wang, W.; Jin, Y.; Chong, D. Dynamic prediction of lifecycle carbon emissions in residential communities: Insights into the role of residential form. *Environ. Impact Assess. Rev.* **2025**, *112*, 107845. [CrossRef]
19. Bedra, K.B.; Zheng, J.; Li, J.; Sun, Z.; Zheng, B. Automating Microclimate Evaluation and Optimization during Urban Design: A Rhino–Grasshopper Workflow. *Sustainability* **2023**, *15*, 16613. [CrossRef]
20. Abdallah, A.S.H.; Mahmoud, R.M.A.; Aloshan, M.A. Optimizing Urban Spaces: A Parametric Approach to Enhancing Outdoor Recreation Between Residential Areas in Riyadh, Saudi Arabia. *Buildings* **2025**, *15*, 1527. [CrossRef]
21. van der Walt, R.E.; Grobbelaar, S.S.; Booysen, M.J. Indoor temperature and $CO_2$ in South African primary school classrooms: Inspecting brick, container, and prefab structures. *J. Clean. Prod.* **2024**, *470*, 143120. [CrossRef]
22. Mahyuddin, N.; Essah, E.A. Spatial distribution of $CO_2$ Impact on the indoor air quality of classrooms within a University. *J. Build. Eng.* **2024**, *89*, 109246. [CrossRef]
23. Summa, S.; Remia GPerna, C.D.; Stazi, F. Mechanically ventilated classrooms in central Italy's heritage school buildings: Proposal of archetypes and $CO_2$ prediction models. *Build. Environ.* **2024**, *265*, 111963. [CrossRef]
24. Wu, B.; Zhao, S.; Liu, Y.; Zhang, C. Do meteorological variables impact air quality differently across urbanization gradients? A case study of Kaohsiung, Taiwan, China. *Heliyon* **2025**, *11*, 41694. [CrossRef] [PubMed]
25. Maiques, M.; Tarragona, J.; Gangolells, M.; Casals, M. Energy implications of meeting indoor air quality and thermal comfort standards in Mediterranean schools using natural and mechanical ventilation strategies. *Energy Build.* **2025**, *328*, 115076. [CrossRef]
26. Li, L.; Mirzabeigi, S.; Soltanian-Zadeh, S.; Dong, B.; Krietemeyer, B.; Gao, P.; Wilson, N.; Zhang, J. A high-performance multi-scale modular-based green design studio platform for building and urban environmental quality and energy simulations. *Sustain. Cities Soc.* **2025**, *119*, 106078. [CrossRef]
27. Moghadam, T.T.; Bruton, K.; O'Sullivan, D.T.J.; Norton, B. Energy efficient achievement of indoor air quality and thermal comfort using mechanical ventilation heat recovery and solar-energy pre-heating. *Energy Convers. Manag.* **2025**, *327*, 119528. [CrossRef]
28. Yao, S.; Li, M.; Yuan, J.; Huo, Q.; Zhao, S.; Wu, Y. Optimization design of layout dimension for residential buildings weighing up daylighting, thermal comfort, and indoor air quality with a low-carbon decision-making. *J. Build. Eng.* **2024**, *98*, 111328. [CrossRef]
29. Gebreyesus, T.; Borgemeister, C.; Herrero-J'auregui, C.; Kelboro, G. Transforming urban air quality: Green infrastructure strategies for the urban centers of Ethiopia. *Environ. Pollut.* **2024**, *363*, 125244. [CrossRef]
30. Abdeen, N.; Rafaat, T. Assessing vertical green walls for indoor corridors in educational buildings and its impact outdoor: A field study at the universities of Canada in Egypt. *Results Eng.* **2024**, *21*, 101838. [CrossRef]
31. Gustafsson, M.S.; Lindén, J.; Johansson, E.M.; Watne, Å.K.; Uddling, J.; Sjölie, D.; Pleijel, H. Well-planned greenery improves urban air quality—Modelling the effect of altered airflow and pollutant deposition. *Atmos. Environ.* **2024**, *338*, 120829. [CrossRef]
32. Kofel, D.; Bourgeois, I.; Paganini, R.; Pulfer, A.; Grossiord, C.; Schmale, J. Quantifying the impact of urban trees on air quality in Geneva, Switzerland. *Urban For. Urban Green.* **2024**, *101*, 128513. [CrossRef]
33. Kaveh, S.; Habibi, A.; Nikkar, M.; Aflaki, A. Optimizing green infrastructure strategies for microclimate regulation and air quality improvement in urban environments: A case study. *Nat.-Based Solut.* **2024**, *6*, 100167. [CrossRef]
34. Islam, A.; Pattnaik, N.; Moula, M.M.; Rötzer, T.; Pauleit, S.; ARahman, M. Impact of urban green spaces on air quality: A study of PM10 reduction across diverse climates. *Sci. Total Environ.* **2024**, *955*, 176770. [CrossRef]
35. Reiminger, N.; Jurado, X.; Maurer, L.; Vazquez, J.; Wemmert, C. Advancing urban air quality modeling with solar radiation-included computational fluid dynamics simulations. *Atmos. Pollut. Res.* **2025**, *16*, 102383. [CrossRef]
36. HBRC. The Egyptian Code for Enhancing Energy Use in Buildings. 2006. Available online: https://www.hbrc.edu.eg/en (accessed on 1 April 2025).
37. Noaa. 2025. Available online: http://www.noaa.gov/ (accessed on 1 April 2025).

38. Meteoblue. 2025. Available online: https://www.meteoblue.com (accessed on 1 April 2025).
39. *ANSI/ASHRAE Standard 62.1-2019*; Ventilation for Acceptable Indoor Air Quality. American Society of Heating, Refrigerating and Air-Conditioning Engineers: Atlanta, GA, USA, 2019.
40. Mahmoud, R.M.A. Multi-objective Approach for Optimizing Buildings' Forms and Kinetic Façade Systems in Office Buildings. *Mansoura Eng. J.* **2024**, *49*, 3. [CrossRef]
41. Muhaisen, A.S. Ventilation performance of courtyards as a passive cooling strategy in hot climates. *Renew. Energy* **2006**, *31*, 2061–2071.
42. Al-Masri, N.; Abu-Hijleh, B. Courtyard housing in midrise buildings: An environmental assessment in hot-arid climate. *Renew. Sustain. Energy Rev.* **2012**, *16*, 1892–1898. [CrossRef]
43. Abdallah, A.S.H.; Mahmoud, R.M.A. Sustainable Mitigation Strategies for Enhancing Student Thermal Comfort in the Educational Buildings of Sohag University. *Buildings* **2025**, *15*, 2048. [CrossRef]
44. Persily, A.; de Jonge, L. Carbon dioxide generation rates for building occupants. *Indoor Air* **2017**, *27*, 868–879. [CrossRef]
45. Xu, X.; Yin, C.; Wang, W.; Xu, N.; Hong, T.; Li, Q. Revealing Urban Morphology and Outdoor Comfort through Genetic Algorithm-Driven Urban Block Design in Dry and Hot Regions of China. *Sustainability* **2019**, *11*, 3683. [CrossRef]
46. Norouziasas, A.; Ha, P.P.; Ahmadi, M.; Rijal, H.B. Evaluation of urban form influence on pedestrians' wind comfort. *Build. Environ.* **2022**, *224*, 109522. [CrossRef]
47. Saura, R.B.D.; Andante, R.J.M. Detection of cyanide in freshwater fishes relative to sex dimorphism using landmark-based geometric morphometrics in Agusan del Sur, Philippines. *Int. J. Biosci.* **2018**, *12*, 2222–5234. [CrossRef]

**Disclaimer/Publisher's Note:** The statements, opinions and data contained in all publications are solely those of the individual author(s) and contributor(s) and not of MDPI and/or the editor(s). MDPI and/or the editor(s) disclaim responsibility for any injury to people or property resulting from any ideas, methods, instructions or products referred to in the content.

*buildings*

*Article*

# Sustainable Mitigation Strategies for Enhancing Student Thermal Comfort in the Educational Buildings of Sohag University

**Amr Sayed Hassan Abdallah [1,*] and Randa Mohamed Ahmed Mahmoud [2]**

1 Department of Architectural Engineering, College of Engineering, Imam Mohammad Ibn Saud Islamic University (IMSIU), Riyadh 11432, Saudi Arabia

2 Department of Architecture, Faculty of Engineering, Assiut University, Assiut 71516, Egypt; randa.m@aun.edu.eg

* Correspondence: asabdullah@imamu.edu.sa

**Abstract:** Improving students' thermal comfort in university courtyards and indoor spaces promotes walkability, enhances livability, and fosters social interaction among students. This study aims to improve students' outdoor thermal comfort in university courtyards, to reduce heat transfer to classrooms, and to accordingly reduce energy consumption in university buildings in hot arid climates. Thus, the proposed coupled methodology for the case study, the Faculty of Agriculture, New Sohag University, Egypt, consists of three stages. First, monitoring and questionnaire surveys were conducted in the open courtyard and the classroom to obtain air temperature, wind speed, thermal image, and $CO_2$ and thermal comfort analysis. Secondly, the Envi-met model was used to investigate the impact of six improvement solutions on improving thermal comfort in the courtyard. Third, retrofitting strategies in the building envelope were evaluated to decrease heat transfer and energy consumption by DesignBuilder software. Consequently, the findings revealed a high outdoor air temperature, which causes discomfort for students. Hence, the simulation results concluded that the significant reduction of physiological equivalent temperature (PET), which ranged between 11.1 °C and 13.9 °C, occurred after applying the hybrid improvement solutions (vegetation area and semi-shading or pergola-shading). Moreover, integrating a combination of retrofitting strategies into the faculty buildings contributed to a 30% reduction in energy consumption. Ultimately, the proposed methodology aims to assist architects and urban designers in the early design stages by providing the appropriate environmental solutions for the universities' courtyards and buildings in hot arid climates.

**Keywords:** outdoor thermal comfort; coupled simulation; Envi-met model; energy consumption; retrofitting strategies

## 1. Introduction

In recent years, educational buildings have gained increasing attention for their environmental design, encompassing both indoor environments and outdoor spaces. On the one hand, improving thermal comfort in the main courtyard encourages students to practice social activities. At the same time, reducing energy consumption in university buildings is a critical step toward sustainable campus development. Hence, improving outdoor and indoor thermal comfort for students in university courtyards is essential for enhancing well-being and promoting the use of open spaces, especially in hot arid climates. Accordingly, the relevant literature is categorized into the following two main

 https://doi.org/10.3390/buildings15122048

areas: (a) improving outdoor thermal comfort in educational buildings and (b) energy conservation methods inside educational buildings. Several studies utilized different survey methods and simulation software to investigate the impact of their proposed mitigation solutions. For example, the author of [1] used Envi-met to study the effect of shading devices on outdoor thermal comfort and indoor illuminance. The research was conducted at a university campus in Guangzhou, China. They concluded that shading could reduce the Universal Thermal Climate Index (UTCI), physiological equivalent temperature (PET), and radiant temperature by 5.9 °C, 12.0 °C, and 24.8 °C, respectively. In addition, the authors of [2] studied the impact of a courtyard's different dimensions and ratios at the Faculties of Agriculture and Education, New Sohag University, Egypt. It was found that the Faculty of Agriculture and Education, with a H/W ratio equal to 1.2, achieved acceptable thermal comfort for students. On the other hand, several studies addressed the indoor thermal comfort inside educational buildings' spaces. For instance, the authors of [3] surveyed university students to assess the indoor thermal comfort of naturally ventilated classrooms in Bangladesh. Results revealed that the mean values of air temperature, wind speed, and relative humidity were 31 °C, 0.8 m/s, and 78%, respectively.

### 1.1. *Background and Literature*

#### 1.1.1. Improving Outdoor Thermal Comfort in Educational Buildings

Furthermore, studies have proposed novel strategies for improving the thermal comfort of outdoor spaces in educational buildings. For example, the authors of [4] studied the campus of George Institute of Technology, USA. They concluded that a highly dense tree canopy has the potential to reduce air temperature by 3.77 °C. Further, the authors of [5] proposed nine passive strategies to improve the outdoor thermal comfort at a public school with an E-shaped style in New Assiut City. The results revealed that the PET value was reduced by 18.6 °C by applying the strategy of hybrid diagonal staggered shading with trees. The authors of [6] studied the effect of vegetation in courtyards of educational buildings on outdoor thermal comfort and energy performance. It was found that the tree configuration could decrease the total cooling energy demand by 7.52 kWh/m$^2$ and discomfort hours by 12.5%. The authors of [7] used Envi-met to evaluate microclimate parameters on the university campus in Kuala Lumpur, Malaysia. It was found that, because of shading and vegetation, the outdoor space can improve the comfort level. Furthermore, the importance of urban geometry and shading in outdoor spaces has been investigated using Envi-met and Rayman for calculating PET by the authors of [8]. The results demonstrated that urban shading could reduce mean radian temperature (MRT) and PET by 34 °C and 17.6 °C, respectively.

Nevertheless, the authors of [9] proposed eight vegetation scenarios to improve outdoor thermal comfort at a public elementary school in Riyadh, Saudi Arabia. Results revealed that applying 10 m high trees could reduce air temperature (Ta), mean radiant temperature (Tmrt), and Universal Thermal Climate Index (UTCI) by 3.2 °C, 7.13 °C, and 17.95 °C, respectively. In addition, the authors of [10] investigated the impact of well-defined boundary spaces on the outdoor thermal comfort of university campuses in Eastern China. The result was a set of design guidelines for variables that impact spatial thermal comfort. The authors of [11] monitored a university campus to find the relationship between PET, the Predicted Mean Vote (PMV) index, and the real thermal sensation in Birjand, Iran. They concluded that PET is significantly correlated with the real thermal sensation, with a comfortable range between 16.4 °C and 25.3 °C. Meanwhile, the authors of [12] conducted a questionnaire and assessed the UTCI. The aim was to evaluate the outdoor thermal comfort at an elementary school campus in Guangzhou, China. They concluded that tree planting is the most efficient strategy for improving outdoor thermal comfort.

#### 1.1.2. Energy Conservation Methods Inside Educational Buildings

Moreover, many studies have addressed passive retrofitting strategies to improve energy efficiency and reduce energy demand inside educational buildings. The authors of [13] proposed energy efficiency strategies for building envelope thermal performance and sun shading to reduce energy consumption at universities in China after field measurements, simulation processes, and questionnaires. It was revealed that, while using insulating sun shading systems, the annual cooling and heating load would decline by 45.1% and 38.6%, respectively. Additionally, the authors of [14] evaluated the energy consumption in three buildings at the Federal University of Itajubá, Brazil, to study the effect of building materials, activities, and weather conditions. It was found that the reduction of annual energy consumption and $CO_2$ was 3.6–17.7 MWh and 0.5 $tCO_2eq$. In addition, the authors of [15] assessed the thermal comfort and energy performance inside the educational building with the shading system with phase change material (PCM). The results revealed that the annual cooling energy consumption declined by 44%. Moreover, the authors of [16] conducted field measurements and surveyed students' views regarding the thermal comfort inside university classrooms. They concluded that the acceptable range of temperature was 21.3–25.4 °C and 29.6 °C in winter and summer, respectively, and the $CO_2$ concentration level was 2500 ppm.

Additionally, the influence of energy rationalization on optimizing the building envelope was investigated in the Mechanical Engineering Department building at Ain Shams University, Egypt, by [17]. It was found that the retrofitting strategy contributed to a decline in the annual energy consumption by 20%. The authors of [18] studied the impact of shading and glazing configuration on indoor thermal comfort and $CO_2$ concentration in educational schools in New Zealand. The results show that using dynamic shading and glazing could decrease Ta by 1.6–3.0 °C and $CO_2$ by 50%. The authors of [19] studied two courtyards as a passive strategy for improving indoor thermal comfort and energy consumption in the school building in Argentina. The results show that energy consumption was reduced by 21%. Furthermore, the authors of [20] proposed a set of façade retrofitting strategies to enhance energy efficiency in Saudi school buildings. The results revealed that the retrofitting strategies could reduce the cooling energy, lighting energy, and annual energy costs by ratios of 17%, 49%, and 18%, respectively. Additionally, the authors of [21] measured the indoor air temperature, relative humidity, $CO_2$, $PM_{10}$, and $PM_{2.5}$ inside the classes of the Technical University, India, to investigate the indoor air quality (IAQ). Consequently, the range of $PM_{10}$ was between 1.8 $\mu g/m^3$ and 159.7 $\mu g/m^3$, and the range of $PM_{2.5}$ was between 18.2 $\mu g/m^3$ and 108 $\mu g/m^3$.

### *1.2. Research Objectives*

However, the critical role played by the university buildings involves shaping students' personalities and providing a comfortable social climate. There is a research gap in investigating and providing methodologies for improving outdoor and indoor thermal comfort together at the university buildings in Egypt. Nevertheless, the discomfort issues, such as high air temperature and high incidence of solar radiation inside the courtyard and high indoor temperatures of educational spaces, have been determined by conducting field measurements and questionnaire surveys inside the Faculty of Agriculture at New Sohag University, Sohag, Egypt. Therefore, this study aims to improve students' outdoor thermal comfort in university courtyards, in addition to reducing energy consumption in university buildings in hot arid climates. This helps to encourage students to practice social activities in outdoor spaces. Hence, there is a need for using Envi-met and DesignBuilder to simulate and investigate the effect of a set of outdoor mitigation solutions and indoor retrofitting strategies. Consequently, a coupled simulation methodology was proposed by using Envi-met and DesignBuilder software programs, which contributed the following:

(a) Improving students' thermal comfort in open courtyards;
(b) Increasing social interaction in open courtyards;
(c) Conserving energy in an educational university building.

Moreover, the novelty of this study lies in integrating field monitoring and conducting a questionnaire with coupled simulation methodology to improve outdoor thermal comfort by applying six different improvement solutions and to reduce indoor energy consumption by applying five retrofitting strategies. Consequently, the optimal improvement solution and the retrofitting strategy could be obtained to achieve the study's main aim. Based on the previous study, there is a need for using Envi-met and DesignBuilder to implement a coupled simulation methodology to simulate and investigate the effect of a set of outdoor mitigation solutions and indoor retrofitting strategies in a university building in a hot arid region. Thus, this study is integrated with some of the previous studies, such as [2,4,7,10,11,13], in studying several mitigation solutions, but this study distinguishes from them by applying new hybrid mitigation solutions, such as a hybrid between the vegetation area and pergola-shading and a hybrid greenery shading (metal mesh shading panels covered with ivy plants). Ultimately, this proposed methodology aims to support the architects and urban designers during the early design stages in providing the appropriate environmental solutions for the university courtyards and buildings in hot arid climates. This paper is structured below, starting with the methodology in the next Section 2 (case study and the proposed methodology), then the results and discussion in Section 3, and finally the conclusion section.

## 2. Methodology

### 2.1. Case Study Description

Sohag City is located in southern Egypt with a longitude of 26.5591° N, 31.6957° E. Sohag City is characterized by a clear sky throughout the year, with direct strong solar radiation [22]. Therefore, the case study of the Faculty of Agriculture at New Sohag University was chosen as a model for university buildings established in hot arid climates. It consists of five floors, with a floor area of 3147 $m^2$ and a ground floor consisting of three courtyards, two closed courtyards and an open middle courtyard. A set of classes, labs, and staff offices overlook these courtyards. Figure 1 shows the location of the study area concerning Sohag City, New Sohag University, and its plan, elevation, and different outside photos. The courtyard ratio (H/W) is 0.7 for two closed courtyards, and the H/W is 1.2 for the open middle courtyard, which is the study area. Spaces with different functions (classes, labs, offices) were used for monitoring. The average dimensions for most of the classes, labs, and offices are 60 $m^2$, 24 $m^2$, and 24 $m^2$ respectively. These spaces have different orientations with 0.4% for WWR for outside windows. The materials used inside the middle courtyards are tile interlook and sand in some places under the sitting area. The dimensions of the sitting area are 3 × 3, and it is made from wood with a steel frame. Very few vegetations with short herbaceous plants are used inside the courtyard.

### 2.2. The Proposed Improvement Methodology

Consequently, this study proposed a coupled simulation methodology to improve the outdoor thermal comfort in the middle open courtyard of the case study and to reduce indoor energy consumption. As shown in Figure 2, the proposed methodology consists of three stages, which are detailed and illustrated in the following sections:

(a) Monitoring the case study;
(b) Simulation the outdoor thermal comfort by ENVI-met Version 5.1, (ENVI-met GmbH, Essen, NW, Germany).
(c) Simulating the indoor energy consumption by DesignBuilder software version (V.5.0.3.007), (DesignBuilder Software Ltd., Stroud, UK).

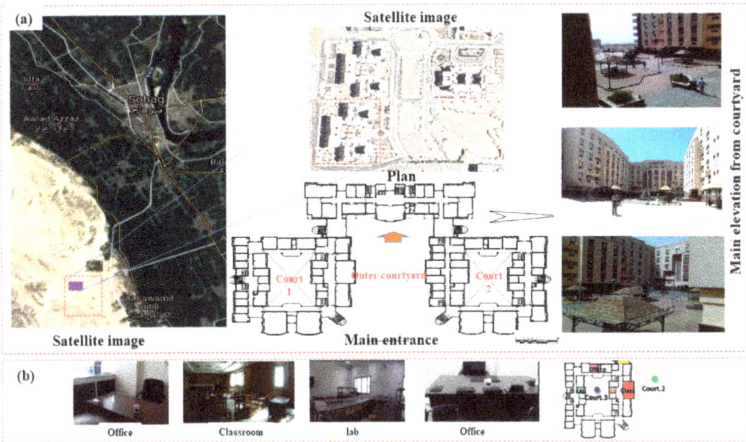


**Figure 1.** The overview of the Faculty of Agriculture building: (**a**) the outside view (main elevation, plan, and satellite image), (**b**) the inner picture (author).

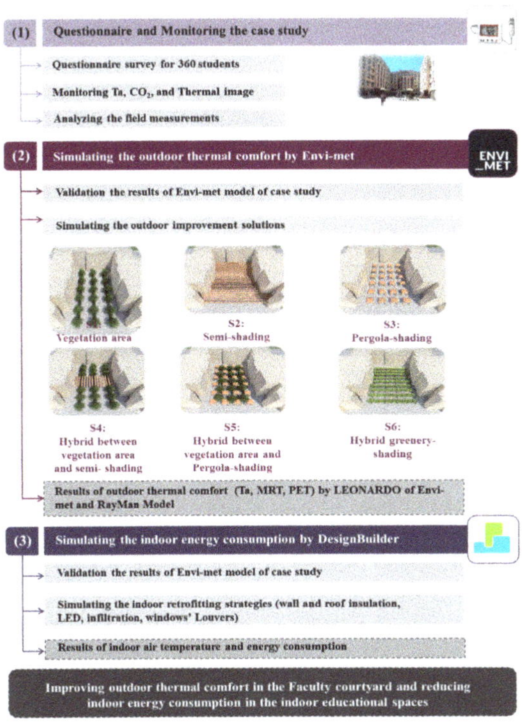


**Figure 2.** The proposed improvement methodology of the Faculty of Agriculture.

2.2.1. Stage 1: In Situ Measurements of the Case Study

In the first stage, monitoring was conducted during the hot days at the end of the second semester (May 2018) from 9:00 to 14:00 inside classes in different orientations and outdoors inside the middle open courtyards with existing students. This period was selected due to the increase of the outdoor temperature, before the end of semester and final exam, and as an extension of past research for one year evaluation and monitoring [2]. Measurements from the timespan from 9:00 to 14:00 inside classes were chosen because most lectures are during that period. Measurement locations: A number of spaces overlooking courtyards with different orientations were selected, representing spaces with different functions (classes, labs, offices) to analyze the effect of different orientation on indoor environment. These spaces used natural ventilation and mechanical fans for ventilation. The detailed evaluations for indoors and outdoors were analyzed for one year with a questionnaire survey using 360 effective questionnaires that were randomly distributed to students staying in university courtyards doing a different activity while measurements were conducted [2]. The structure of the questionnaire was based on ASHRAE Standard 55, 2019. The first part contains demographic and student characteristic information. The second section asks about the thermal sensation and their sensation of air humidity and wind speed. The thermal sensation vote (TSV) scale was the traditional ASHRAE 7-point scale (−3 very cold, −2 cold, −1 slightly cold, 0 neutral, 1 slightly hot, 2 hot, 3 very hot). For air humidity, estimation was based on a 7-point scale humidity sensation vote (HSV) (−3 very dry, −2 dry, −1 slightly dry, 0 just right, 1 slightly humid, 2 humid, 3 very humid). For the wind speed, estimation was based on a 5-point scale air velocity vote (AV) (−2 very low, −1 low, 0 enough, 1 high, 2 very high). The third part addresses the source of feeling uncomfortable, such as the reason for discomfort inside the courtyard according to temperature. Questionnaires were distributed for subjective assessment between 1:00 and 2:00 p.m. during the hot period, aligning with the time when most students take their break in the courtyard. Measurements were conducted for different outdoor parameters using the measurement devices to calculate the index of physiological equivalent temperature (PET) that could define the thermal acceptability ranges of students inside the courtyards, taking into account the effect of shading and radiation flux [23]. Additionally, the effects of courtyard design and ratio (H/W) on airspeed and carbon dioxide concentration ($CO_2$) were taken into consideration. Thermal monitoring for the outer environment using thermal images was conducted, and a calculation for the sky view factor was obtained based on the Rayman software 1.2. Thermal image evaluation was conducted to focus on the main problem for student thermal comfort inside the courtyards concerning the material used in the courtyard grounds and the material used for the building façade with heat transfer.

2.2.2. Stage 2: Simulating the Outdoor Thermal Comfort by ENVI-met

In the second stage, Envi-met software version 5.0.3 was used to build a model of the case study and to conduct the simulation process for the middle open courtyard. The validation point was selected in the middle of the courtyard to be exposed to all ambient conditions from the building shade, courtyard components such as shading units, and prevailing wind direction in the courtyard. Thus, the results of the Envi-met model were compared with the field measurements to validate the simulation model during 7 h of the usual study day from 9:00 to 15:00 on 28 May 2018. Therefore, Figure 3 illustrates the coefficient of determination ($R^2$) as equal to 0.93 for air temperature, which indicates successful validation of the model with a strong correlation close to the real conditions based on the past [24]. Based on the questionnaire and field survey, the middle open courtyard includes small shrubs, four small shaded places, and a dark-colored floor, which caused students thermal discomfort. Hence, the middle open courtyard was selected for studying rather than the closed courtyards because of its superior ability to facilitate natural ventilation and, thus, to improve outdoor thermal comfort. Consequently, a set of six improvement solutions was proposed to apply to the middle open courtyard. Additionally, their impact on improving the students' thermal comfort in the middle open courtyard was investigated by Envi-met. The simulation process was conducted on 1 May 2024 (from 9:00 to 15:00) to be as close as possible to the measurement date in May 2018. Figure 4 elaborates on the characteristics of the six improvement solutions to increase the shading area in the middle open courtyard. There are the following three improvement solutions: adding a vegetation area, adding semi-shading, and adding pergola shading. In addition, the three hybrid improvement solutions are the following: a hybrid between the vegetation area and semi-shading, a hybrid between the vegetation area and pergola shading, and a hybrid greenery shading (metal mesh shading panels covered with ivy plants).

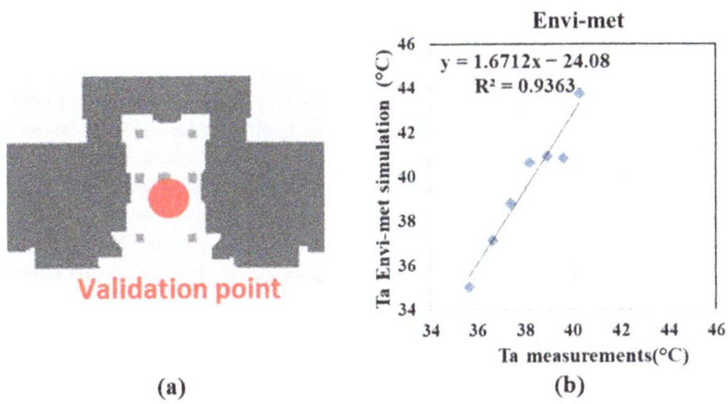


**Figure 3.** The validation process in Envi-met: (**a**) location of the validation point in the courtyard, and (**b**) linear regression of air temperature in Envi-met with relation to the measurement.

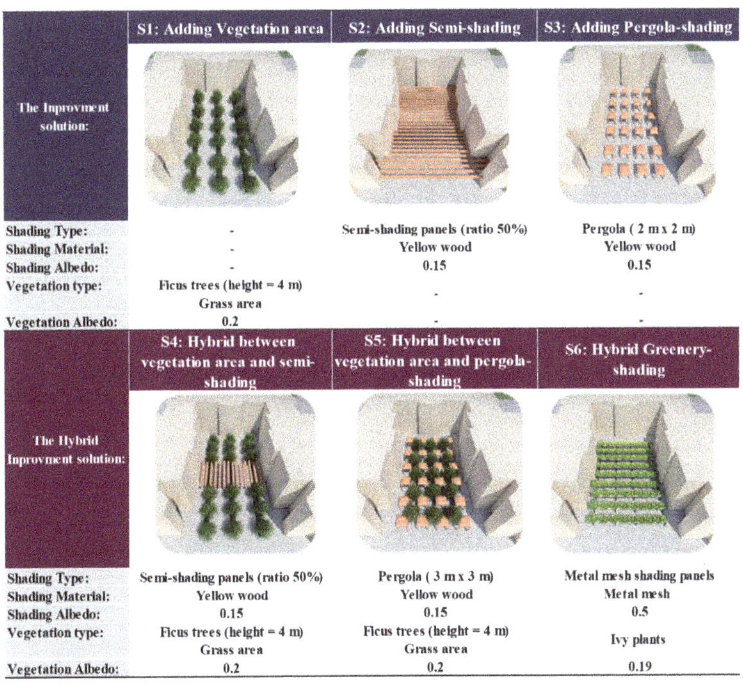


**Figure 4.** The six proposed improvement solutions.

#### 2.2.3. Stage 3: Simulating the Indoor Energy Consumption by DesignBuilder

In the third stage, the model of the Faculty of Agriculture was built using DesignBuilder software version (V.5.0.3.007) based on the material described in Table 1. The study used the specification of material from the real field model, and the U-value was calculated in DesignBuilder simulation based on the real model and material specification. Figure 5a shows the external view of the building in the DesignBuilder software. The model was validated based on the measurement of air temperature in one class on the second floor for one day measurement overlooking the middle courtyards, as shown in Figure 5b. This classroom was taken in the north direction, as most of the classrooms take the north orientation in Sohag University. The calibrated model results in a strong coefficient of determination ($R^2$) equal to 0.96, which has been accepted in the past literature [24]. This produced an actual environment of the educational building to investigate different retrofitting strategies for improving energy consumption. The building model was calibrated using indoor measurements from 9:00 to 14:00 inside classes overlooking courtyards with different orientations, representing spaces with different functions (classes, labs, offices). The five retrofitting strategies were proposed for the DesignBuilder model to reduce building annual energy consumption and heat transfer to the indoor environment. The five strategies are the following: (a) adding wall insulation, (b) adding wall and roof insulation, (c) adding window louvers for different orientations, (d) adding air infiltration, and (e) replacing the existing lighting system with LED units, besides combining all strategies.

**Table 1.** Description of building materials of the Faculty of Agriculture.

| Building Part | Material | U-Value ($W/m^2K$) | R-Value ($W/m^2K$) | Thickness (m) | Properties | |
|---|---|---|---|---|---|---|
| Glass windows | Single glass | 5.7 | | 0.006 | Total solar transmission (SHGC)<br>Direct solar transmission<br>Light transmission | 0.623<br>0.487<br>0.749 |
| External and internal walls | Red brick (with 2 cm cement plaster on every side) with a thickness of 25 cm without thermal insulation | 0.944 | 1.059 | 0.29 | **Inner surface**<br>Convective heat transfer coefficient ($W/m^2$-K)<br>Radiative heat transfer oefficient ($W/m^2$-K)<br>Surface resistance ($m^2$-K/W)<br>**Outer surface**<br>Convective heat transfer coefficient ($W/m^2$-K)<br>Radiative heat transfer oefficient ($W/m^2$-K)<br>Surface resistance ($m^2$-K/W) | <br>2.152<br>5.540<br>0.130<br><br>19.8<br>5.130<br>0.040 |
| | Ceiling tile<br>Cement mortar | | 2.881 | 0.43 | **Inner surface**<br>Convective heat transfer coefficient ($W/m^2$-K)<br>Radiative heat transfer oefficient ($W/m^2$-K)<br>Surface resistance ($m^2$-K/W) | <br>4.46<br>5.540<br>0.100 |
| Roof | Sand | 0.353 | | | | |
| | Concrete (lightweight)<br>Heat insulation (polyurethane)<br>Water insulation (rubber)<br>Concrete<br>Cement plaster | | | | **Outer surface**<br>Convective heat transfer coefficient ($W/m^2$-K)<br>Radiative heat transfer oefficient ($W/m^2$-K)<br>Surface resistance ($m^2$-K/W) | <br>19.8<br>5.130<br>0.040 |

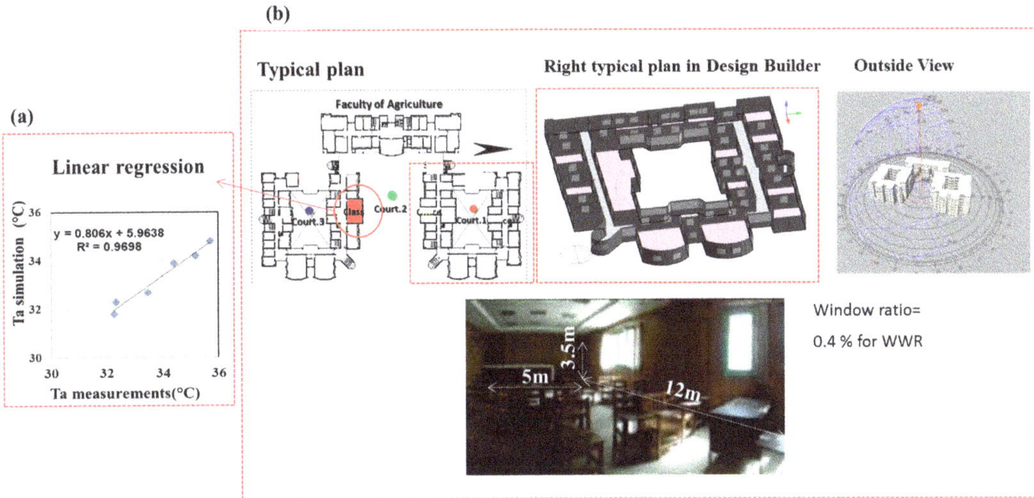


**Figure 5.** The validation process in DesignBuilder: (**a**) linear regression of indoor air temperature in DesignBuilder with relation to measurement, (**b**) the model of the case study (plan, outside view) and classroom dimension in the Faculty of Agriculture.

## 3. Results and Discussion

### *3.1. Evaluation of In Situ Measurements of Air Temperature, Questionnaire, and PET Thermal Comfort Index*

Figure 6 shows the following temperature and questionnaire survey results of the agriculture building: (a) the indoor temperature inside classes, the air temperature inside the outer open courtyards, and the PET pattern inside the open courtyard, (b) the results of the thermal sensation vote and air velocity vote inside the courtyards of buildings. It was concluded that air temperature inside different classes of the second floor has nearly the same pattern, and most indoor temperatures are high and away from the 90% acceptability limits of the Adaptive Comfort Standard (ACS) of [25] using natural ventilation. Additionally, the thermal performance inside the middle courtyard shows warm and strong heat stress, especially after 12:00 pm, due to less shading and trees in the outer open courtyards. High outdoor air temperature strongly affects students' outdoor thermal comfort while practicing different social activities inside the courtyards. Additionally, during the hot period, thermal satisfaction in the Faculty of Agriculture courtyard was not achieved for the majority of students. Only 20% reported feeling "neutral" regarding the temperature, while 73% perceived it as "slightly to very hot." This dissatisfaction is primarily attributed to the courtyard's H/W ratio of 0.7, leading to a low sky view factor (SVF) and a confined layout. These factors contribute to increased solar radiation, lack of vegetation, and limited air movement, which are exacerbated by the closure of shared courtyard doors. Consequently, 30% of students perceived the air speed as "low," while just 23% found it "appropriate," indicating insufficient natural ventilation. To investigate the urban morphology and outdoor thermal comfort, SVF was used as an indicator based on the calculation of Rayman. The high value of Sky View Factor (SVF) equal to 0.84 maximizes the solar radiation coming into the open courtyard, which affects the increase of air temperature and heat stress for students.

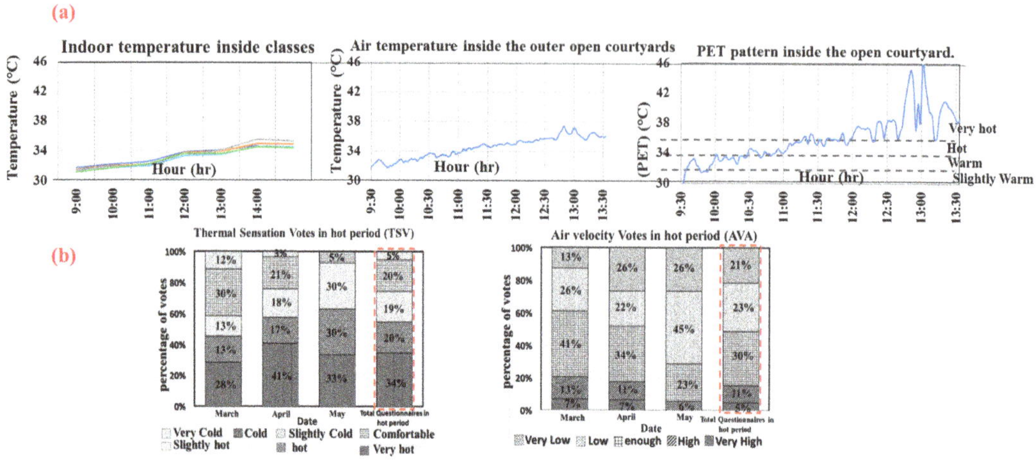

Figure 6. Temperature and questionnaire survey results of the agriculture building: (a) indoor temperature inside classes, air temperature inside the outer open courtyards, and PET pattern inside the open courtyard, and (b) results of the thermal sensation vote and air velocity vote inside the courtyards of buildings.

### 3.2. Analysis of In Situ Measurements of Airspeed and $CO_2$ Concentration

The air speed inside the courtyard was measured concerning its carbon dioxide concentration and its effect on student thermal comfort. Figure 7 shows the values of $CO_2$ concentration (a) inside three courtyards with different aspect ratios, and (b) two classes overlooking the middle courtyard of the Faculty of Agriculture. It was concluded that the level of $CO_2$ concentration decreases in different courtyards when opening the entrance doors with cross ventilation, and this is equivalent to an air speed equal to 3.0 m/s. It was observed that the courtyard with an aspect ratio equal to 1.2 achieved high air speed with low $CO_2$ concentration, as shown in Figure 7a. The middle courtyard achieved the lowest average $CO_2$ concentration, with an average of 550 ppm and 600 ppm during the opening and closing of the connection with other courtyards, respectively, and this is higher than the standard according to [26]. According to ASHRAE Standard 62.2-2019 [26], the baseline outdoor $CO_2$ concentration is assumed to be 1000 ppm, which is typically used as a reference value for outdoor air in ventilation calculations. This causes the rise of $CO_2$ concentration (960 ppm) in the class overlooking the middle courtyard with a H/W ratio equal to 0.7 due to not opening windows in the presence of students.

### 3.3. Evaluation of In Situ Measurements of Thermal Images in the Courtyard

The thermal image was used to analyze the surface temperature for different materials in the building and urban canyons, including the ground with different materials and vegetation. Additionally, it was used for the diagnosis of the current situation concerning the material used and the effect of heat transfer from material used inside the courtyards and the material used for the building façade inside the courtyards. The evaluation was conducted using thermal imaging to improve the weak point using simulation software. Figure 8 shows a thermal image inside the middle courtyard. The ground surface temperature of the outer courtyards was very high, with an average temperature of 41 °C, while the surface temperature under the shaded area (pergola) decreased to 27 °C. High surface temperature for the courtyard ground increases air temperature and reflects heat to students causing students' discomfort. This is due to the use of dark tiles in most ground areas that absorb and release heat to the surroundings. Additionally, low air speed in the courtyard, while closing the connection with other courtyards, causes more heat release to the air temperature inside the courtyard and adds to student discomfort. High outdoor surface temperature causes high heat release and transfer to the indoor environment and discomfort to students. Based on a questionnaire survey for students staying inside the middle courtyard, more than 73% felt discomfort, "slightly too hot". This caused students' dissatisfaction, for most of the students, regarding high solar radiation and no shading for the courtyard. SVF in the middle courtyard is equal to 0.84. This increases the solar radiation received in the courtyard, which increases outdoor air temperature. The dark tile inside the courtyard emits heat to the students inside the courtyard, as does the absence of trees. Based on the monitoring, questionnaire survey, and thermal analysis of the courtyard and classrooms, the increase of outdoors with low tree density and shading was correlated with student thermal comfort. Additionally, high indoor air temperature correlated with student thermal comfort and high heat transfer. The simulation sections help to solve the problem that appears in the evaluation of the thermal image. Therefore, two numerical simulations were used. First numerical simulation using the Envi-met tool was used to improve student outdoor thermal comfort to increase social interaction based on three strategies inside the middle courtyards. Second, the DesignBuilder model was used to simulate retrofitting strategies to reduce heat transfer inside the classroom and achieve a low-energy building.

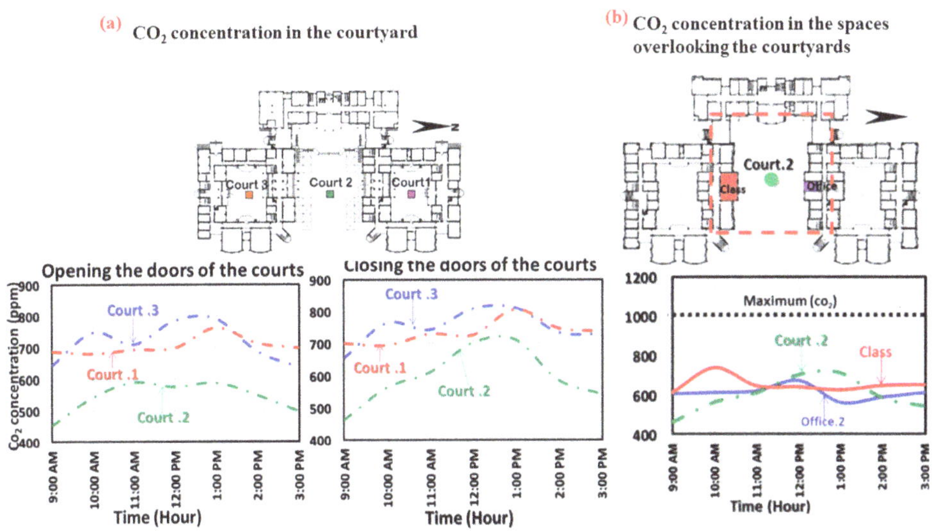


**Figure 7.** $CO_2$ concentration for the Faculty of Agriculture building: (**a**) $CO_2$ concentration in the three courtyards, and (**b**) $CO_2$ concentration for two classes overlooking the middle courtyard.

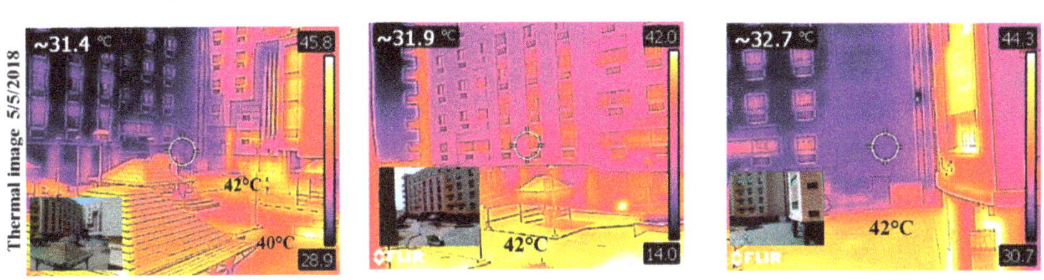


**Figure 8.** Thermal image of the outer surface temperature of the outer courtyards at 1 pm.

### 3.4. Envi-met Simulation Results

The results of the six improvement solutions were obtained using Envi-met software. Figure 9 shows the air temperature pattern during the workday from 9 a.m. to 3 p.m. in two points, with point 1 in the deep of the middle open courtyards and point 2 in the opening side. In point 1, the average of Ta in the base case was 37.07 °C based on the simulation results in Envi-met, but its reduction, as a result of applying the improvement solutions, ranged between 2.69 °C and 3.91 °C. Thus, the highest reduction of the average of Ta was obtained by applying the improvement solution S4 of a hybrid between the vegetation area and semi-shading, while the average of Ta was 33.16 °C. Additionally, the solution of S5 of a hybrid between the vegetation area and pergola shading contributed to the reduction of the Ta average by 3.23 °C. Conversely, the lowest reduction was obtained by applying S3, which was adding pergola shading, whereas the average of Ta reached 34.48 °C, which is still a beneficial improvement.

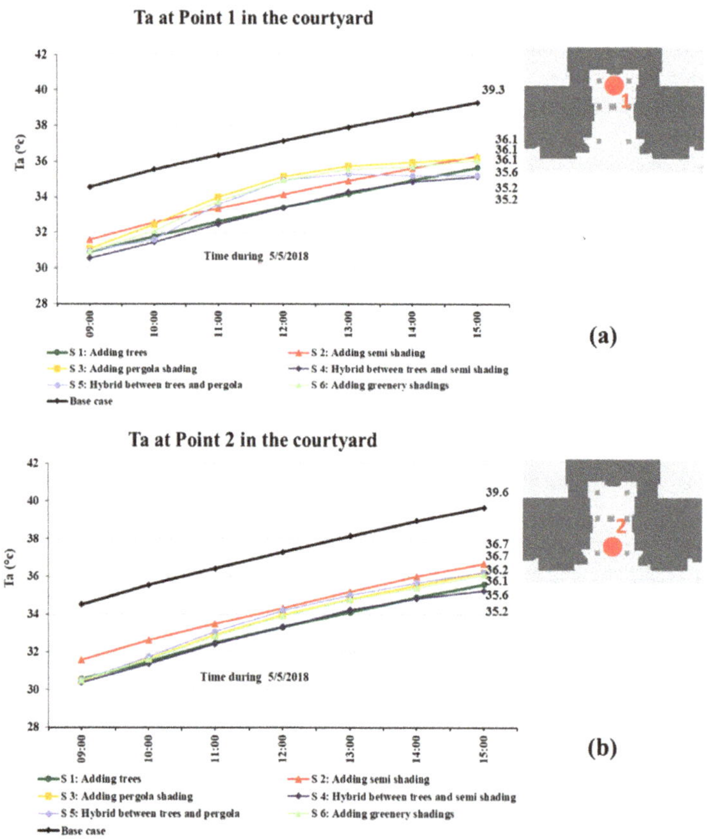


**Figure 9.** Air temperature simulation results of the base case and six improvement solutions at (**a**) point 1 in the deep of the courtyard and (**b**) point 2 in the opening side of the courtyard.

On the other hand, in point 2, the average of Ta in the base case was 37.2 °C, but its reduction as a result of applying the improvement solutions ranged between 3.44 °C and 4.10 °C. Hence, the highest reduction of the average of Ta was obtained by applying the improvement solution S4 of a hybrid between the vegetation area and semi-shading, while the average of Ta was 33.11 °C. Additionally, the solutions of S1 (adding vegetation area) and S2 (adding semi-shading) assisted in the reduction of the Ta average by 4.01 °C.

However, the lowest reduction was obtained by applying S5 of a hybrid between the vegetation area and pergola shading, whereas the average of Ta reached 33.78 °C, which is still a beneficial improvement. Moreover, these results are compatible with the results of [27], who found that adding trees in the educational building courtyard reduces the air temperature by 4.9 °C. Additionally, the results of a reduction of Ta using hybrid solutions (from 2.9 °C to 4.10 °C) agree with the results of [5,28], who reduced Ta by 2.1 °C and 5.8 °C by using hybrid solutions. It is concluded that a significant reduction of Ta occurred by applying the hybrid improvement solutions at the two points, which ranged between 3.23 °C and 4.1 °C.

Figure 10 shows the hourly mean radiant temperature for the six improvement solutions compared to the base case. MRT is very sensitive to different urban strategies. In point 1, the average of MRT in the base case was very high and reached 62.3 °C, although its reduction as a result of applying the improvement solutions ranged between 5.61 °C and 21.02 °C. Hence, the highest reduction of the average of MRT was obtained by applying the improvement solution S4 of a hybrid between the vegetation area and semi-shading, while the average of MRT reaching only 41.3 °C. Additionally the solution of S5 (a hybrid between the vegetation area and pergola shading) and S6 (a hybrid of greenery shading) achieved reductions of the average of MRT by 18.37 °C and 14.22 °C, respectively. The other three improvement solutions, S1, S2, and S3, lightly reduced the average of MRT. This is due to the strong impact of hybrid solutions in providing a large shaded area and moderate air temperature as a result of increasing the vegetation and shading area together.

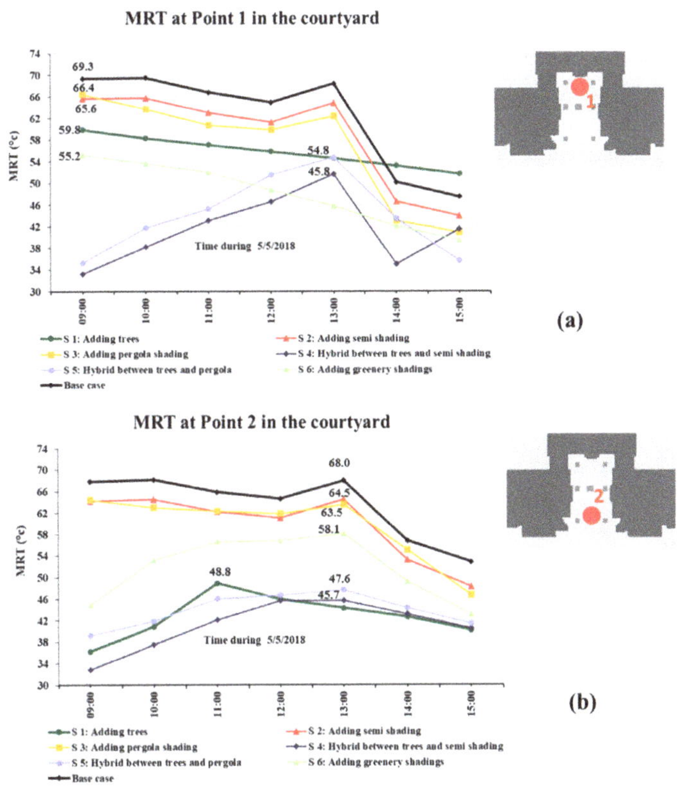


**Figure 10.** The mean radiant temperature simulation results of the base case and six improvement solutions at (**a**) point 1 in the deep of the courtyard and (**b**) point 2 in the opening side of the courtyard.

On the other side, in point 2, the average of MRT in the base case was 63.4 °C, but its reduction as a result of applying the improvement solutions ranged between 3.88 °C and 22.3 °C. Moreover, the highest reduction of the average of MRT was obtained by applying the improvement solution S5 of a hybrid between the vegetation area and pergola shading, while the average of MRT was 41.08 °C. Although, adding pergola shading only, as in solution S3, achieved the lowest reduction of the average of MRT. Therefore, it confirmed the necessity of hybridizing the improvement solutions between shading units and vegetation elements. These results are consistent with the previous researcher [29], who clarified the impact of the hybrid solution in reducing the MRT value by 9.2 °C. Additionally, the results of [30] indicated that the hybrid solution contributed to the reduction of MRT by a range from 12 °C to 15.3 °C, which agrees with our results (11.7 °C to 22.3 °C). In addition, the results are consistent with the results of [5,28]. It is concluded that a significant reduction of MRT occurred by applying the hybrid improvement solutions at the two points, which ranged between 14.2 °C and 22.3 °C.

Eventually, PET values were obtained using LEONARDO outputs and RayMan outputs. LEONARDO is a data visualization and analysis tool within Envi-met that allows users to analyze the simulation results, such as Ta, MRT, etc. Figure 11 shows the hourly PET values for the six improvement solutions at two points inside the courtyards. In point 1, the average of PET in the base case was very high and reached 51.09 °C (very hot grade), which caused thermal discomfort and difficulty in doing activities for students. Nevertheless, the average PET reduction as a result of applying the improvement solutions ranged between 6.34 °C and 13.9 °C. Hence, the highest reduction of the average of PET was obtained by applying the improvement solution S4 of a hybrid between the vegetation area and semi-shading, while the average of PET reached only 37.1 °C. Additionally, the solution of S5 (a hybrid between the vegetation area and pergola shading) and S6 (hybrid greenery-shading) achieved reductions of the average of PET by 13.46 °C and 11.23 °C, respectively. Accordingly, the grade of outdoor thermal comfort based on PET improved from a very hot grade (above 42 °C) in the base case to a warm grade (38.1–42 °C) by applying S4 and S5, which is a significant improvement [5].

Conversely, in point 2, the average of PET in the base case was 49.7 °C, but its reduction as a result of applying the improvement solutions ranged between 8.1 °C and 13.1 °C. Moreover, the highest reduction of the average PET was obtained by applying the improvement solution S5 of a hybrid between the vegetation area and pergola shading, while the average PET was 36.6 °C. However, adding pergola shading only, as in solution S3, achieved the lowest reduction of the average of PET (8.1 °C). Indeed, almost all the improvement solutions achieved a high reduction in PET in point 2 because of its location at the opening side of the courtyard and the availability of higher wind speeds than at point 1. Thus, these results are compatible with the results of [5], which indicated that the reduction of PET reached 15.9 °C by applying hybrid solutions in the school courtyard. Additionally, the reduction of PET based on adding vegetation area (reached 11.9 °C) agrees with the results of [6], which indicated the reduction of PET reached 14.4 °C by adding trees in the courtyard. Although the Ta, MRT, and PET reduction resulting from applying a single type of shading, such as semi-shading (S2) and pergola shading (S3), was the least due to the small shading area provided. The solution of adding trees (S1) assisted in the reduction of Ta, MRT, and PET, which reduced due to the trees' properties of providing shade and air humidity, which cool the outdoor air. However, the hybrid improvement solutions were the most efficient in reducing Ta, MRT, and PET. It was concluded that a significant reduction of PET occurred by applying the hybrid improvement solutions at point 1 and by applying almost all the improvement solutions at point 2. In addition, using greenery shading panels covered with ivy plants is more effective than using regular wooden shading (semi-shading panels and pergola units) in reducing Ta, MRT, and PET because of the hanging plants' effect, but still, other hybrid solutions (S4, S5) are the most efficient solutions due to the large trees effect, wide vegetation, and wide shading area. These reduce the heat transfer from shaded ground to the surrounding area, with significant thermal comfort for students.

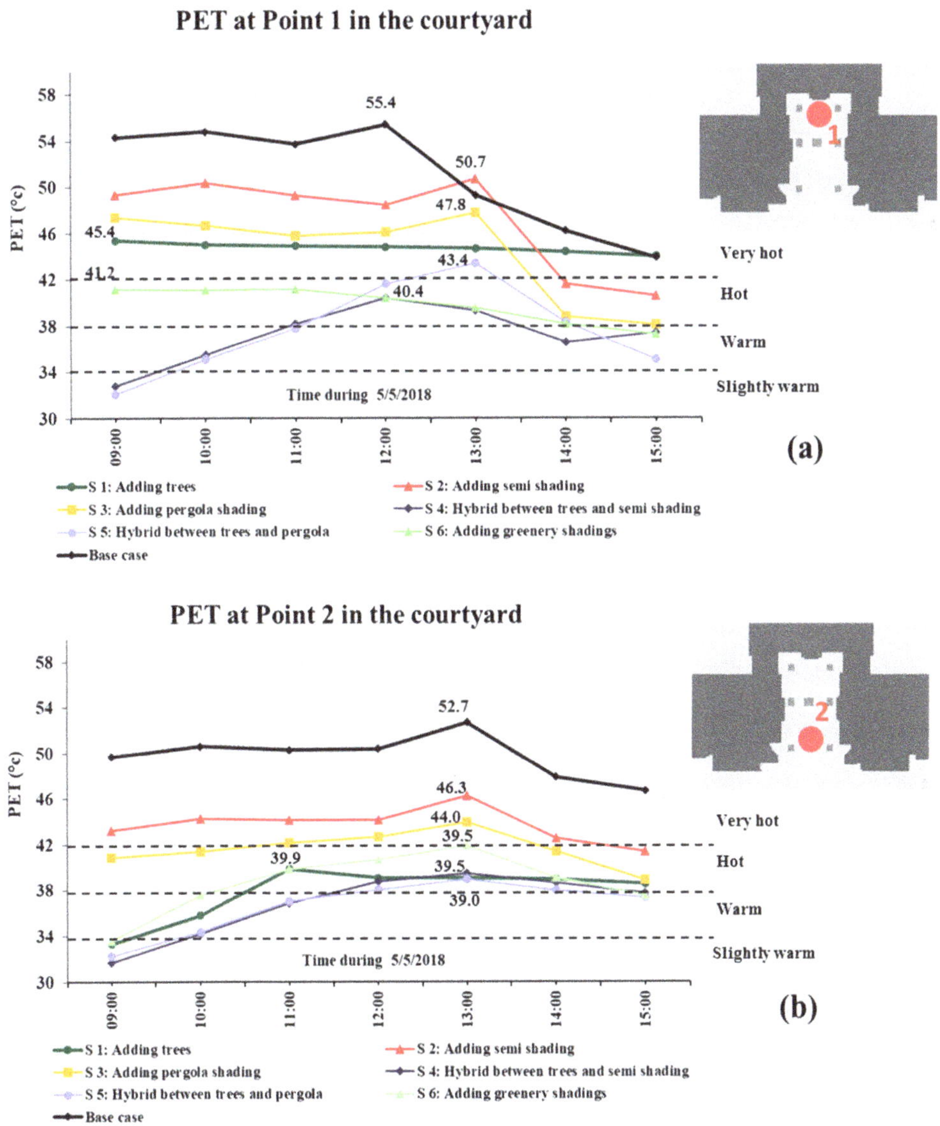


**Figure 11.** The PET hourly values of the base case and six improvement solutions at (**a**) point 1 in the deep of the courtyard and (**b**) point 2 on the opening side of the courtyard.

Figure 12 shows the contour map of $T_{MRT}$ of different improvement solutions at two periods of time (11:00 a.m. and 2:00 p.m.). The results of the contour map emphasize the significant achievement of hybrid solutions compared to other solutions.

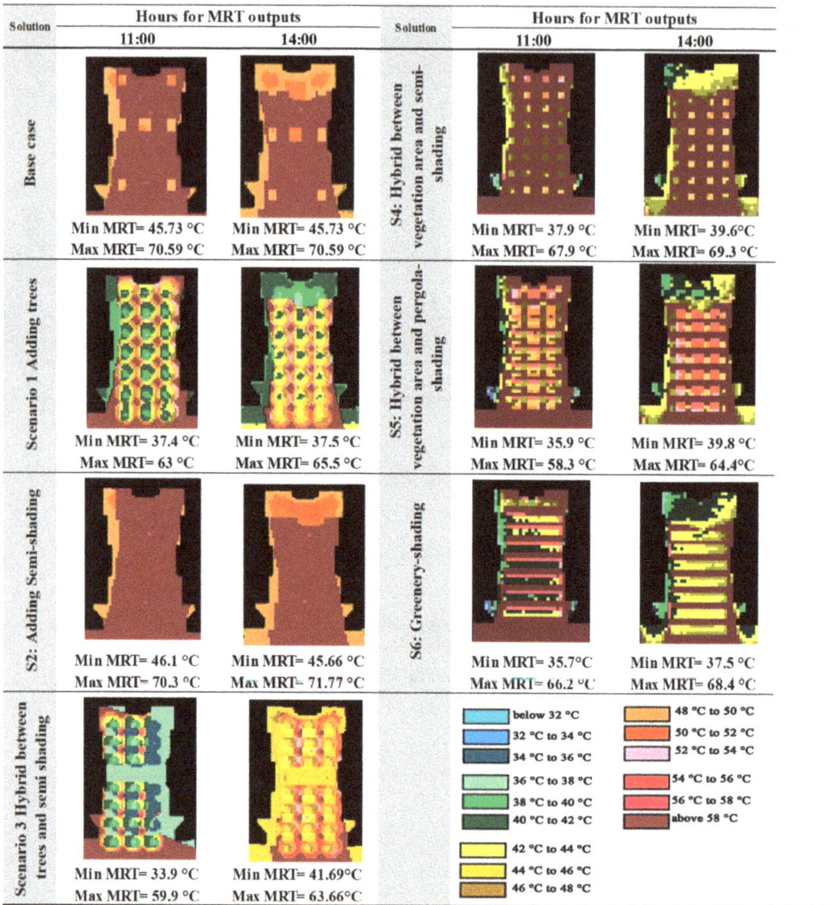


**Figure 12.** Contour map for MRT distribution with max and min temperatures inside the open courtyards for the base case and improvement solution.

### 3.5. DesignBuilder Simulation Results

The DesignBuilder simulation model was used to reduce heat transfer to an indoor environment that strongly affects student thermal behavior and thermal comfort. Thus, four retrofitting strategies were adopted for the building envelope to achieve student comfort. Additionally, replacing LED units was used to increase energy consumption compared to another strategy that aimed to achieve thermal comfort and a low-energy building. Figure 13 shows the average indoor temperature difference between the different retrofitting strategies in one of the classrooms where measurement was conducted. It was concluded that using a passive design strategy and other mechanical ventilation achieved the lowest indoor temperature of 33.22 °C. Then, by integrating the window louver, the indoor air temperature was reduced by 0.04 °C, while the average indoor temperature for the wall insulation was nearly the same as the base case due to the existing case study using a wall with a thickness of 25 cm and a double wall in the same part of the façade of the classroom. This causes low heat transfer to the indoor environment.

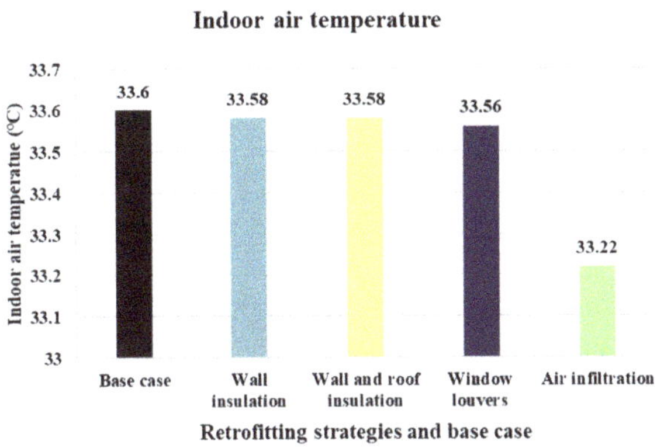


**Figure 13.** Indoor air temperature for the validated classroom based on different retrofitting strategies.

Nevertheless, the impact of different retrofitting strategies on total annual energy consumption was studied and is shown in Figure 14. It was observed that the impact of wall and roof insulation, air infiltration (enhanced natural ventilation), and using LED lighting systems was the most significant on the annual energy consumption, while the percentage of energy consumption reduction reached 4%, 5%, and 25%, respectively. Additionally, combining all the significant strategies achieved a reduction in energy consumption of 30%. Hence, these results are compatible with the results of [31], which indicated that the energy saving of using wall insulation in residential buildings was about 2.2%. Additionally, the results agree with the previous research that concluded the reduction of energy consumption based on combining retrofitting strategies was 31.1%. However, using the LED lighting retrofitting strategy in this study achieved only 25% energy savings compared to 30% in the previous work, due to the shorter time of using artificial lighting at universities than in residential buildings in [31].

### *3.6. Abilities and Limitations*

The proposed methodology has been discussed, and its abilities are listed as follows:

(a) Relying on a coupled simulation methodology to improve the outdoor thermal comfort in the middle open courtyard of the case study and to reduce indoor energy consumption;
(b) Providing results of outdoor and indoor thermal comfort;
(c) Studying six improvement solutions for outdoor spaces and five retrofitting strategies for indoor spaces at the same time;
(d) Investigating the impact of three different types of shading: semi-shading panels, pergola shading, and greenery shading (metal mesh shading panels covered with ivy plants);
(e) Investigating the impact of two hybrid improvement solutions (between the vegetation area and semi-shading, and between the vegetation area and pergola shading);
(f) Ability to apply the proposed methodology in several universities' courtyards in hot climate cities;
(g) The credibility of the proposed methodology is due to the agreement between the field measurements and the model results;
(h) The flexibility of upgrading the proposed methodology to include new improvement solutions to contribute to solving global warming issues and confronting climate change issues.

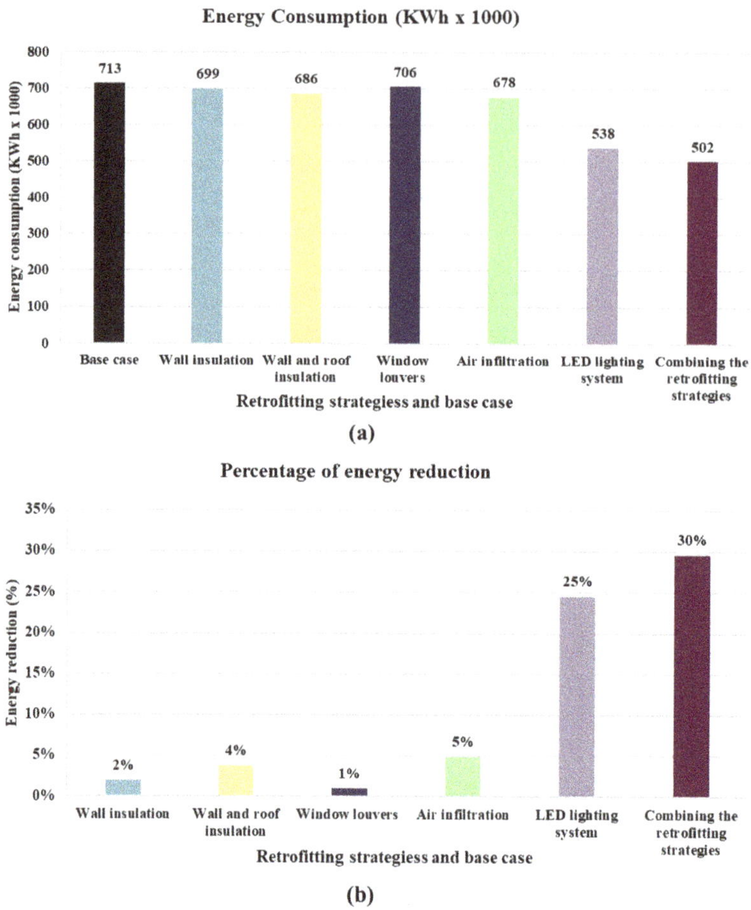


**Figure 14.** The indoor energy after applying the retrofitting strategies for the whole building: (**a**) energy consumption and (**b**) the percentage of energy reduction.

On the other side, the limitations of the proposed methodology are listed as follows:

(a) The limitation of shading types and ignoring other types, such as tent shading and arcades;

(b) The limitation of the building height is equal to 20 m, and the courtyard ratio (H/W) is 1.2;

(c) The limitation of ignoring other improvement solutions, such as ground material, surface albedo, shading ratio, and shading height;

(d) The limitation of ignoring the impact of the geometric building parameters on outdoor thermal comfort, such as self-shading, H/W, and the opening side of the courtyard.

## 4. Conclusions and Recommendations

Improving the courtyard condition strengthens social relations for students and reduces energy consumption for classes overlooking the courtyard. Thus, this study aims to improve the outdoor students' thermal comfort in university courtyards, in addition to reducing energy consumption in university buildings in hot arid climates, especially at Sohag University. Moreover, this study integrated field monitoring and a questionnaire with coupled simulation methodology to improve outdoor thermal comfort by applying six different improvement solutions (using the Envi-met model) and to reduce indoor energy consumption by applying five retrofitting strategies (using DesignBuilder). The results from the monitoring, questionnaire, thermal image analysis, and simulation can be summarized as follows:

- Through monitoring, the high outdoor air temperature inside the middle courtyard, with a high value of SVF equal to 0.84, causes discomfort for students and high heat stress due to the lack of shading and low tree density. The PET values exceed 38 °C, from 10:00 to 13:00 only.
- Based on thermal image analysis, the ground surface temperature of the black tiles reaches 41 °C in most areas of the open courtyard with no trees and shade. This causes high heat reflection on a student staying in the outdoor area, making them uncomfortable. The surface temperature decreases to 27 °C under the shaded area.
- Closing the connection between different courtyards, the effect strongly affects $CO_2$ concentration. The middle courtyard achieves the lowest average $CO_2$ concentration, with an average of 550 ppm and 600 ppm during the opening and closing of the connection with other courtyards, respectively, compared to the closed courtyard. This causes the rise of $CO_2$ concentration (960 ppm) in the class overlooking the middle courtyard with a H/W ratio equal to 0.7 due to not opening windows in the presence of students.
- To achieve student comfort in the courtyard between buildings, a significant reduction of PET ranges between 11.1 °C and 13.9 °C, which occurs by applying the hybrid improvement solutions (vegetation area and semi-shading or pergola shading). Accordingly, the grade of outdoor thermal comfort based on PET improves to become located in the warm grade, which ranges from 34.1 °C to 38 °C during almost all day hours.
- The reduction of the average of Ta ranges between 2.69 °C and 4.10 °C, and the reduction of the average of MRT ranges between 3.88 °C and 22.3 °C, as a result of applying the improvement solutions.
- In addition, using greenery shading panels covered with ivy plants is more effective than using regular wooden shading (semi-shading panels and pergola units) in reducing Ta, MRT, and PET because of the hanging plants' effect, but still other hybrid solutions (S4, S5) are the most efficient solutions due to the large trees effect, wide vegetation, and wide shading area.
- The integration of louvers, air infiltration, and wall and roof insulation, using an LED lighting system, achieves a significant percentage reduction in annual energy consumption equal to 30% with low heat transfer using DesignBuilder.

Therefore, it is recommended to apply hybrid improvement solutions (vegetation area and semi-shading or pergola shading), especially for shallow canyons and courtyards with an aspect ratio (H/W) equal to 1.2 in the existing university building of a hot arid climate or a new building. This achieves student thermal comfort in open spaces more than using shading only or increasing tree density only. The results help designers to integrate the optimum solutions in the current university buildings and the early design of new education buildings. Moreover, it is proposed that the future work of this methodology should be expanded to include different shapes of open and closed courtyards and to apply diverse mitigation solutions such as phase change material.

**Author Contributions:** Conceptualization, A.S.H.A. and R.M.A.M.; methodology, A.S.H.A. and R.M.A.M.; software, R.M.A.M.; validation, R.M.A.M. formal analysis, A.S.H.A. and R.M.A.M.; investigation, A.S.H.A. and R.M.A.M.; resources, A.S.H.A. and R.M.A.M.; data curation, A.S.H.A. and R.M.A.M.; writing—original draft preparation, A.S.H.A. and R.M.A.M.; writing—review and editing,

A.S.H.A. and R.M.A.M.; visualization, R.M.A.M.; supervision, A.S.H.A.; project administration; funding acquisition. All authors have read and agreed to the published version of the manuscript.

**Funding:** This work was supported and funded by the Deanship of Scientific Research at Imam Mohammad Ibn Saud Islamic University (IMSIU) (grant number IMSIU-DDRSP2502).

**Data Availability Statement:** Data are contained within the article.

**Conflicts of Interest:** The authors declare that they have no known competing financial interests or personal relationships that could have appeared to influence the work reported in this paper.

## References


1. Lam, C.K.C.; Weng, J.; Liu, K.; Hang, J. The effects of shading devices on outdoor thermal and visual comfort in Southern China during summer. *Build. Environ.* **2023**, *228*, 109743. [CrossRef]
2. Abdallah, A.S.H.; Mohammad, D.A.; Ali, A.-M.M. The influence of different courtyard ratios in university buildings on their thermal performance during the hot period: (Faculties of Agriculture and Education, New Sohag University, Egypt as a case study), Simulation for a Sustainable Built Environment. In *IOP Conference Series Earth and Environmental Science 397*; IOP Publishing Ltd.: Bristol, UK, 2019. [CrossRef]
3. Talukdar, M.S.J.; Talukdar, T.H.; Singh, M.K.; Abdul Baten, M.; Hossen, M.S. Status of thermal comfort in naturally ventilated university classrooms of Bangladesh in hot and humid summer season. *J. Build. Eng.* **2020**, *32*, 101700. [CrossRef]
4. Mallen, E.; Bakin, J.; Stone, B.; Sivakumar, R.; Lanza, K. Thermal impacts of built and vegetated environments on local microclimates in an Urban University campus. *Urban Clim.* **2020**, *32*, 100640. [CrossRef]
5. Mahmoud, R.M.A.; Abdallah, A.S.H. Assessment of outdoor shading strategies to improve outdoor thermal comfort in school courtyards in hot and arid climates. *Sustain. Cities Soc.* **2022**, *86*, 104147. [CrossRef]
6. Darvish, A.; Eghbali, G.; Eghbali, S.R. Tree-configuration and species effects on the indoor and outdoor thermal condition and energy performance of courtyard buildings. *Urban Clim.* **2021**, *37*, 100861. [CrossRef]
7. Ghaffarianhoseini, A.; Berardi, U.; Al-Obaidi, K. Analyzing the thermal comfort conditions of outdoor spaces in a university campus in Kuala Lumpur, Malaysia. *Sci. Total Environ.* **2019**, *666*, 1327–1345. [CrossRef]
8. Nasrollahi, N.; Namazi, Y.; Taleghani, M. The effect of urban shading and canyon geometry on outdoor thermal comfort in hot climates: A case study of Ahvaz, Iran. *Sustain. Cities Soc.* **2021**, *65*, 102638. [CrossRef]
9. Binabid, J.; Anteet, Q. Numerical study of vegetation effects on thermal comfort for outdoor spaces at a public school in hot and arid climate. *Environ. Adv.* **2024**, *15*, 100482. [CrossRef]
10. Qi, Y.; Chen, L.; Xu, J.; Liu, C.; Gao, W.; Miao, S. Influence of university campus spatial morphology on outdoor thermal environment: A case study from Eastern China. *Energy Built Environ.* **2025**, *6*, 43–56. [CrossRef]
11. Khalili, S.; Fayaz, R.; Zolfaghari, S.A. Analyzing outdoor thermal comfort conditions in a university campus in hot-arid climate: A case study in Birjand, Iran. *Urban Clim.* **2022**, *43*, 101128. [CrossRef]
12. Guo, T.; Lin, Z.; Zhao, Y.; Fang, Z.; Fan, Y.; Zhang, X.; Yang, J.; Li, Y. Investigation and optimization of outdoor thermal comfort in elementary school campuses: Example from a humid-hot area in China. *Build. Environ.* **2024**, *248*, 111055. [CrossRef]
13. Ge, J.; Wu, J.; Chen, S.; Wu, J. Energy efficiency optimization strategies for university research buildings with hot summer and cold winter climate of China based on the adaptive thermal comfort. *J. Build. Eng.* **2018**, *18*, 321–330. [CrossRef]
14. Batlle, E.A.O.; Palacio, J.C.E.; Lora, E.E.S.; Reyes, A.M.M.; Moreno, M.M.; Morejón, M.B. A methodology to estimate baseline energy use and quantify savings in electrical energy consumption in higher education institution buildings: Case study, Federal University of Itajubá (UNIFEI). *J. Clean. Prod.* **2020**, *244*, 118551. [CrossRef]
15. Park, J.H.; Yun, B.Y.; Chang, S.J.; Wi, S.; Jeon, J.; Kim, S. Impact of a passive retrofit shading system on educational building to improve thermal comfort and energy consumption. *Energy Build.* **2020**, *216*, 109930. [CrossRef]
16. Wang, X.; Yang, L.; Gao, S.; Zhao, S.; Zhai, Y. Thermal comfort in naturally ventilated university classrooms: A seasonal field study in Xi'an, China. *Energy Build.* **2021**, *247*, 111126. [CrossRef]
17. Emil, F.; Diab, A. Energy rationalization for an educational building in Egypt: Towards a zero energy building. *J. Build. Eng.* **2021**, *44*, 103247. [CrossRef]
18. Amini, R.; Ghaffarianhoseini, A.; Ghaffarianhoseini, A.; Berardi, U. Numerical investigation of indoor thermal comfort and air quality for a multi-purpose hall with various shading and glazing ratios. *Therm. Sci. Eng. Prog.* **2021**, *22*, 100812. [CrossRef]
19. Cantón, M.A.; Ganem, C.; Barea, G.; Llano, J.F. Courtyards as a passive strategy in semi dry areas. Assessment of summer energy and thermal conditions in a refurbished school building. *Renew. Energy* **2014**, *69*, 437–446. [CrossRef]
20. Aloshan, M.; Aldali, K. Empirical study of facade retrofits for optimizing energy efficiency and cooling in school buildings in Saudi Arabia. *Energy Rep.* **2024**, *12*, 4105–4128. [CrossRef]

21. Sahu, V.; Gurjar, B.R. Spatial and seasonal variation of air quality in different microenvironments of a technical university in India. *Build. Environ.* **2020**, *185*, 107310. [CrossRef]
22. Meteoblue. 2018. Available online: https://www.meteoblue.com (accessed on 1 December 2019).
23. Ali-Toudert, F.; Djenane, M.; Bensalem, R.; Mayer, H. Outdoor thermal comfort in the old desert city of Beni-Isguen, Algeria. *Clim. Res.* **2005**, *28*, 243–256. [CrossRef]
24. Nasrollahi, N.; Hatami, Z.; Taleghani, M. Development of outdoor thermal comfort model for tourists in urban historical areas; A case study in Isfahan. *Build. Environ.* **2017**, *125*, 356–372. [CrossRef]
25. *ASHRAE Standard 55*; Thermal Environmental Conditions for Human Occupancy. America Society of Heating Refrigerating and Air-Conditioning Engineers, Inc.: Atlanta, GA, USA, 2017.
26. *ASHRAE Standard 62-2*; Ventilation and Acceptable Indoor Air Quality in Residential Buildings. America Society of Heating Refrigerating and Air-Conditioning Engineers, Inc.: Atlanta, GA, USA, 2019.
27. Zolch, T.; Mohammad, A.R.; Pfleiderer, E.; Wagner, G.; Pauleit, S. Designing public squares with green infrastructure to optimize human thermal comfort. *Build. Environ.* **2019**, *149*, 640–654. [CrossRef]
28. Mahmoud, H.; Ragab, A. Urban Geometry Optimization to Mitigate Climate Change: Towards Energy-Efficient Buildings. *Sustainability* **2021**, *13*, 27. [CrossRef]
29. Chen, C.; Zhang, Y.; Han, J.; Li, X. An investigation of the influence of ground surface properties and shading on outdoor thermal comfort in a high-altitude residential area. *Front. Arch. Res.* **2021**, *10*, 432–446. [CrossRef]
30. Abdallah, A.S.H.; Mahmoud, R.M.A.; Aloshan, M.A. Optimizing Urban Spaces: A Parametric Approach to Enhancing Outdoor Recreation Between Residential Areas in Riyadh, Saudi Arabia. *Buildings* **2025**, *15*, 1527. [CrossRef]
31. Abdallah, A.S.H.; Mahmoud, R.M.A. Investigation of Greening Façade and Retrofitting strategies on Outdoor Thermal Comfort and Indoor Energy Consumption in New Assiut City, Egypt. *Mansoura Eng. J.* **2023**, *48*, 9. [CrossRef]

**Disclaimer/Publisher's Note:** The statements, opinions and data contained in all publications are solely those of the individual author(s) and contributor(s) and not of MDPI and/or the editor(s). MDPI and/or the editor(s) disclaim responsibility for any injury to people or property resulting from any ideas, methods, instructions or products referred to in the content.

buildings

MDPI

Article

# Consideration of Thermal Comfort, Daylighting Comfort, and Life-Cycle Decarbonization in the Retrofit of Kindergarten Buildings in China: A Case Study

**Kai Hu [1,2], Chao Xu [1], Wenjun Li [1], Jing Ye [1], Yankai Yang [2] and Yizhe Xu [1,*]**

1 China Construction Yipin Investment and Development Co., Ltd., Wuhan 430070, China; kaihu2024@outlook.com (K.H.); chaoxu2024@outlook.com (C.X.); liwenjun2025@outlook.com (W.L.); jingye2025@outlook.com (J.Y.)

2 College of Urban Construction, Nanjing Tech University, No. 200, North Zhongshan Road, Nanjing 210009, China; yankaiyang2024@outlook.com

* Correspondence: yizhe.xu@njtech.edu.cn

**Abstract:** Kindergartens play a crucial role in nurturing the physical, cognitive, and social development of children. Hence, designing kindergarten buildings requires the consideration of the unique requirements and behavior of children. Considering the rapid urbanization of China and its commitment to achieving the 3060 carbon goal, in this study, we examine the retrofitting of kindergarten buildings in China and propose a retrofit optimization method for kindergarten buildings that considers thermal comfort, daylighting, and life-cycle carbon emissions. Through this method, information on the thermal and daylighting comfort of occupants, weather data, occupant scheduling, and envelope and energy system of the kindergarten building to be retrofitted can be obtained through various approaches, such as video playback, field investigation, literature research, and consult drawings. On this basis, optimization variables are selected, and a physical model is established to guide the retrofit process. Afterward, a rapid comprehensive optimization framework based on parallel computing is adopted to obtain the comprehensive optimal design scheme for the building to be retrofitted. The proposed method is applied to a kindergarten building retrofit case in Nanjing, China, and the results show that the optimal comprehensive scheme results in a reduction in carbon emissions of 34,158.3 kg, an increase in the thermal comfort period of 2.7%, and an improvement in daylighting comfort of 79.7% over the benchmark scheme. The significance of this study extends beyond its potential for widespread application in kindergarten building retrofits. It contributes to advancing sustainable building design and environmental stewardship, creating healthier and more comfortable learning environments for children while mitigating the environmental impact of buildings. Moreover, it emphasizes the importance of considering children's unique needs and behaviors in building design, ultimately leading to better outcomes for their overall development.

**Keywords:** kindergarten building retrofit; children comfort; life-cycle decarbonization; design optimization

## 1. Introduction

### 1.1. Background

Human-centered and sustainable development are two essential themes in the realm of building design and are crucial concepts that must be considered in the building design process. The kindergarten, a quintessential educational building, has received much attention from society due to its primary occupants, namely, children. With the urbanization rate of China surpassing 60% [1], urban renewal and retrofitting will serve as a primary focus in the field of building design in the future, with kindergartens as a key area of interest among designers and researchers. In traditional retrofitting of kindergartens, emphasis is placed on seismic reinforcement, thermal insulation, and energy conservation, while the comfort of occupants has been considered less [2,3].

Kindergartens are occupied primarily by children and teachers, with children constituting 90% of the total occupants. As children are at the stage of growth and development, the comfort of the indoor environment greatly impacts their physical and mental health. It has been shown in previous studies that a reasonable lighting environment can enhance the learning efficiency and decrease the incidence of myopia among children [4], while a comfortable thermal environment is beneficial to brain development and the mental health of children [5,6]. While trade-offs between thermal comfort, daylighting, and energy have been considered in some optimization studies of educational buildings [7], few studies have targeted design optimization of kindergarten building environments by fully considering the physiological and behavioral differences between children and adults.

Presently, the carbon peaking and carbon neutrality policy of the Chinese government, released in 2020, imposes higher requirements on energy conservation and emission reduction for educational buildings. Traditional energy-saving retrofitting only targets the operational energy consumption of buildings and fails to consider their carbon emissions from a life-cycle perspective. Therefore, achieving life-cycle decarbonization retrofitting of kindergarten buildings while satisfying the comfort requirements of occupants is a crucial challenge. This is very important for promoting the healthy growth of children and realizing building decarbonization.

### *1.2. Literature Review*

#### 1.2.1. Study on the Thermal Comfort of Children

Children are the primary occupants of kindergartens, and the indoor thermal environment is closely linked to their growth and development. However, the thermal comfort of children is often neglected by adults. Existing research on indoor thermal comfort primarily focuses on office and residential spaces for adults, with relatively few studies addressing the thermal environment of educational buildings. The main indicators of thermal comfort currently include the air temperature, operative temperature, predicted mean vote (PMV), and predicted percentage dissatisfied (PPD). In several studies, thermal comfort in educational buildings has been explored. For instance, Wang et al. investigated thermal comfort in a school classroom in Germany, with the air temperature as an evaluation indicator, and they targeted the thermal comfort range between 20 °C and 26 °C [8]. Xu et al. studied thermal comfort optimization of primary and secondary school classrooms in China, utilizing the satisfaction of the temperature set point and PMV to measure thermal comfort in classrooms [9,10]. Al-Rashidi et al. calculated clothing insulation for Kuwaiti boys and girls aged 6 to 17 in winter and summer, highlighting the importance of studying the clothing insulation characteristics of different age groups in PMV calculations [11]. Jindal studied the thermal environment in a classroom of a government residential school in the composite climate zone of Ambala, India, noting that students exhibit favorable heat resistance in terms of the operative temperature, with their thermal comfort range exceeding Indian and international adult standards [12]. Zomorodian et al. reviewed the research on thermal comfort in different educational buildings and examined different thermal comfort evaluation indicators and their applicability [13]. Li et al. investigated thermal comfort in kindergartens in Chongqing and Wuhan, China, using the PMV and thermal sensation vote (TSV) indicators to quantify the thermal comfort of children. They found that the main factors affecting the thermal comfort of children were the activity level and subjective regulation of the indoor thermal environment by teachers [5].

The above clearly indicates that the most widely used evaluation indicator for the thermal comfort of children is the PMV, which is suitable for steady-state thermal comfort evaluation in air-conditioned rooms. However, only a few researchers have fully considered the adaptability of the PMV indicator to the evaluation of thermal comfort among children of different ages. Additionally, there are few studies on the thermal comfort of children in kindergartens, where children are typically 4–6 years old (corresponding to grades 1, 2, and 3). Their physiological and activity characteristics significantly differ from those of primary and secondary school students. Therefore, further study is needed to determine the annual hourly thermal comfort of children in kindergarten buildings.

In contrast to office and commercial buildings, kindergarten buildings exhibit notable regional characteristics due to the influence of region, climate, customs, and teaching arrangements. In most existing research on thermal comfort optimization of kindergartens in China, certain measures have been adopted to improve the thermal environment in kindergartens (such as strengthening natural ventilation and local space adjustment), but the thermal environment in kindergarten buildings has been optimized from a whole-building perspective in few studies.

#### 1.2.2. Study on the Daylighting Comfort of Children

Daylight irradiation has been shown to promote bone development, prevent rickets, osteoporosis, and other diseases, and enhance the metabolism and blood circulation in children [14]. The optimization of daylighting comfort in kindergarten buildings has been widely studied. For example, Pagliolico et al. investigated the application of a photobio screen (PBS) shading system in a kindergarten in San Marcel, Italy, using various daylighting indicators to measure visual comfort. They found that the PBS shading system effectively facilitated an improvement in visual comfort in the kindergarten classroom [15]. Samiou et al. optimized daylighting in Greek kindergarten classrooms using spatial daylight autonomy (sDA), annual sun exposure (ASE), and daylight glare probability (DGP) as evaluation indicators for daylighting comfort. They optimized the size, position, and shading equipment of external windows and found that ASE and glare effects in the classroom were effectively alleviated after optimization [16]. Vásquez et al. proposed a multimethod approach to determine the preferences of children for the light environment and window landscape in kindergarten classrooms, and found that children can distinguish daylighting needs according to their activities [17]. Salleh et al. investigated the indoor environment in kindergarten buildings in Malaysia and found that glare in the classroom is the main factor affecting visual comfort [18]. Gao et al. studied the daylighting optimization of Chinese kindergartens using the daylight factor (DF) and illuminance, analyzing the effects of various local daylighting optimization measures for kindergarten classrooms, including orientation, interior decoration, classroom size, side window height, skylight position, and reflector features [19].

Although the daylighting of kindergartens has been extensively researched, kindergarten buildings often have distinct regional characteristics. Notably, the building scale, number of stories, classroom size, and number of students are closely related to regional characteristics. Chinese kindergartens are generally multistory buildings with internal corridors, and their classrooms typically serve multiple functions, such as teaching, activities, lunch breaks, and dining, resulting in larger classroom areas than those found in primary and secondary schools. The architectural features of large classrooms and inner corridors (external windows cannot be set on both the northern and southern sides) make it difficult to optimize the daylighting of kindergarten classrooms, and further research is needed.

#### 1.2.3. Energy Conservation and Emission Reduction of Kindergarten Buildings

Carbon emissions have always been a research hotspot in the field of building design, and many researchers have explored this topic [20–23]. In 2019, the Ministry of Housing and Urban–Rural Development of China issued a standard for calculating building carbon emissions, which clearly specifies the carbon emission calculation methods for the production, construction, operation, and final demolition stages [24]. In 2020, the Chinese government issued a carbon peaking and carbon neutrality policy, which greatly promoted research in this field [25,26]. Therefore, the optimization of kindergarten design must account for the issue of carbon emissions. Hammad et al. investigated energy conservation and emission reduction of kindergartens in Jordan, achieving an annual reduction in carbon dioxide emissions of 11.7 t through increased wall insulation, solar hot water systems, photovoltaic systems, and fresh air heat recovery [27]. Zhang et al. proposed a green design method for kindergartens based on renewable energy, reducing fossil energy consumption through the integration of renewable energy at the site and surrounding environment [28]. Causone et al. studied the retrofitting of a zero-energy kindergarten in a smart district renovation project, significantly reducing HVAC energy consumption and improving the indoor environmental quality through optimization [3]. Gajic et al. proposed an energy performance indicator by designing and measuring the parameters of representative samples of old and new kindergartens located in the temperate climate zone of Banja Luka, suggesting that kindergartens should be regarded as a unique type of building in building energy audits and certifications [2].

While energy conservation and emission reduction of kindergartens have been investigated in the above studies, the focus was on operating energy consumption and carbon emissions, with few studies analyzing kindergarten carbon emissions from a life-cycle perspective. Additionally, there are significant differences between the optimization of new buildings and retrofit optimization of existing buildings. In retrofitting existing buildings, greater attention is given to the application of envelope thermal insulation materials and renewable energy, such as photovoltaic systems [29]. Thus, the embodied carbon emissions of materials must be considered in the optimization process.

In references [8,30,31], it was noted that maintaining a comfortable indoor environment often requires high energy consumption, which could also lead to an increase in building carbon emissions. Therefore, multi-objective optimization methods are often applied in design optimization studies that balance comfort and carbon emissions to obtain suitable design schemes (including the weight method and Pareto method [32]). However, existing research indicates that there are significant differences in comfort needs among different populations [33,34]. If building design optimization can be further tailored to the comfort range of a specific group of people, it is clear that a better balance between comfort and carbon emissions can be attained. Children and kindergarten buildings are typical representatives of specific population groups and building types for retrofitting, respectively, and there is currently a lack of research on the correlation between the comfort of children and energy consumption in kindergarten buildings.

### 1.3. Research Gaps and Main Contributions

Drawing from the aforementioned review, it is evident that additional research is imperative to explore retrofit optimization strategies for kindergarten buildings in China, with a particular emphasis on assessing the annual hourly thermal and daylighting comfort levels of 4-to-6-year-old children as the primary occupants. Furthermore, the trade-off between occupant comfort and life-cycle decarbonization of these buildings warrants further investigation. Therefore, this paper proposes a design optimization method for retrofitting kindergarten buildings, which holistically integrates occupants' comfort evaluation and life-cycle decarbonization considerations.

The remainder of this paper is arranged as follows: in Section 2, the overall method of life-cycle decarbonization of comfortable kindergarten buildings is introduced in detail. In Section 3, the method is applied to the retrofitting of a real kindergarten project, and the results are analyzed. In Section 4, the study findings are examined, and prospects for future research are provided. A summary is presented in the last section of this paper.

## 2. Methodology

The fundamental process of this study, as depicted in Figure 1, can be primarily segmented into three components: theoretical research on children's comfort, information gathering and evaluation prior to retrofit design, and the optimization and analysis of design schemes. Given that the comfort of indoor thermal and daylight environments is primarily influenced by the building envelope and energy systems, the primary focus of this study is on optimizing the envelope and energy system of kindergarten buildings.

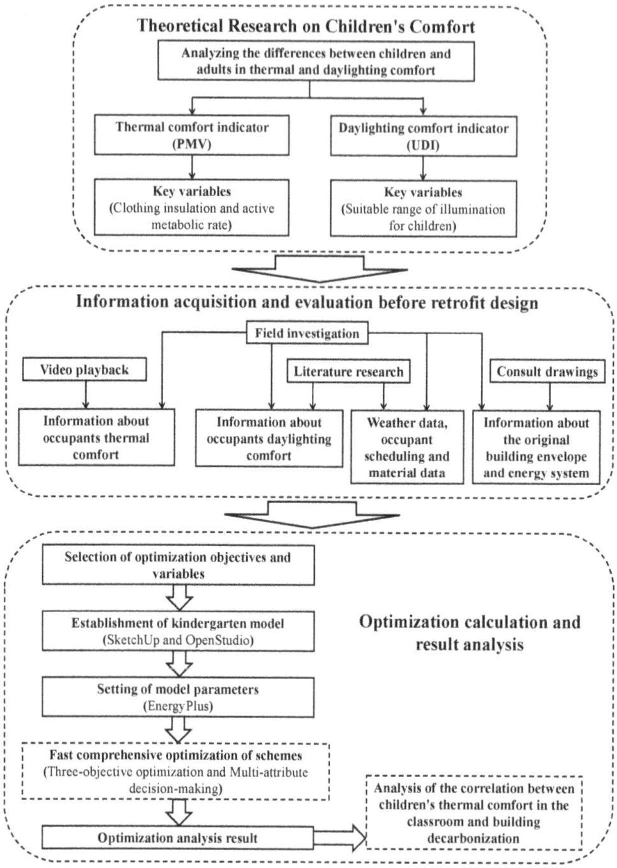


**Figure 1.** Basic process of retrofit optimization of kindergarten buildings.

## 2.1. *Information Acquisition and Evaluation before Retrofit Design*

### 2.1.1. Information about the Thermal Comfort of Occupants

Based on the literature review provided in Section 1.2.1, the PMV is the most commonly used indicator for evaluating the thermal comfort of children. It comprehensively considers the effects of ambient air temperature, radiation temperature, air humidity, air flow rate, human activity level, and clothing insulation on thermal comfort. In this study, the PMV is utilized as an indicator for evaluating the indoor thermal comfort in kindergarten buildings. Notably, in addition to environmental variables, the PMV is significantly influenced by the clothing insulation and metabolic rate of occupants [35]. It is important to recognize that the primary occupants of kindergartens are children. Therefore, when evaluating thermal comfort, it is crucial to consider the clothing insulation and active metabolic rate of children.

Clothing insulation is influenced by various factors, including the outdoor temperature, living habits, and activities. Although simple clothing insulation estimation methods have been provided in standards such as ASHRAE Standard 55-2013 [36] and ASHRAE Handbook 2009 [37], these standards are primarily applicable to adults. In Chinese kindergartens, clothing for children is often managed by nursing teachers to facilitate teaching and management. These teachers require children to dress or undress according to real-time activities and the indoor temperature. To ensure the practicality of retrofit optimization, we employ field investigations and video reviews in this study to determine the clothing worn by children during typical periods in different seasons. Equation (1) can then be used to calculate the insulation of the clothing worn by children during different periods [36].

$$I_{cl} = 0.835\sum_i I_{clu,i} + 0.161 \tag{1}$$

where $I_{cl}$ is the insulation of the clothing as a whole in clo and $I_{clu,i}$ is the insulation of a single piece of clothing in clo. Numerous standards and studies have provided metabolic rate values for adults engaged in activities of different levels [38]. However, the metabolic rate of children significantly differs from that of adults, and there are considerable variations in the metabolic rate among children of different ages [5]. In this study, Equations (2)–(5) can be used to calculate the basic metabolic rate of children, which involves determining the resting metabolic rate and skin surface area of children [39]. These parameters depend on the child's weight and height, as well as other factors such as age, region, and eating habits. It should be noted that Equations (2) and (4) only apply to children aged 3–10 years [39]. To obtain the metabolic rate of children engaged in different activities, field investigation should be conducted to obtain physical information of children in the kindergarten to be retrofitted. Afterward, various activity-related metabolic rates can be calculated based on the activity level conversion specified in ISO 8996 [40].

$$RMR = 0.082W_c + 0.545H_c + 1.736 \tag{2}$$

$$Q = \frac{RMR \times 10^6}{3600 \times 24} \tag{3}$$

$$A_D = 0.00659H_c + 0.0126AW_c - 0.1603 \tag{4}$$

$$M_b = \frac{Q}{A_D} \tag{5}$$

where $RMR$ is the resting metabolic rate in MJ/d; $W_c$ is the child's weight in kg; $H_c$ is the child's height in m; $Q$ is the metabolic rate per unit time in W; $A_D$ is the child's skin surface area of the child in m$^2$; and $M_b$ is the basic metabolic rate of the child in W/m$^2$.

#### 2.1.2. Information on Occupant Daylighting Comfort

Based on Section 1.2.2, an excessively high illuminance can cause glare, while insufficient illuminance can result in visual impairment and other problems for children. In previous studies, values of 100 lux and 2000/3000 lux have been utilized as the lower and upper limits of illuminance, respectively, as per references [7,41,42]. However, these values are primarily applicable to adults, and the acceptable range of illuminance for the visual comfort of children differs significantly from that for adults. Before retrofitting, the lower and upper limits of illuminance should be determined based on the age of the children [19].

#### 2.1.3. Information on Weather Data, Occupant Scheduling Data, and Material Data

Weather data, occupant scheduling data, and material data involved in the retrofit are also important contents of the information acquisition process before retrofit design. Weather data primarily comprise the annual hourly outdoor dry bulb temperature, humidity, wind speed, wind direction, and illuminance, which can be obtained through literature investigation and research. Occupant scheduling data primarily involve hourly usage patterns in different rooms, including timing control strategies for equipment in rooms, and can typically be obtained through field investigation. Material data involved in retrofitting can generally be obtained through literature reviews and research.

#### 2.1.4. Information on the Original Building Envelope and Energy System

During pre-retrofitting information acquisition, evaluating the envelope and energy system of the original building is essential. While some original buildings may contain a building energy management system (BAS) that can be used to perform a comprehensive and detailed evaluation, the equipment performance in most original buildings is relatively outdated, and a BAS is not commonly configured. Therefore, in this study, the original drawings and field investigation data are used to evaluate the building envelope and energy system to determine the parts that require retrofitting. During envelope investigation, it is crucial to obtain information on the thermal insulation performance of the external wall, external windows, roof, and ground of the original building. Similarly, during energy system investigation, it is essential to evaluate the performance of existing lighting lamps, domestic hot water equipment, elevator equipment, and HVAC equipment.

### 2.2. *Optimization Calculation for the Comprehensive Optimal Scheme*

After information acquisition and evaluation before retrofit design, the design optimization of the scheme can be targeted.

#### 2.2.1. Selection of Optimization Objectives and Variables

Section 1.3 indicates that we focus on the trade-off between thermal comfort, daylighting, and life-cycle decarbonization through multi-objective optimization. The optimization objectives are defined in Equations (6)–(8), where larger $TC$ and $UDI$ values indicate better performance, whereas smaller $C_{lc}$ values indicate better performance.

$$\begin{cases} TC = \frac{\sum_i (tf_i t_i)}{\sum_i t_i} \in [0,1] \\ tf_i = \begin{cases} 1\ if -1 \leq PMV \leq 1 \\ 0\ if\ PMV < -1 \vee PMV > 1 \end{cases} \end{cases} \quad (6)$$

$$\begin{cases} UDI = \frac{\sum_i (wf_i l_i)}{\sum_i t_i} \in [0,1] \\ wf_i = \begin{cases} 1 \text{ if } E_{Lower\ lim} \leq E_{Daylight} \leq E_{Upper\ lim} \\ 0 \text{ if } E_{Daylight} < E_{Lower\ lim} \vee E_{Daylight} > E_{Upper\ lim} \end{cases} \end{cases} \quad (7)$$

$$\begin{cases} C_{lc} = C_m + C_o + C_r \\ C_m = C_{m1} + C_{m2} \\ C_{m1} = \sum_{l=1}^{n} P_{ml} \times K_{ml} \\ C_{m2} = \sum_{j=1}^{n} P_{cj} \times K_{cj} \times A_{cj} \\ C_o = TES \times K_d \times Y \\ C_r = C_{r1} + C_{r2} \\ C_{r1} = C_m \times cof_r \\ C_{r2} = \sum_{l=1}^{n} P_{ml} \times K_{ml} \times cof_l \end{cases} \quad (8)$$

where $TC$ denotes the annual occupied time fraction of thermal comfort hours, $i$ denotes the number of occupied hours in a year, $t_i$ denotes each occupied hour in a year, $tf_i$ denotes a thermal comfort weighting factor, $UDI$ denotes the annual occupied time fraction of the indoor horizontal daylight illuminance observed at a given test point in a given domain, $wf_i$ denotes a daylighting weighting factor, $E_{Daylight}$ denotes the horizontal illuminance at a given point in lux, $E_{Upper\ lim}$ denotes the preset upper limit value of the horizontal illuminance at a given point in lux, and $E_{Lower\ lim}$ denotes the preset lower limit value of the horizontal illuminance at a given point in lux. Moreover $C_{lc}$ denotes the total carbon emissions during the building life cycle in kg; $C_m$ denotes the carbon emissions at the building manufacturing stage in kg; $C_o$ denotes the carbon emissions at the building operation stage in kg; $C_r$ denotes the carbon emissions at the building end-of-life recycling stage in kg; $C_{m1}$ denotes the embodied carbon emissions of materials in kg; $C_{m2}$ denotes the construction carbon emission in kg; $P_{ml}$ denotes the quantity of various building materials used in t; $K_{ml}$ denotes the carbon emission factors of different materials in $tCO_2/t$; $l$ denotes the types of materials; $P_{cj}$ denotes the quantities of a certain construction process in $m^3$; $K_{cj}$ denotes the carbon emission factors of a certain construction process in $tCO_2/m^3$; $A_{cj}$ denotes the area of a certain construction process in $m^2$; $j$ denotes the construction process types; $TES$ denotes the annual power consumption in kWh; $K_d$ denotes the carbon emission factor of electricity obtained from the grid (generation using fossil fuels) in $kgCO_2/kWh$; $Y$ denotes the theoretical building life in years; $C_{r1}$ denotes the carbon emission during building demolition in kg; $C_{r2}$ denotes the decarbonization amount recovered from building materials in kg; $cof_r$ denotes the carbon emission coefficient of building demolition (0.1 in this study according to reference [43]); and $cof_l$ denotes the recovery factor of various building materials.

Given the optimization objectives outlined and the research focus of this paper (i.e., retrofitting kindergarten buildings), the optimization variables can be classified into four primary categories: external wall-related variables, external window-related variables, external shading-related variables, and energy system-related variables. These variables have been described in detail in our team's previous studies [9,44], and will not be repeated here.

#### 2.2.2. Model Establishment and Parameter Setting

In this study, we used SketchUp to create a building model, and then imported EnergyPlus through OpenStudio for parameter setting and simulation calculation purposes. The model parameters primarily include weather, scheduling, material, thermal comfort, and daylighting comfort-related parameters. Weather parameters were set using data from the EnergyPlus website. Scheduling parameters were based on hourly usage patterns from field investigations. Material parameters were set considering performance and embodied carbon emissions of building materials from literature research. Thermal comfort parameters were determined by hourly clothing insulation and active metabolic rate from video playback and field investigation. Daylighting comfort parameters include appropriate illuminance value and sensor grid density [16].

#### 2.2.3. Fast Optimization Calculation

After setting the model parameters, fast comprehensive optimization of the scheme can be conducted. Meta-models represented by artificial neural networks (ANNs) are currently commonly used in design optimization research, and their adoption can effectively increase the efficiency of design optimization by replacing physical models. However, our team has noted in previous studies that generating sample spaces and optimizing ANN models typically require a significant amount of time [10,44]. To address this issue, we propose a scheme for a fast, comprehensive optimization framework based on parallel computing, as shown in Figure 2. The parallel computation process of this framework involves two main components: generating the sample space and optimizing the ANN model. In the calculation process, an initial sample with a size of 100 was generated by coupling EnergyPlus with Python language, and the sample was used for ANN model training and testing (divided into a training set and a test set at a ratio of 4:1). According to our team's reference [10], the SEGA-Q algorithm was used in this study for ANN model hyperparameter optimization, and the specific process will not be repeated here. When the ANN model is optimized, the sample space generation part is continued. If the mean relative error (MRE) of the optimized ANN model meets the requirement (MRE $\leq$ 1%), the model is adopted in the subsequent analysis. If not, the generated samples are added to the initial sample set to form a new sample, which is then used to optimize the ANN model. After obtaining the optimal ANN model, it can be combined with optimization algorithms for multi-objective optimization calculations to generate a Pareto solution set. Our team has compared the performance of the above three algorithms in building design optimization in previous studies [9,10], and the results show that the NSGA-II provides significant advantages. However, in reference [45], the performance of the NSGA-II and NSGA-III models was further compared in building design optimization, and the results showed that the NSGA-III model is more suitable for solving multi-objective optimization problems with more than two objectives. Therefore, the NSGA-III model was used as a multi-objective optimization algorithm in this study. On this basis, this study uses the entropy weight method to screen the Pareto set, in order to select a comprehensive optimization scheme that takes into account thermal comfort, daylighting, and life-cycle decarbonization [46].

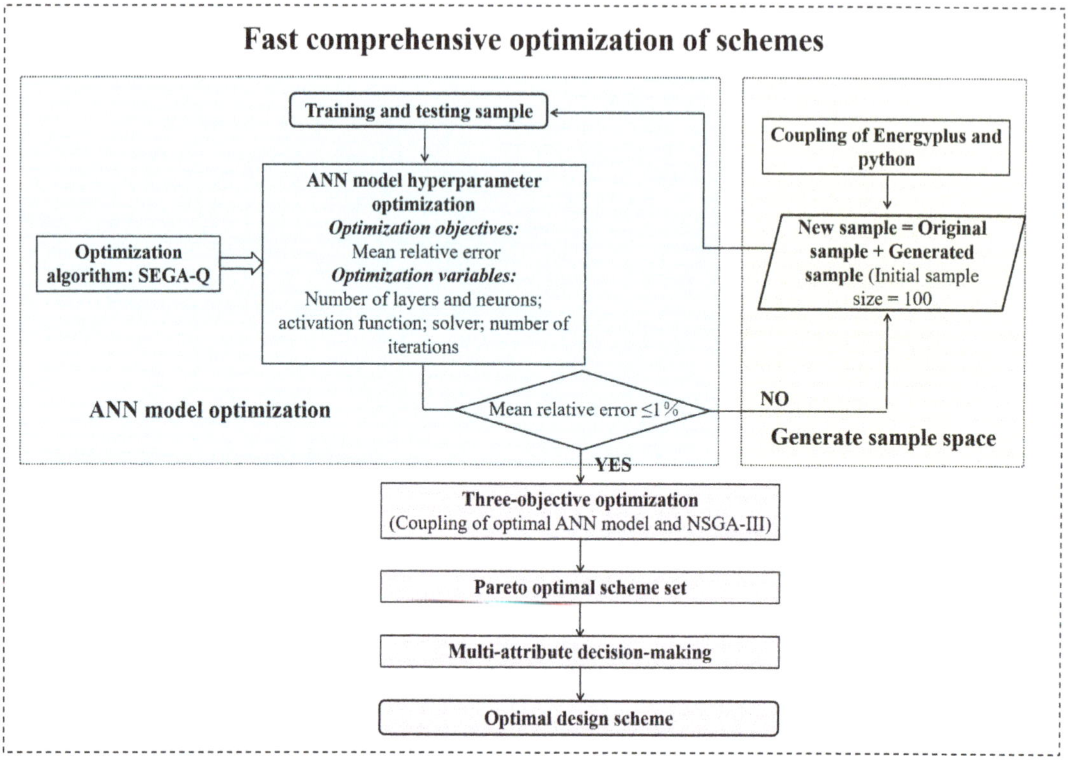


**Figure 2.** Comprehensive optimization framework based on parallel computing.

## 3. Case Study

### *3.1. Case Information*

To demonstrate the effectiveness of the proposed approach for retrofitting kindergarten buildings, we conducted a case study of a kindergarten in Nanjing from the 1990s and optimized its retrofit design scheme. The geographical coordinates of Nanjing are 31°14″ to 32°37″ N and 118°22″ to 119°14″ E. It is situated in a subtropical monsoon climate zone and is classified as a hot-summer and cold-winter area in the Chinese construction thermal zoning classification. Throughout the year, there is a demand for both cooling and heating. The typical annual outdoor air dry bulb temperature change in Nanjing is shown in Figure 3.

Detailed information on the kindergarten is presented in Table 1, and its appearance (before retrofit) and overall layout are shown in Figure 4. Notably, the kindergarten comprises two buildings, with Building 1 serving as the primary teaching facility, featuring a total of nine standard classrooms and one multifunctional classroom. The first grade encompasses four classrooms, while the second and third grades encompass three classrooms, with approximately the same number of students in each class. Building 2 is an office building that includes offices, canteens, and teacher activity rooms. Given that the primary focus of this study is to ensure a comfortable environment for children, design scheme optimization is solely confined to the main teaching building (Building 1). In addition, there are multiple residential buildings around the kindergarten, and in practical terms, the obstruction of these surrounding buildings will have a significant impact on the daylighting of the kindergarten. However, considering that the focus of this study is on the design optimization of kindergarten buildings, the occlusion of these surrounding buildings was ignored.

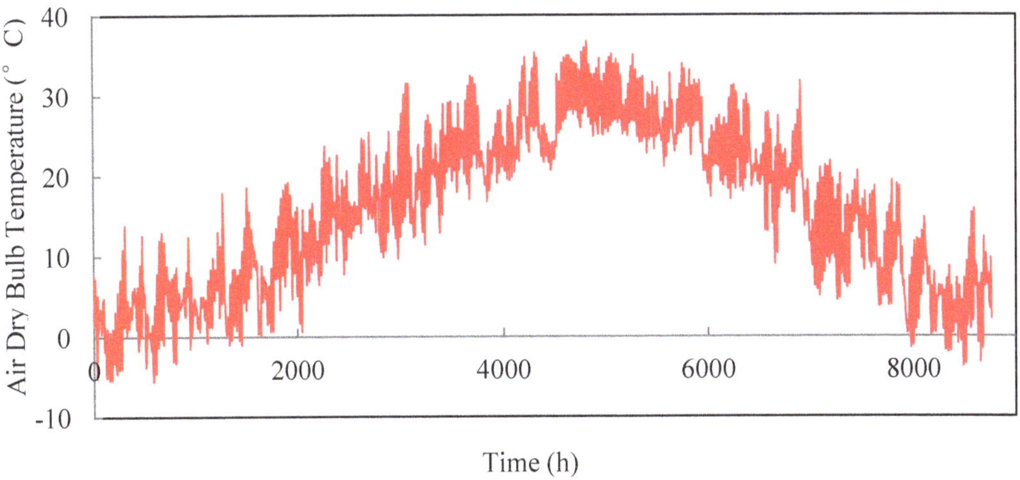


**Figure 3.** Annual variation in the outdoor air dry bulb temperature in Nanjing.

**Table 1.** Detailed information on the case kindergarten.

| Parameters | Values (Building 1) |
|---|---|
| Number of floors | 2 |
| Number of classrooms per floor | 5 |
| Floor height (m) | 3.5 |
| Building area ($m^2$) | 321 |
| Total floor area ($m^2$) | 664.8 |
| Classroom length and width (m) | 6.4 |
| Total number of children | 183 |
| Number of teachers | 32 |
| Vacation | Winter vacation: 31 January to 26 February<br>Summer vacation: 1 July to 31 August |
| Occupied period (weekdays) | 8:00–16:00 |

### *3.2. Information Acquisition and Evaluation of the Case Kindergarten*

Based on the content in Section 2, the first step involves conducting a survey to evaluate the thermal comfort of children in the kindergarten. To this end, we employed field investigation and video playback techniques (as illustrated in Figure 5) to gather information on the annual clothing conditions of children across three grades in the kindergarten. Subsequently, we calculated the clothing insulation for each period using Equation (2), and the results are listed in Table 2. Additionally, we collected information on the height and weight of the children in these three grades. With the use of these data and Equations (3)–(6), we calculated the metabolic rate of the children under various activities, as presented in Table 3. Notably, there is no essential difference in the activity metabolic rate among the students of different grades, and there is no need to treat this factor differently during optimization.

Subsequently, it is crucial to ascertain the required daylighting threshold for children. As per reference [19], the suitable range of illuminance for children varies between 100 and 1000 lux. Furthermore, according to reference [47], the recommended upper limit of the glare index in classrooms, laboratories, and offices varies between 20 and 22. In this study we selected a value of 22 as the upper limit value of the glare index. Then, through field investigation, the occupant scheduling data in this case can be collected, as listed in Table 4. Weather data were obtained directly from the EnergyPlus website.

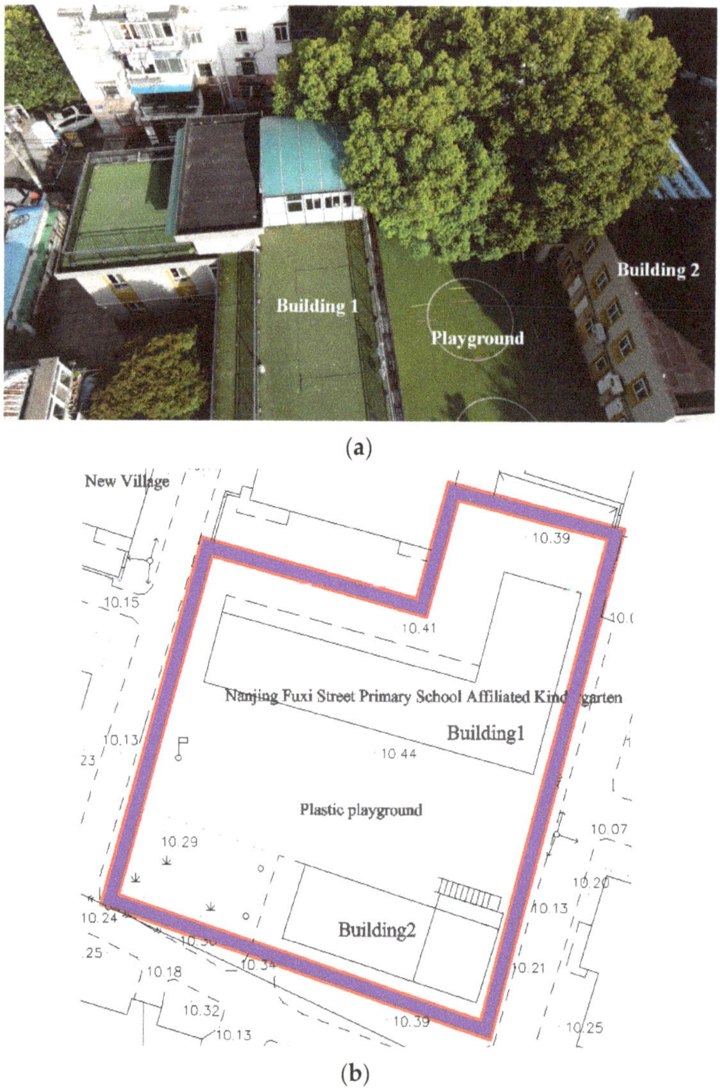


**Figure 4.** Appearance (before retrofit) and overall layout of the kindergarten: (**a**) appearance of the kindergarten area (obtained by an unmanned aerial vehicle); (**b**) overall layout of the kindergarten (derived from original drawings).

As stated in Section 2.1.4, the sole focus of this paper is the retrofitting of kindergarten building envelopes and energy systems. Hence, it is crucial to evaluate the performance of the building's envelope and energy system before retrofitting in this case. Based on the original drawings, the performance parameters of the case building envelope can be obtained, as listed in Table 5. The table indicates that the external walls and roofs of the building lack insulation materials, and the structure of the external windows is not rational. These factors were treated as variables in the subsequent design optimization. The artificial lighting equipment of the kindergarten was updated several years ago and energy-saving eye protection lights are now used, rendering it unnecessary to make any adjustments as part of the retrofit. The HVAC system currently uses split air conditioners, which have been in use for many years and include different equipment brands (purchased gradually), making it difficult to estimate their operational energy efficiency. Therefore, to improve the energy efficiency of the building, a VRV system with fresh air will be uniformly installed as part of the retrofit scheme.

Play activities (summer) — Classes (summer)

Meals (summer) — Noon break (summer)

Play activities (winter) — Classes (winter)

Meals (winter) — Noon break (winter)

**Figure 5.** Acquisition of annual clothing condition information for children through video playback).

**Table 2.** Average clothing insulation for each period.

| Seasons | Activity Types | Values [clo] |
| --- | --- | --- |
| Summer | Play activities<br>Classes<br>Meals | 0.31 |
| | Noon break | 0.33 |
| Winter | Play activities<br>Classes<br>Meals | 1.52 |
| | Noon break | 1.80 |

**Table 3.** Average height, weight, and metabolic rate of the children in each grade.

| Grades Parameters | 1 | 2 | 3 |
|---|---|---|---|
| Height [cm] | 101.3 | 107.6 | 115.1 |
| Weight [kg] | 18.2 | 19.4 | 22.5 |
| Basic metabolic rate [W/m$^2$] | 62.5 | 60.2 | 58.0 |
| Metabolic rate during the various activities [W/m$^2$] | Play activities: 187.5<br>Classes: 90.6<br>Meals: 154.4<br>Noon break: 63.1 | Play activities: 180.6<br>Classes: 87.3<br>Meals: 148.7<br>Noon break: 60.8 | Play activities: 174<br>Classes: 84.1<br>Meals: 143.3<br>Noon break: 58.6 |

Note: The conversion relationship between the above basic metabolic rate and the metabolic rate during the various activities is based on [34].

**Table 4.** Occupant scheduling data for the case kindergarten (weekdays).

| Grades | Average Number of People in Each Class | Play Activities | Classes | Meals | Noon Break |
|---|---|---|---|---|---|
| 1 | 18 | 8:00–9:00; 10:00–11:30; | 9:00–10:00; 14:00–15:00 | 11:30–12:30 | 12:30–14:00 |
| 2 | 24 | 9:00–10:00; | 8:00–9:00; 10:00–11:30; | | |
| 3 | 24 | 15:00–16:00 | 14:00–15:00 | | |

**Table 5.** Performance parameters of the case building envelope (before the retrofit).

| Parameters | Main Materials and Thickness | Performance Indicator Value |
|---|---|---|
| External wall | Main material: concrete bricks + cement mortar + ceramic tile<br>Total thickness: 240 mm | U value = 1.636 |
| Floor | Main material: floated coat + concrete + plasterboard<br>Total thickness: 150 mm | U value = 2.322 |
| Roof | Main material: concrete + cement mortar + waterproof roll + artificial turf<br>Total thickness: 300 mm | U value = 0.510 |
| External window (sliding window) | Outside layer: clear_3 mm; middle layer: air, 6 mm; inside layer: clear, 3 mm | U value = 3.159; SHGC = 0.762; VT = 0.812 |

*3.3. Design Optimization of the Case Kindergarten Building*

In Section 3.2, an investigation was conducted of the personnel, envelope, and equipment information for retrofitting the kindergarten building. Based on the acquired data, design optimization calculations are conducted in this section. In this paper, three optimization objectives are introduced in Section 2.2.1, namely, TC, UDI, and $C_{lc}$. Considering the particular context of the kindergarten under investigation, the optimization variables of this case study are presented in Table 6.

As per Section 2.2.2, this study established a physical model using SketchUp, as depicted in Figure 6. Subsequently, the physical model is imported into EnergyPlus for parameter settings. The weather parameters, scheduling-related parameters, and thermal-comfort-related parameters are introduced in Section 3.2, while the material-related parameters (carbon-emission-related parameters) involved in this study are shown in Table 7. Moreover, to ensure the rationality of parameter settings related to daylighting comfort, this study used a classroom as an example to verify the grid independence of the illuminance sensor.

**Table 6.** Optimization variables of the kindergarten building retrofit.

| No. | Optimization Variables | Value Range |
|---|---|---|
| X1 | External wall insulation | EPS board: 20–50 mm (interval: 5 mm) |
| X2 | External roof insulation | EPS board: 30–70 mm (interval: 5 mm) |
| X3 | Glass types of the different facade windows (double-paned window; filling gas: air, 6 mm) | Outer/Inner glass: clear, 3 mm; clear, 6 mm; bronze, 3 mm; bronze, 6 mm; gray, 3 mm; gray, 6 mm; green, 3 mm; green, 6 mm |
| X4 | WWR of the different facades | South: 0.2–0.8 (interval: 0.1)<br>North: 0.2–0.8 (interval: 0.1)<br>East: 0.2–0.8 (interval: 0.1)<br>West: 0.2–0.8 (interval: 0.1) |
| X5 | Fixed sunshades for the windows in the facades | South (overhanging depth): 0–2.0 m (interval: 0.1 m) |
| X6 | Cooling set point in the classrooms during each activity period | Play activities: 24 °C–30 °C (interval: 0.5 °C)<br>Classes: 24 °C–30 °C (interval: 0.5 °C)<br>Meals: 24 °C–30 °C (interval: 0.5 °C)<br>Noon break: 24 °C–30 °C (interval: 0.5 °C) |
| X7 | Heating set point in the classrooms during each activity period | Play activities: 16 °C–22 °C (interval: 0.5 °C)<br>Classes: 16 °C–22 °C (interval: 0.5 °C)<br>Meals: 16 °C–22 °C (interval: 0.5 °C)<br>Noon break: 16 °C–22 °C (interval: 0.5 °C) |
| X8 | Proportion of the roof photovoltaic area | 0–1 (interval: 0.001) |

**Figure 6.** Physical model of the kindergarten building to be retrofitted.

Upon completing the parameter settings, the Python programming language can be utilized to construct the rapid comprehensive optimization framework, as discussed in Section 2.2.3. The optimal ANN model can be generated first, as presented in Table 8. When the sample size reaches 1700, the mean relative error of the optimized ANN model satisfies the requirement ($\leq$1%). Compared to references [10,44], this method significantly reduces computational costs and does not necessitate generating a vast sample set before conducting metamodel training.

On this basis, the optimal ANN model is combined with the NSGA-III to perform a three-objective optimization calculation (with a population of 50), and the outcomes are demonstrated in Figure 7. It can be seen that there are a total of 50 Pareto solutions in the Pareto solution set, among which the range of $C_{lc}$ is −77,336.1–785,327.2 kg; the range of $TC$ is 38.5–66.4%; and the range of $UDI$ is 1.2–93.3%. As all the aforementioned solutions are nondominated, a multi-attribute decision-making method is necessary to further select the optimal design scheme. Based on the relevant content in Section 2.2.3, the comprehensive optimal scheme attained in this research is presented in Table 9 and Figure 8. Additionally, considering that most of the retrofit schemes for buildings in actual engineering are designed according to codes, this study also generated the benchmark scheme based on the code [48] (as shown in Table 9) and compared it with the comprehensive optimal scheme.

**Table 7.** Material-related parameters (carbon-emission-related parameters).

| Parameters | Values |
| --- | --- |
| $K_{mEPS}$ | 5220 kg $CO_2$/t |
| $K_{mconcrete}$ | 135 kg $CO_2$/t |
| $K_{mglassclear}$ | 500 kg $CO_2$/t |
| $K_{mglassbronze}$ | 600 kg $CO_2$/t |
| $K_{mglassgrey}$ | 600 kg $CO_2$/t |
| $K_{mglassgreen}$ | 600 kg $CO_2$/t |
| $K_{mphotovoltaic}$ | 158 kg $CO_2/m^2$ |
| $K_d$ | 0.59 kg$CO_2$/kWh |
| Y | 50 |
| $cof_r$ | 0.1 |
| $cof_{EPS}$ | |
| $cof_{concrete}$ | |
| $cof_{glass}$ | 0.3 |
| $cof_{photovoltaic}$ | 0.3 |

Note: This study assumes that the service life of wall-related materials is 50 years, while the service life of windows and photovoltaic-related materials is only 25 years.

**Table 8.** Hyperparameter settings of the ANN model.

| Hyperparameters | Optimal Values |
| --- | --- |
| Sample size | 1700 |
| Solver | lbfgs |
| Activation function | tanh |
| Hidden layer sizes | (18, 12, 6) |
| Hidden layer numbers | 3 |
| Initial learning rate | 0.1 |
| Maximum number of iterations | 20,000 |
| MRE (%) | 0.98 |
| $R^2$ | 0.991 |

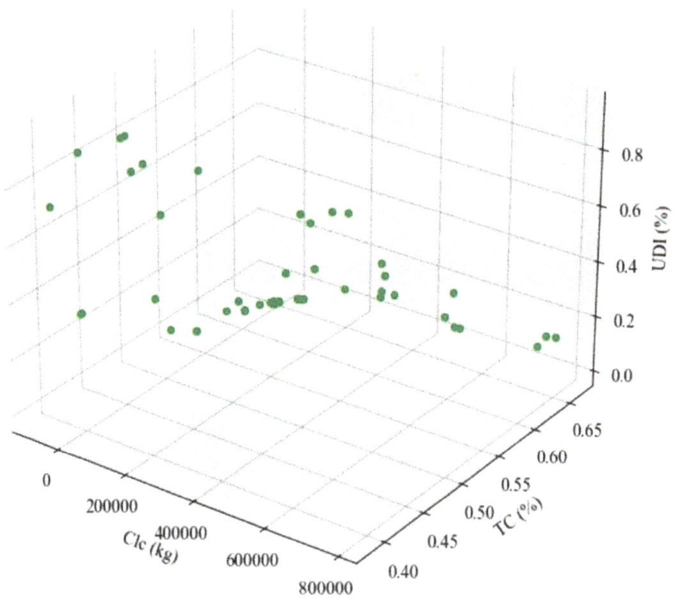


**Figure 7.** Pareto solution set obtained by three-objective optimization.

**Table 9.** Comprehensive optimal design scheme for kindergarten building retrofit.

| No. | Optimization Variables | Comprehensive Optimal Scheme | Benchmark Scheme |
|---|---|---|---|
| X1 | External insulation of wall | EPS board: 30 mm<br>(Total U value of the wall = 0.670) | EPS board: 30 mm<br>(Total U value of the wall = 0.670) |
| X2 | External insulation of roof | EPS board: 30 mm<br>(Total U value of the roof = 0.362) | EPS board: 30 mm<br>(Total U value of the roof = 0.362) |
| X3 | Glass types of different facade windows<br>(double-paned window; filling gas: Air_6 mm) | Outer glass: bronze, 6 mm;<br>Inner glass: gray, 6 mm<br>(U value = 3.111; SHGC = 0.442; VT = 0.526) | Outer glass: clear, 3 mm;<br>Inner glass: clear, 3 mm<br>(U value = 3.159; SHGC = 0.762; VT = 0.812) |
| X4 | WWR of different facades | South: 0.4<br>North: 0.4<br>East: 0.2<br>West: 0.2 | South: 0.6<br>North: 0.6<br>East: 0.4<br>West: 0.4 |
| X5 | Fixed sunshades for windows on south facade | Overhanging depth: 2.0 m | Overhanging depth: 2.0 m |
| X6 | Cooling set point in classrooms of each activity period | Play activities: 25.5 °C<br>Classes: 30 °C<br>Meals: 25.5 °C<br>Noon break: 30 °C | 26 °C |
| X7 | Heating set point in classrooms of each activity period | Play activities: 16.5 °C<br>Classes: 22 °C<br>Meals: 16.5 °C<br>Noon break: 17.5 °C | 18 °C |
| X8 | Proportion of the roof photovoltaic area | 1 | 1 |
| | $TC$ | 41.2% | 38.5% |
| | $UDI$ | 92.6% | 12.9% |
| | $TES$ | 4205.89 kWh | −2155.73 kWh |
| | $C_o$ | −2481.48 kg | −1271.88 kg |
| | $C_{lc}$ | −9647.4 kg | 24,510.9 kg |

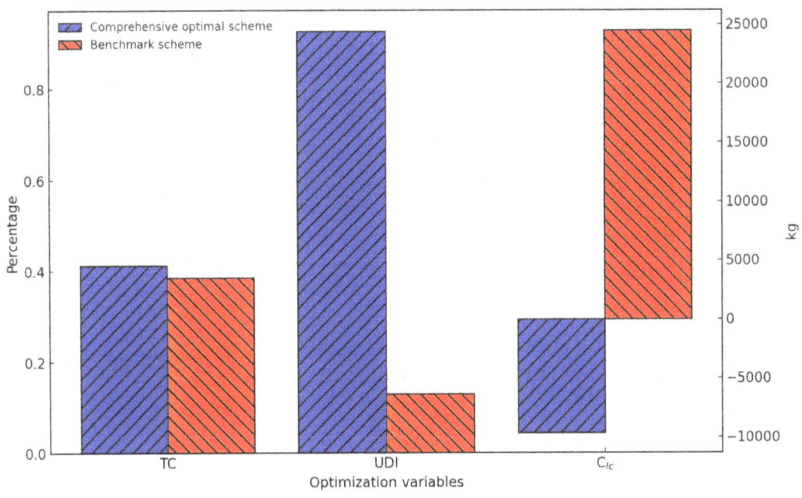


**Figure 8.** Comprehensive optimization results for kindergarten building retrofit.

Upon comparing the Pareto scheme set obtained from three-objective optimization and the benchmark scheme, it is evident that three-objective optimization can achieve a maximum carbon emission reduction of 101,847 kg (i.e., 77,336.1 kg of carbon emissions can be saved for the region on the basis of achieving zero emissions during the life cycle). Moreover, it can increase the thermal comfort period by up to 27.9% and the daylighting comfort period by up to 80.4%. Upon comparing the comprehensive optimal scheme and the benchmark scheme, it is evident that the annual operating carbon emissions in both schemes are negative. Notably, the annual production energy is higher than the energy consumption. However, the comprehensive optimal scheme can also achieve savings of

9647.4 kg of carbon emissions based on achieving zero emissions during the life cycle. Additionally, it can facilitate increases in the thermal comfort period of 2.7% and the daylighting comfort period of 79.7%.

## 4. Results and Discussion

### 4.1. Analysis of the Research Results

The above case study indicates that the approach adopted in this study for setting the cooling/heating temperature set points during different periods is rational and better aligned with the thermal comfort requirements of children in the kindergarten building. While increasing the thermal comfort period, it can also yield a reduction in the energy consumption (i.e., reducing operating carbon emissions). Additionally, the current design codes fail to consider the differences in daylighting comfort between children and adults. This leads to the illuminance values on the working surface of the classroom often exceeding the comfort range of children, resulting in a low $UDI$ value in the benchmark scheme. However, by optimizing the window glass type, WWR, and shading, it is possible to effectively address this issue.

In this paper, we examine life-cycle decarbonization retrofitting of comfortable kindergarten buildings in China (considering children in particular). Through multi-objective optimization combined with multi-attribute decision-making, a balance is achieved between thermal comfort, daylighting comfort, and life-cycle decarbonization. In traditional design concepts, maintaining a comfortable indoor environment often requires high energy consumption. However, we found that, while fully considering the metabolic rate and clothing insulation of children during different periods, the comfort of children and building decarbonization are not completely opposite goals, and can be balanced within a certain range (as demonstrated by the comparison between the comprehensive optimal scheme and the benchmark scheme in Table 9).

In addition, the parallel computing framework proposed in this study can significantly reduce the sample set required for meta-model training, thereby significantly improving the efficiency of generating high-precision meta-models.

### 4.2. Comparison with Similar Studies

The optimization of educational buildings has always been a hot topic in the field of building research. In reference [49], the impact of an egg-crate shading device on the indoor environmental quality of classrooms in southern Spain was studied. The results indicated that the implementation of the egg-crate shading device resulted in improvements in the operative temperature and illuminance in the classroom. In reference [50], the adaptive thermal comfort of students aged 10–11 years in summer classrooms in the Seville region was studied. The researchers compared the PMV and thermal sensation vote (TSV) results, and the results showed that the PMV model is not suitable for adaptive thermal comfort evaluation of children aged 10–11 in the Seville region. In reference [7], three-objective optimization of primary school classrooms in Tehran was studied, aiming to simultaneously improve the energy, thermal comfort, and daylighting performance of primary school classrooms. In reference [51], design optimization of natural ventilation in educational buildings was studied. In the research process, a multi-objective optimization algorithm was used to achieve the optimal balance between the indoor air quality, thermal comfort, energy consumption, and life-cycle cost.

Compared with the above studies, the innovation of this study encompasses: (1) proposing a design optimization method that comprehensively considers children's thermal comfort, daylighting comfort, and life-cycle decarbonization; (2) proposing a rapid design optimization framework based on parallel computing; and (3) exploring the correlation between the thermal comfort of children in the classroom and building decarbonization on the basis of considering their metabolic rate and clothing insulation. Overall, this study has important reference value for thermal comfort, daylighting, and decarbonization optimization of kindergarten buildings in various regions

### 4.3. Analysis of Research Limitations and Prospects for Future Directions

This study is not without limitations. Firstly, in the simulation process, we utilized annual data generated from meteorological parameters in 2002 to account for the inherent uncertainty in these parameters. However, these simulated data deviate from actual weather data, which may introduce a certain degree of deviation in the practical application of the optimization results. Secondly, as the building in this case study is still undergoing renovation, actual test data from the renovated building were not obtainable for this research. This lack of data precluded further verification of the reliability of the simulation results. Future research will entail following up on the progress of this project to obtain such data. Thirdly, the focus of this study was on the overall optimization effect of buildings. Therefore, in the simulation calculation process for thermal comfort, it was assumed that the distributions of wind speed, temperature, and humidity within the classroom were uniform. In reality, however, there are variations in the distribution of thermal comfort within classrooms.

Furthermore, it is important to note that children are in a stage of rapid growth and development, with significant physiological differences among different age groups, and correspondingly, distinct needs regarding thermal and light environments. Unfortunately, existing design codes lack relevant content on daylighting comfort for children and do not account for the differences in the range of daylighting comfort between children and adults. Given that moderate daylighting is beneficial for stimulating the development of children's visual nerves, further research is necessary to determine the appropriate range of illuminance for children of different ages. Additionally, climate change will inevitably impact the indoor environmental comfort and operational energy consumption of buildings. Therefore, subsequent research could further investigate the impact mechanism of climate uncertainty on various types of educational buildings and quantify the degree of uncertainty. This would facilitate the development of optimization methods for educational building design that consider climate uncertainty.

## 5. Conclusions

In this paper, we propose an optimization methodology for retrofitting kindergarten buildings, considering thermal comfort, daylighting, and life-cycle carbon emissions. After conducting a comprehensive investigation of the parameters pertaining to children's comfort, we calculated relevant indicators of thermal and daylighting comfort, ensuring that the retrofitted kindergarten building promotes optimal child growth. In the course of our research, necessary retrofit parameters were obtained through a combination of video playback, field investigation, literature review, and examination of drawings. Based on this rigorous approach, optimization variables were selected, and a physical model was established. Ultimately, a rapid comprehensive optimization framework was employed for optimization calculations to derive the optimal scheme for kindergarten buildings. By utilizing a typical kindergarten building in Nanjing as a case study, we validated the effectiveness of the proposed method. The results revealed that: (1) in comparison to the benchmark scheme, multi-objective optimization could reduce carbon emissions by up to 101,847 kg, increase the thermal comfort period by up to 27.9%, and enhance the daylighting comfort period by up to 80.4%. (2) When compared to the benchmark scheme, the comprehensive optimal scheme resulted in a reduction in carbon emissions of 34,158.3 kg, an extension of the thermal comfort period by 2.7%, and an improvement in daylighting comfort by 79.7%. (3) The determined cooling/heating temperature set points are highly suitable for kindergarten buildings, as they fulfill children's thermal comfort needs while also mitigating carbon emissions.

Compared with other similar studies, the main innovation of this research lies in: (1) proposing a design optimization method that comprehensively considers children's thermal comfort, daylighting comfort, and life-cycle decarbonization; (2) proposing a rapid design optimization framework based on parallel computing; and (3) exploring the correlation between the thermal comfort of children in the classroom and building decarbonization on the basis of considering their metabolic rate and clothing insulation.

**Author Contributions:** Conceptualization, K.H. and Y.X.; methodology, Y.X. and W.L.; software, J.Y.; validation, K.H., C.X. and Y.Y.; formal analysis, C.X.; investigation, Y.X.; writing—original draft preparation, K.H.; writing—review and editing, Y.X.; visualization, C.X.; funding acquisition, Y.X. All authors have read and agreed to the published version of the manuscript.

**Funding:** The authors gratefully acknowledge the support of the National Natural Science Foundation of China (No. 51708287), a grant from the Jiangsu Provincial Department of Housing and Urban Rural Development (No. 2023ZD026), a grant from the China Construction Yipin Investment & Development Co., Ltd and a grant from the Nanjing Construction Industry Science and Technology Plan Project (No. Ks2415).

**Data Availability Statement:** The data presented in this study are available on request from the corresponding author. The data are not publicly available due to privacy.

**Conflicts of Interest:** Kai Hu, Chao Xu, Wenjun Li, Jing Ye were employed by the company China Construction Yipin Investment and Development Co., Ltd. The remaining authors declare that the research was conducted in the absence of any commercial or financial relationships that could be construed as a potential conflict of interest.

## References

1. State Statistical Bureau. *Statistical Bulletin of the People's Republic of China on National Economic and Social Development in 2021*; State Statistical Bureau: Hong Kong, China, 2022.
2. Gajić, D.; Stupar, D.; Antunović, B.; Janković, A. Determination of the energy performance indicator of kindergartens through design, measured and recommended parameters. *Energy Build.* **2019**, *204*, 109511. [CrossRef]
3. Causone, F.; Carlucci, S.; Moazami, A.; Cattarin, G.; Pagliano, L. Retrofit of a Kindergarten Targeting Zero Energy Balance. *Energy Procedia* **2015**, *78*, 991–996. [CrossRef]
4. Sanchez-Tocino, H.; Villanueva Gomez, A.; Gordon Bolanos, C.; Alonso Alonso, I.; Vallelado Alvarez, A.; Garcia Zamora, M.; Frances Caballero, E.; Marcos-Fernandez, M.A.; Schellini, S.; Galindo-Ferreiro, A. The effect of light and outdoor activity in natural lighting on the progression of myopia in children. *J. Fr. Ophtalmol.* **2019**, *42*, 2–10. [CrossRef] [PubMed]
5. Li, S.; Wang, Y.; Lining, Y. Children's thermal comfort in kindergarten buildings of Chongqing and Wuhan in winter. *J. HVAC* **2017**, *47*, 136–142.
6. Fonseca Gabriel, M.; Paciência, I.; Felgueiras, F.; Cavaleiro Rufo, J.; Castro Mendes, F.; Farraia, M.; Mourão, Z.; Moreira, A.; de Oliveira Fernandes, E. Environmental quality in primary schools and related health effects in children. An overview of assessments conducted in the Northern Portugal. *Energy Build.* **2021**, *250*, 111305. [CrossRef]
7. Bakmohammadi, P.; Noorzai, E. Optimization of the design of the primary school classrooms in terms of energy and daylight performance considering occupants' thermal and visual comfort. *Energy Rep.* **2020**, *6*, 1590–1607. [CrossRef]
8. Wang, Y.; Kuckelkorn, J.; Zhao, F.-Y.; Liu, D.; Kirschbaum, A.; Zhang, J.-L. Evaluation on classroom thermal comfort and energy performance of passive school building by optimizing HVAC control systems. *Build. Environ.* **2015**, *89*, 86–106. [CrossRef]
9. Xu, Y.; Zhang, G.; Yan, C.; Wang, G.; Jiang, Y.; Zhao, K. A two-stage multi-objective optimization method for envelope and energy generation systems of primary and secondary school teaching buildings in China. *Build. Environ.* **2021**, *204*, 108142. [CrossRef]
10. Xu, Y.; Yan, C.; Pan, Y.; Zhao, K.; Li, M.; Zhu, F.; Jiang, Y. A three-stage optimization method for the classroom envelope in primary and secondary schools in China. *J. Build. Eng.* **2022**, *52*, 104487. [CrossRef]
11. Al-Rashidi, K.; Loveday, D.; Al-Mutawa, N. Investigating the applicability of different thermal comfort models in air-conditioned classrooms in Kuwait. In Proceedings of the 10th REHVA World Congress on Sustainable Energy Use in Buildings, Clima 2010, Antalya, Turkey, 9–12 May 2010.
12. Jindal, A. Thermal comfort study in naturally ventilated school classrooms in composite climate of India. *Build. Environ.* **2018**, *142*, 34–46. [CrossRef]
13. Zomorodian, Z.S.; Tahsildoost, M.; Hafezi, M. Thermal comfort in educational buildings: A review article. *Renew. Sustain. Energy Rev.* **2016**, *59*, 895–906. [CrossRef]
14. Wirz-Justice, A.; Skene, D.J.; Munch, M. The relevance of daylight for humans. *Biochem. Pharmacol.* **2020**, *191*, 114304. [CrossRef]
15. Pagliolico, S.L.; Lo Verso, V.R.M.; Zublena, M.; Giovannini, L. Preliminary results on a novel photo-bio-screen as a shading system in a kindergarten: Visible transmittance, visual comfort and energy demand for lighting. *Sol. Energy* **2019**, *185*, 41–58. [CrossRef]
16. Samiou, A.I.; Doulos, L.T.; Zerefos, S. Daylighting and artificial lighting criteria that promote performance and optical comfort in preschool classrooms. *Energy Build.* **2022**, *258*, 111819. [CrossRef]
17. Vásquez, N.G.; Felippe, M.L.; Pereira, F.O.R.; Kuhnen, A. Luminous and visual preferences of young children in their classrooms: Curtain use, artificial lighting and window views. *Build. Environ.* **2019**, *152*, 59–73. [CrossRef]

18. Salleh, N.M.; Kamaruzzaman, S.N.; Riley, M.; Ahmad Zawawi, E.M.; Sulaiman, R. A quantitative evaluation of indoor environmental quality in refurbished kindergarten buildings: A Malaysian case study. *Build. Environ.* **2015**, *94*, 723–733. [CrossRef]
19. Gao, S. Kindergarten Design Strategy and Simulation Research Guided by Building a Healthy Natural Light Environment for Children's Growth. Master's Thesis, Beijing University of Civil Engineering and Architecture, Beijing, China, 2020.
20. Hammad, A.W.A.; Akbarnezhad, A.; Oldfield, P. Optimising embodied carbon and U-value in load bearing walls: A mathematical bi-objective mixed integer programming approach. *Energy Build.* **2018**, *174*, 657–671. [CrossRef]
21. Cho, H.; Mago, P.J.; Luck, R.; Chamra, L.M. Evaluation of CCHP systems performance based on operational cost, primary energy consumption, and carbon dioxide emission by utilizing an optimal operation scheme. *Appl. Energy* **2009**, *86*, 2540–2549. [CrossRef]
22. Vettorato, D.; Geneletti, D.; Zambelli, P. Spatial comparison of renewable energy supply and energy demand for low-carbon settlements. *Cities* **2011**, *28*, 557–566. [CrossRef]
23. Ho, Y.-F.; Chang, C.-C.; Wei, C.-C.; Wang, H.-L. Multi-objective programming model for energy conservation and renewable energy structure of a low carbon campus. *Energy Build.* **2014**, *80*, 461–468. [CrossRef]
24. *GB/T51366-2019*; Standard for Calculating Carbon Emissions from Buildings. National Standards of the People's Republic of China: Beijing, China, 2019.
25. Piccardo, C.; Dodoo, A.; Gustavsson, L. Retrofitting a building to passive house level: A life cycle carbon balance. *Energy Build.* **2020**, *223*, 110135. [CrossRef]
26. Lai Huang, W.; Li, J. Optimizing the Roadmap to Carbon Neutralization with a New Paradigm. *Engineering* **2021**, *7*, 1678–1679. [CrossRef]
27. Hammad, M.; Ebaid, M.S.Y.; Al-Hyari, L. Green building design solution for a kindergarten in Amman. *Energy Build.* **2014**, *76*, 524–537. [CrossRef]
28. Zhang, Y.; Wang, W.; Wang, Z.; Gao, M.; Zhu, L.; Song, J. Green building design based on solar energy utilization: Take a kindergarten competition design as an example. *Energy Rep.* **2021**, *7*, 1297–1307. [CrossRef]
29. Luo, X.J.; Oyedele, L.O. A data-driven life-cycle optimisation approach for building retrofitting: A comprehensive assessment on economy, energy and environment. *J. Build. Eng.* **2021**, *43*, 102934. [CrossRef]
30. Li, B.; You, L.; Zheng, M.; Wang, Y.; Wang, Z. Energy consumption pattern and indoor thermal environment of residential building in rural China. *Energy Built Environ.* **2020**, *1*, 327–336. [CrossRef]
31. Wang, R.; Lu, S.; Feng, W. A three-stage optimization methodology for envelope design of passive house considering energy demand, thermal comfort and cost. *Energy* **2020**, *192*, 116723. [CrossRef]
32. Xu, Y.; Yan, C.; Liu, H.; Wang, J.; Yang, Z.; Jiang, Y. Smart energy systems: A critical review on design and operation optimization. *Sustain. Cities Soc.* **2020**, *62*, 102369. [CrossRef]
33. Day, J.K.; McIlvennie, C.; Brackley, C.; Tarantini, M.; Piselli, C.; Hahn, J.; O'Brien, W.; Rajus, V.S.; De Simone, M.; Kjærgaard, M.B.; et al. A review of select human-building interfaces and their relationship to human behavior, energy use and occupant comfort. *Build. Environ.* **2020**, *178*, 106920. [CrossRef]
34. Yun, H.; Nam, I.; Kim, J.; Yang, J.; Lee, K.; Sohn, J. A field study of thermal comfort for kindergarten children in Korea: An assessment of existing models and preferences of children. *Build. Environ.* **2014**, *75*, 182–189. [CrossRef]
35. Zheng, Z.; Zhang, Y.; Mao, Y.; Yang, Y.; Fu, C.; Fang, Z. Analysis of SET* and PMV to evaluate thermal comfort in prefab construction site offices: Case study in South China. *Case Stud. Therm. Eng.* **2021**, *26*, 101137. [CrossRef]
36. *ASHRAE Standard 55*; ASHRAE. Thermal Environmental Conditions for Human Occupancy. American Society of Heating, Refrigerating and Air-Conditioning Engineers: Atlanta, GA, USA, 2013.
37. American Society of Heating, Refrigerating and Air-Conditioning Engineers. *ASHRAE Handbook Fundamentals*; American Society of Heating, Refrigerating and Air-Conditioning Engineers: Atlanta, GA, USA, 2013.
38. Zhang, S.; He, W.; Chen, D.; Chu, J.; Fan, H.; Duan, X. Thermal comfort analysis based on PMV/PPD in cabins of manned submersibles. *Build. Environ.* **2019**, *148*, 668–676. [CrossRef]
39. He, J. Testing Thermal Environment Parameters in the Kindergarten and Study of Children Thermal Comfort. Master's Thesis, Chongqing University, Chongqing, China, 2016.
40. *ISO. EN ISO 8996*; Ergonomics of the Thermal Environment—Determination of Metabolic Rate. International Organization for Standardization: Geneva, Switzerland, 2004.
41. Dussault, J.-M.; Gosselin, L. Office buildings with electrochromic windows: A sensitivity analysis of design parameters on energy performance, and thermal and visual comfort. *Energy Build.* **2017**, *153*, 50–62. [CrossRef]
42. Reffat, R.M.; Ahmad, R.M. Determination of optimal energy-efficient integrated daylighting systems into building windows. *Sol. Energy* **2020**, *209*, 258–277. [CrossRef]
43. Zou, Y. Research on the Calculation of the Whole Life Cycle Carbon Emission and the Carbon Reduction Strategy of Chaoyang Wanda Plaza. Ph.D. Thesis, Shenyang Jianzhu University, Shenyang, China, 2020.
44. Xu, Y.; Yan, C.; Yan, S.; Liu, H.; Pan, Y.; Zhu, F.; Jiang, Y. A multi-objective optimization method based on an adaptive meta-model for classroom design with smart electrochromic windows. *Energy* **2021**, *243*, 122777. [CrossRef]
45. Razmi, A.; Rahbar, M.; Bemanian, M. PCA-ANN integrated NSGA-III framework for dormitory building design optimization: Energy efficiency, daylight, and thermal comfort. *Appl. Energy* **2022**, *305*, 117828. [CrossRef]

46. Xu, Y.; Yan, C.; Wang, D.; Li, J.; Shi, J.; Lu, Z.; Lu, Q.; Jiang, Y. Coordinated optimal design of school building envelope and energy system. *Sol. Energy* **2022**, *244*, 19–30. [CrossRef]
47. U.S. Department of Energy. *EnergyPlus*™; version 9.0.1; Input Output Reference. Documentation; U.S. Department of Energy: Washington, DC, USA, 2018.
48. *GB 50099-2011*; Code for Design of School. National Standards of the People's Republic of China: Beijing, China, 2011.
49. Calama-González, C.; Suárez, R.; León-Rodríguez, Á.; Ferrari, S. Assessment of Indoor Environmental Quality for Retrofitting Classrooms with an Egg-Crate Shading Device in A Hot Climate. *Sustainability* **2019**, *11*, 1078. [CrossRef]
50. Aparicio-Ruiz, P.; Barbadilla-Martín, E.; Guadix, J.; Muñuzuri, J. A field study on adaptive thermal comfort in Spanish primary classrooms during summer season. *Build. Environ.* **2021**, *203*, 108089. [CrossRef]
51. Acosta-Acosta, D.F.; El-Rayes, K. Optimal design of classroom spaces in naturally-ventilated buildings to maximize occupant satisfaction with human bioeffluents/body odor levels. *Build. Environ.* **2020**, *169*, 106543. [CrossRef]

**Disclaimer/Publisher's Note:** The statements, opinions and data contained in all publications are solely those of the individual author(s) and contributor(s) and not of MDPI and/or the editor(s). MDPI and/or the editor(s) disclaim responsibility for any injury to people or property resulting from any ideas, methods, instructions or products referred to in the content.

buildings

MDPI

Article

# Building Energy Efficiency Enhancement through Thermochromic Powder-Based Temperature-Adaptive Radiative Cooling Roofs

Ge Song, Kai Zhang *, Fei Xiao, Zihao Zhang, Siying Jiao and Yanfeng Gong

College of Urban Construction, Nanjing Tech University, Nanjing 211816, China
* Correspondence: kai.zhang.ch@njtech.edu.cn

**Abstract:** This paper proposes a temperature-adaptive radiative cooling (TARC) coating with simple preparation, cost effectiveness, and large-scale application based on a thermochromic powder. To determine the energy efficiency of the proposed TARC coating, the heat transfer on the surface of the TARC coating was analyzed. Then, a typical two-story residential building with a roof area of 258.43 m$^2$ was modeled using EnergyPlus. Finally, the energy-saving potential and carbon emission reduction resulting from the application of the proposed TARC roof in buildings under different climates in China were discussed. The results showed that the average solar reflectivity under visible light wavelengths (0.38–0.78 μm) decreases from 0.71 to 0.37 when the TARC coating changes from cooling mode to heating mode. Furthermore, energy consumption can be reduced by approximately 17.8–43.0 MJ/m$^2$ and 2.0–32.6 MJ/m$^2$ for buildings with TARC roofs compared to those with asphalt shingle roofs and passive daytime radiative cooling (PDRC) roofs, respectively. This also leads to reductions in carbon emissions of 9.4–38.0 kgCO$_2$/m$^2$ and 1.0–28.9 kgCO$_2$/m$^2$ for the buildings located in the selected cities. To enhance building energy efficiency, TARC roofs and PDRC roofs are more suitable for use on buildings located in zones with high heating demands and high cooling demands, respectively.

**Keywords:** temperature-adaptive roof; radiative cooling; building energy efficiency

## 1. Introduction

Global warming not only endangers the balance of natural ecosystems but also threatens the survival of humans and animals. Reducing carbon emissions can effectively alleviate global warming, especially carbon emissions in the building sector, which account for 30–40% of total global carbon emissions [1–3]. As a passive cooling technique, radiative cooling can be achieved by exchanging heat with outer space through an atmospheric window (8–13 μm) without extra energy input, which effectively reduces building carbon emissions [4,5].

Radiative cooling materials are most convenient when employed as cool roofs in buildings, and they can reduce building energy consumption by 30–40% under different climates in China [6–8]. Many efforts have been made to prepare radiative cooling materials and to enhance the cooling energy efficiency for their application as cool roofs in buildings. Lin et al. [9] prepared a radiative cooling coating composed of a titanium dioxide–polydimethylsiloxane ($TiO_2$-PDMS) layer at the bottom and an aluminum oxide–polydimethylsiloxane ($Al_2O_3$-PDMS) layer at the top. Their study showed that the reflectivity in the solar spectrum and the emissivity in the atmospheric window of the coating are 92.2% and 95.3%, respectively. Furthermore, approximately 9.8 GJ per year of cooling energy is reduced by applying the proposed coating as the cool roof of a single-story office building in Mumbai. A sturdy aerogel with nano/microporous structures using cross-linked silica-hybridized cellulose acetate was proposed by Liu et al. [10] for achieving radiative cooling. Their investigation indicated that the solar reflectivity and emissivity in the atmospheric window of the proposed aerogel are 96% and 97%, respectively, which can produce an energy reduction of 13.87 kW/m$^2$ when applying the aerogel to a modeled baseline building in China. Chen et al. [11] investigated the energy savings potential of porous polymer radiative cooling coatings as rooftops for concrete-based roofs, galvanized steel-based roofs, and commercial multilayered roofs in Hong Kong. The results showed that the annual cooling electricity savings of applying the porous polymer radiative cooling coating-based cool roofs are 3.23–47.71 kWh/m$^2$ compared to those of typical white coating-based roofs. Yang et al. [12] prepared a phosphate-activated geopolymer-based coating with a solar reflectivity of 0.9471 and a mid-infrared emissivity of 0.9634. Their experiments demonstrated that the surface temperature decreases by 3.8 °C below the ambient temperature under direct sunlight in Hong Kong. Compared to those of the budlings with a commercial cool roof and an ordinary cement roof, the energy consumption of the building with the proposed coating-based cool roof decreased by 266.61–627.75 kWh and 1240.87–2742.51 kWh, respectively. A cellulose nanocrystal aerogel grating (CAG) prepared by Cai et al. [13] achieved a solar reflectivity of 97.4% and an atmospheric window emissivity of 94%. Their study showed that 10.5 °C of sub-ambient cooling is achieved under a solar irradiance of 620 W/m$^2$. Furthermore, their simulation using EnergyPlus also indicated that 47% of the global cooling energy consumption occurs every year. A radiative cooling coating with a reflectivity of 97.4% and an emissivity of 94.4% was proposed by Yang et al. [14] based on calcined zeolite and polyurea. Their results indicated that the surface temperature of the proposed coating can be reduced by 4.4 °C under direct sunlight in summer. Furthermore, an average efficiency of 13.3% can be achieved for the application of a coating-based cool roof in a five-story building in China. A metamaterial-based radiative cooling film with a mid-infrared emissivity of 0.91 and a solar reflectivity of 0.92 was adopted by Wang et al. [15] as a cool roof of warehouses in Greenville, Cairo, Jimma, and Mexico. Their study showed that the annual energy consumption of a warehouse can be reduced by 22.4% in Greenville, 21.2% in Cairo, 53.0% in Jimma, and 65.2% in Mexico.

Although radiative cooling-based cool roofs can effectively reduce building energy consumption, radiative cooling remains in a cooling mode throughout the year and increases the heating load during the heating season. This also means that some of the cooling benefits will be offset by the heating penalty during the heating season.

Recently proposed temperature-adaptive radiative cooling (TARC) materials can adaptively change the solar reflectivity and/or the atmospheric window emissivity, accounting for changes in the ambient temperature and providing a possible solution to the increased heating load in the heating season caused by continuous daytime radiative cooling [16]. A similar method of applying phase change thermal storage material-based windows in buildings by Michał et al. [17] has demonstrated a 29.4% reduction in heating energy through windows in summer compared to that of ordinary windows. To alleviate the increased heating load caused by daytime radiative cooling, a passive cooling/heating double-effect material was proposed by our group in 2020 [18] based on polymers embedded with vanadium dioxide ($VO_2$) and dielectric particles. Then, two $VO_2$-based TARC materials were reported on 17 December 2021: (1) Tang et al. [19] developed a $VO_2$-based TARC coating in which the reflectivity and emissivity vary adaptively with changes in the ambient temperature. Their investigation indicated that the emissivity of the proposed TARC coating can change from 0.20 to 0.90, accounting for a phase transition temperature of 22 °C. The annual energy consumption of a typical residential building with TARC coating-based roofs in Maryland, U.S., can be reduced by 22.4 MJ/m$^2$ compared to that of a building with asphalt shingle (AS) roofs. (2) A thermochromic glass with passive radiative cooling regulation based on $VO_2$ was proposed by Wang et al. [20]. They experimentally demonstrated that the emissivity changes from 0.21 to 0.61 at a phase transition temperature of 60 °C. Furthermore, their simulation indicated that a maximum annual energy savings of 324.6 MJ/m$^2$ can be achieved by replacing low-E glass with the proposed thermochromic glass for a twelve-floor office building in seven different climates in the U.S. Since then, many researchers have devoted themselves to the study of TRAC.

Zhang et al. [21] fabricated a $VO_2$-based smart window whose mid-infrared emissivity could be adaptively switched from 0.35–0.68, accounting for a phase transition temperature of 20 °C. The experimental results showed that the indoor temperature of buildings with smart windows can decrease by 3.8 °C under a solar irradiance of 500 $W/m^2$ compared to that of buildings with regular glass windows. A scalable self-adaptive radiative cooling film was fabricated by Huang et al. [22] by incorporating core–shell nanoparticles of $VO_2$ into a polyethylene (PE) matrix. The results showed that 0.09 $kgCO_2/m^2$ of carbon emissions can be achieved by applying the proposed film on the roof of an eighteen-story residential building in Jan. Jinan, China. Wang et al. [23] proposed a thermochromic glass with multilayer anti-reflective coating/vanadium dioxide/anti-reflective coating ($AIN/VO_2/AIN$). Their results indicated that the proposed multilayer thermochromic glass works for heating and cooling when the temperature is below 20 °C or above 40 °C, respectively. This also means that the proposed multilayer thermochromic glass will change from heating to cooling when the temperature is increased from 20 °C (or lower) to 40 °C (or higher). Furthermore, a maximum energy savings of 12.13% can be achieved by applying the proposed multilayer thermochromic glass-based windows in an office building in Nanjing, China.

In addition to $VO_2$-based TARC materials, several TARC materials have also been proposed in existing studies, including hydrogel-based TARC materials [24,25], perovskite-based TARC materials [26,27], and other solution-based TARC materials [28]. However, most of these studies have focused on the preparation of TARC materials and their ability to change the temperature adaptively, with little attention given to the energy savings potential of these materials for their application in buildings. Although the energy-saving potential of $VO_2$-based TARC materials has been investigated, all $VO_2$-based TARC materials require doping with W and annealing. This not only increases the difficulty of fabricating $VO_2$-based TARC materials but also makes the performance of the annealed $VO_2$ extremely unstable [29]. To promote the application of TARC materials for improving building energy conservation, more attention should be given to the following two concerns: (1) the development of a simple-preparation TARC for large scalable application in buildings; and (2) a detailed analysis of energy performance for applying TARC in buildings.

To develop a TARC material with simple preparation, cost effectiveness, and large-scale application that can be widely used for improving building energy efficiency, this paper proposes a TARC coating based on a thermochromic powder. The process of preparing the proposed TARC coating was briefly introduced, and the performance parameters of the TARC coating (e.g., solar reflectivity and mid-infrared emissivity) were measured. Then, the heat transfer on the surface of the TARC coating was analyzed. Based on the measured performance parameters of the TARC coating, a typical two-story residential building with a roof area of 258.43 $m^2$ was modeled using EnergyPlus. Finally, the energy performance of applying the proposed TARC coating in buildings under different climates in China was discussed in detail. The flowchart of this study is shown in Figure 1.

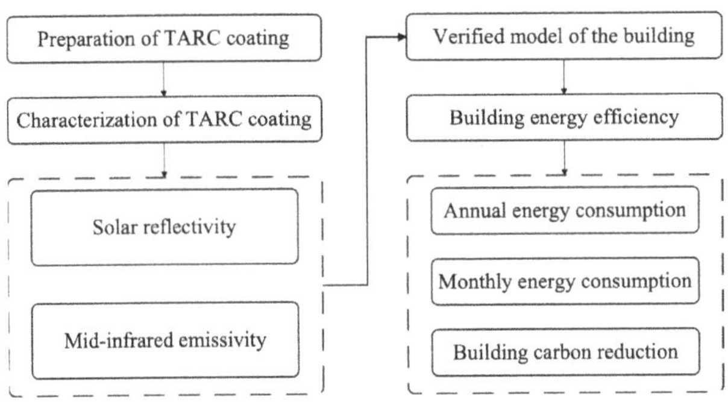


**Figure 1.** Flowchart of this study.

## 2. Methodology

### 2.1. Preparation of the TARC Coating

We prepared a TARC coating with a multilayer structure by doping a thermochromic powder (Shenzhen Tianjinli New Materials Technology Co., Ltd., Shenzhen, China) in ploymethylpentene (TPX) (Dongguan Huijie New Materials Co., Ltd., Dongguan, China) embedded with silicon dioxide ($SO_2$) (Dongguan Xinweijin Industrial Co., Ltd., Dongguan, China). The density, transmittance, and refractive index of the TPX are 0.84 g/cm$^3$, 0.93, and 1.46, respectively. $SiO_2$ is a colorless transparent crystal, whose reflectivity is 2–10% in the visible band. The specific process of preparation included the following steps: (1) $SiO_2$ particles with a diameter of 8 μm were dried in a vacuum oven for 8 h; (2) TPX was mixed with cyclohexane (Nanjing Wanqing Chemical Glass Instrument Co., Ltd., Nanjing, China) at a mass ratio of 0.08, and the mixture was stirred at 60 °C for 2.5 h until the TPX was completely dissolved (TPX dissolved solution). The cyclohexane is a colorless, pungent-smelling liquid, which dissolves polymers very well. (3) The TPX dissolved solution was allowed to stand until the temperature decreased to room temperature and was then mixed with the dried $SO_2$ particles at a volumetric ratio of 0.07 (TPX mixture dissolved solution). The TPX mixture dissolved solution was put into two beakers after 5 min and treated with a cell crusher. (4) The emission layer consisting of the TPX mixture dissolved solution was coated onto reflective layer of the aluminum foil using a coating machine (MSK-AFA-IIID-G, Shenyang Kejing Auto-instrument Co., Ltd., Shenyang, China). (5) The thermochromic power was added into the TPX mixture dissolved solution in another beaker and stirred continuously until the thermochromic powder was fully incorporated. Then, the solution was allowed to stand until all surface bubbles disappeared (thermochromic mixed solution). (6) The prepared thermochromic mixed solution was coated onto the emission layer, leading to a TARC coating with a multilayer structure. A schematic of the prepared TARC coating is shown in Figure 2.

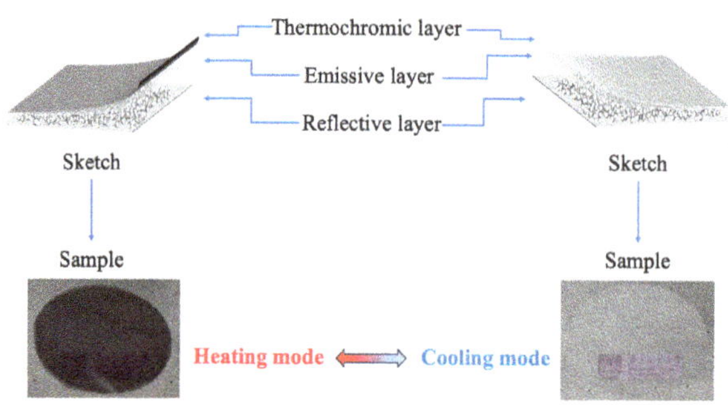


**Figure 2.** TARC coating.

As shown in Figure 2, the TARC coating appears transparent when the surface temperature is higher than the phase transition temperature. The TARC coating has high reflectivity in the solar spectrum and mid-infrared emissivity in this state. This also means that the TARC coating works in a cooling mode, reflecting solar irradiance while also exchanging heat with outer space through the transparent window to obtain cooling energy, thereby achieving sub-ambient radiative cooling. However, the TARC coating works in heating mode once the surface temperature is lower than the phase change temperature (25 °C), and it changes to a semitransparent coating with high solar absorptivity. This increases the absorption of solar irradiance and achieves heat collection.

### 2.2. Characterization of the TARC Coating

To determine the cooling and heating performance of the TARC coating, the solar reflectivity and mid-infrared emissivity were measured at the National Institute of Measurement and Testing Technology and Bohai University, respectively. The changes in the reflectivity and emissivity of the TARC coating are shown in Figure 3. As shown in Figure 3, the average solar reflectivity in the solar spectrum (0.25–2.5 μm) is 0.73 in cooling mode. However, the average solar reflectivity decreases to 0.65 when the TARC coating changes to the heating mode. Especially for visible light wavelengths (0.38–0.78 μm), the average solar reflectivity decreases from 0.71 to 0.37 when the TARC coating changes from cooling mode to heating mode. This means that the absorption of solar irradiance at the visible light wavelength, which accounts for approximately 50% of the total solar irradiance, is significantly suppressed. Furthermore, the average emissivity of 0.93 in the atmospheric window is almost unchanged.

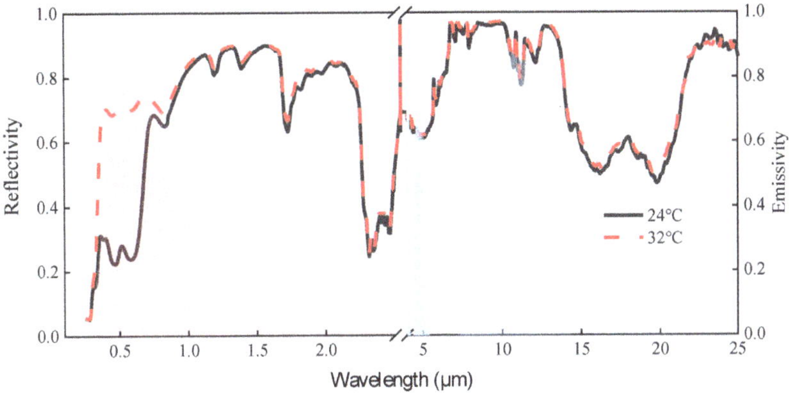


**Figure 3.** Reflectivity and emissivity of the TARC coating.

## 2.3. Modeling of the Building

To derive the energy efficiency potential for the application of the proposed TARC coating in buildings, a typical two-story residential building originating from the U.S. Department of Energy (DOE) building energy codes program [30] is adopted in this study. As shown in Figure 4, the modeled building with a roof area of 258.43 m$^2$ has dimensions of 12.13 m (length) × 9.10 m (width) × 5.79 m (height). A TARC coating/passive daytime radiative cooling (PDRC) coating is directly laid on the original roof (AS roof) to serve as the TARC roof/PDRC roof, and a split air conditioner is configured to supply supplemental cooling/heating to the building during the cooling/heating season. For the application of the TARC roof, the cooling and heating modes change adaptively with the roof surface temperature. Furthermore, other parameters of the modeled typical residential building can be found in Table 1.

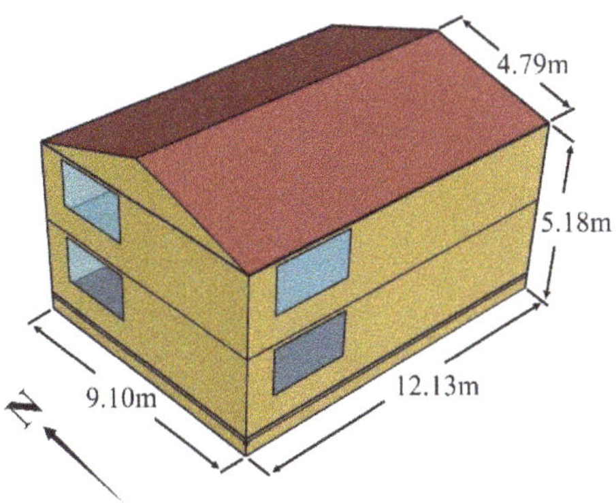


**Figure 4.** Sketch of the modeled building.

**Table 1.** Detailed parameters of the modeled building.

| Item | Details | | |
|---|---|---|---|
| Window–wall ratio | 12.78% | | |
| Occupants | 3 | | |
| Fresh air volume | 30 $m^3/h$ per person | | |
| Lighting density | 8.5 $W/m^2$ | | |
| Equipment power density | 6.0 $W/m^2$ | | |
| Interior design temperatures | Winter: | 20 °C | |
| | Summer: | 27 °C | |
| Roof | TARC | Cooling mode | |
| | | Reflectivity: | 0.65 |
| | | Emissivity: | 0.92 |
| | | Heating mode | |
| | | Reflectivity: | 0.73 |
| | | Emissivity: | 0.93 |
| | PDRC [5] | Reflectivity: | 0.92 |
| | | Emissivity: | 0.91 |
| | AS [6] | Reflectivity: | 0.25 |
| | | Emissivity: | 0.90 |

Since climate significantly impacts the energy performance of the TARC roof as well as changes in cooling and heating modes, five typical cities from all five climatic zones in China are adopted for this study, including Changchun, Beijing, Kunming, Chongqing, and Guangzhou. The criteria for the division of the different climatic zones are given in Ref. [31]. The TMY3 weather data provided by the DOE [32] are adopted in this investigation, and the envelopes of the modeled building are configured according to Ref. [33]. The detailed thermal performances of the envelopes can be found in Table 2.

**Table 2.** Required thermal performances of the envelopes for buildings in China [33].

| City | Climatic Zone | Heat Transfer Coefficient ($W/m^2{\cdot}K$) | |
|---|---|---|---|
| | | Exterior Wall | Roof |
| Changchun | Severe cold zone | 0.2 | 0.1 |
| Beijing | Cold zone | 0.25 | 0.2 |
| Kunming | Moderate zone | 0.4 | 0.25 |
| Chongqing | Hot summer and cold winter zone | 0.4 | 0.25 |
| Guangzhou | Hot summer and warm winter zone | 0.45 | 0.3 |

### 2.4. Heat transfer of the TARC Coating

The heat transfer of the TARC coating is shown in Figure 5, which includes the thermal radiation from the TARC coating ($P_{\text{rad}}$). the absorbed atmospheric radiation of the TARC coating ($P_{\text{atm}}$), the absorbed solar irradiance of the TARC coating ($P_{\text{sun}}$), the convection heat transfer between the TARC coating surface and the ambient air ($P_{\text{conv}}$), and the conduction heat transfer through the TARC coating ($P_{\text{cond}}$).

Then, the heat balance of the TARC coating can be expressed as follows:

$$P_{\text{net}} = P_{\text{rad}} - P_{\text{atm}} - P_{\text{sun}} - P_{\text{conv}} - P_{\text{cond}} \tag{1}$$

where $P_{\text{net}}$ ($W/m^2$) is the net radiative power of the TARC coating.

Although the form of Equation (1) is the same as the heat balance of radiative cooling materials [34], there are differences in their physical meanings. This accounts for the change in the solar absorption under different operation modes. As shown in Figure 5, the solar absorption increases significantly when the TARC coating switches from cooling mode to heating mode, thereby affecting the net radiative power.

The thermal radiation from the TARC coating ($P_{rad}$) can be derived from the following [5]:

$$P_{rad} = \int_0^{\frac{\pi}{2}} \pi \sin 2\theta d\theta \int_0^{\infty} I_B(T_{TARC}, \lambda) d\lambda \tag{2}$$

where $\theta$ (°) is the zenith angle; $T_{TARC}$ (K) is the temperature of the TARC coating surface; $\lambda$ (μm) is the wavelength; and $I_B$ ($T_{TARC}$, $\lambda$) is the spectral radiance of a blackbody at temperature $T_{TARC}$, which is given as follows [6]:

$$I_B(T_{TARC}, \lambda) = 2hc^2 / \left(\lambda^5 (\exp(hc/\lambda k_B T_{TARC}) - 1)\right) \tag{3}$$

where $h$ (J·s) is Planck's constant, $6.626 \times 10^{-34}$; $c$ (m/s) is the speed of light, 299792458; and $k_B$ (J/K) is Boltzmann's constant, $1.3806505 \times 10^{-23}$.

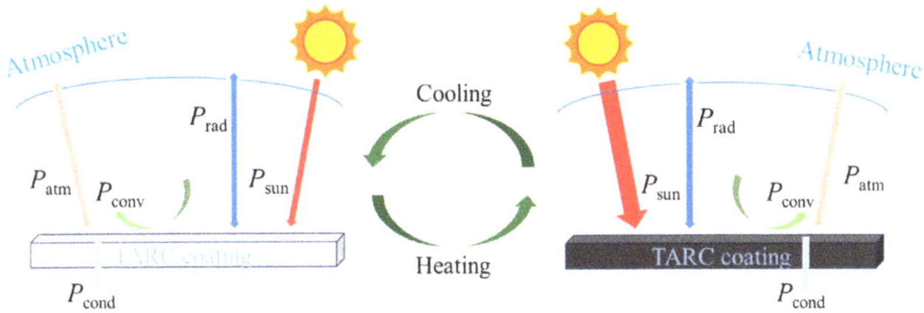


**Figure 5.** Heat transfer of the TARC coating.

The atmospheric radiation absorbed by the TARC coating ($P_{atm}$) can be calculated by the following equation [35]:

$$P_{atm} = \int_0^{\frac{\pi}{2}} \pi \sin 2\theta d\theta \int_0^{\infty} I_B(T_a, \lambda) \varepsilon_r(\lambda, \theta) \varepsilon_a(\lambda, \theta) d\lambda \tag{4}$$

where $T_a$ (K) is the ambient temperature; $\varepsilon_r(\lambda, \theta)$ is the directional emissivity of the TARC coating; and $\varepsilon_a(\lambda, \theta)$ is the angle-dependent emissivity of the atmosphere.

The absorbed solar irradiance of the TARC coating ($P_{sun}$) can be derived from the following [36]:

$$P_{sun} = \int_0^{\frac{\pi}{2}} I_{AM1.5}(\lambda) \varepsilon(\lambda, \theta_{sun}) d\lambda \tag{5}$$

where $I_{AM1 \cdot 5}$ is the solar irradiance relative to the normal solar spectrum; $\theta_{sun}$ (°) is the solar incidence angle; and $\varepsilon(\lambda, \theta_{sun})$ is the absorptivity at solar incidence angle $\theta_{sun}$ of the TARC coating.

The convective heat transfer between the TARC coating surface and the ambient air ($P_{conv}$) is determined by the following [37]:

$$P_{conv} = h_c(T_a - T_{TARC}) \tag{6}$$

where $h_c$ (W/m$^2$·K) is the comprehensive heat transfer coefficient between the TARC coating and ambient air, which can be calculated according to Ref. [38].

The conduction heat transfer through the TARC coating ($P_{cond}$) is determined by the following [6]:

$$P_{cond} = \frac{\lambda_{roof} \Delta T}{\delta} \tag{7}$$

where $\lambda_{roof}$ (W/m$^2$·K) is the thermal conductivity of the TARC coating and the original roof; $\Delta T$ (K) is the difference between the TARC coating and roof; and $\delta$ (m) is the thickness of the TARC coating and roof system.

### 2.5. Model Validation

To ensure the accuracy, the data in Ref. [19] are adopted to verify the model. Although the TARC coating used in Ref. [19] is a $VO_2$-based coating, the building in this study is the same as that in Ref. [19]. Therefore, the only work that needs to be carried out is to modify the properties (i.e., solar reflectivity and mid-infrared emissivity) of the TARC coating in this study to the same parameters as in Ref. [19]. This modification involves only the configuration of the emissivity and reflectivity of the TARC material. Thus, the accuracy of the model can be verified with the work in Ref. [19]. As shown in Figure 6, the relative error between the calculated annual cooling/heating energy consumption and the results in Ref. [19] is lower than 12%/17% for different cities in the U.S., which indicates that the established model can be used in the following study.

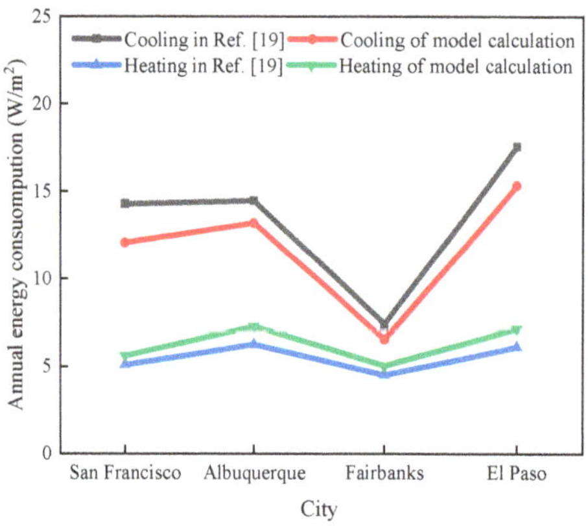


**Figure 6.** Comparison of annual energy consumption.

## 3. Results and Discussion

### 3.1. Effect of the TARC Roof on Annual Building Energy

To determine the energy-saving potential of applying TARC roofs to buildings, the energy consumption of buildings with TARC roofs is compared to that of buildings with AS roofs and PDRC roofs. A comparison of the energy consumption for buildings with the abovementioned roofs is shown in Figure 7.

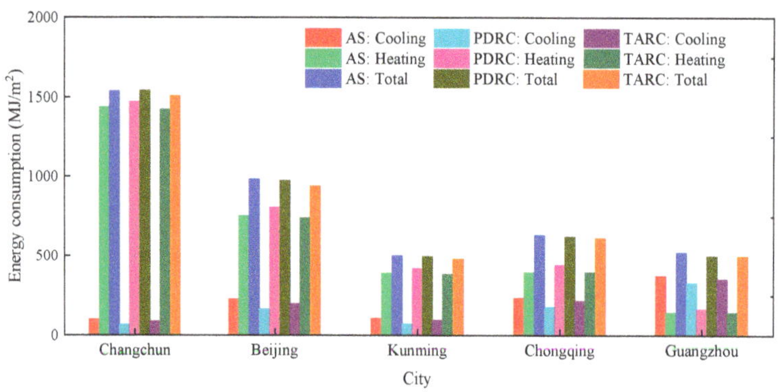


**Figure 7.** Energy consumption of buildings with different roofs.

As shown in Figure 7, the annual total energy consumption for buildings with TARC roofs decreased by 30.6 MJ/m$^2$ in Changchun, 43.0 MJ/m$^2$ in Beijing, 20.3 MJ/m$^2$ in Kunming, 17.8 MJ/m$^2$ in Chongqing, and 23.7 MJ/m$^2$ in Guangzhou compared to that for buildings with AS roofs. However, the annual total energy consumptions of buildings with TARC roofs are 31.8 MJ/m$^2$ lower than those with PDRC roofs in Changchun, 32.6 MJ/m$^2$ lower in Beijing, 14.4 MJ/m$^2$ lower in Kunming, 7.7 MJ/m$^2$ lower in Chongqing, and 2.0 MJ/m$^2$ lower in Guangzhou. This means that the TARC roof can further reduce building energy consumption compared to the AS roof or PDRC roof. In particular, energy savings are greatly improved when TARC roofs are used instead of PDRC roofs for cities with high heating demands in the winter (e.g., Changchun and Beijing). This is mainly because the continuous cooling of the PDRC roof all year leads to an increase in the heating load in winter. However, it should be noted that annual total building energy savings have deteriorated after the use of TARC roofs for cities with high cooling demands (e.g., Chongqing and Guangzhou), even though a minimum energy savings of approximately 7.7 MJ/m$^2$ and 2.2 MJ/m$^2$ can still be achieved in Chongqing and Guangzhou, respectively. This is mainly because the TARC coating is designed to achieve temperature adaptive changes, which requires considering the balance of cooling and heating power. This results in the solar reflectivity of the TARC coating in the cooling mode being lower than that of the PDRC coating, thereby reducing the net radiative cooling power. Thus, the energy savings potential decreases in cities with high cooling demands due to the decrease in reflectivity.

The annual space-conditioned electricity savings are also summarized in Table 3. Accounting for energy savings, the building electricity savings of applying TARC roofs in buildings are 8.6 kWh/m$^2$ in Changchun, 12.0 kWh/m$^2$ in Beijing, 5.7 kWh/m$^2$ in Kunming, 5.0 kWh/m$^2$ in Chongqing, and 6.6 kWh/m$^2$ in Guangzhou compared to those of the AS roof. Furthermore, the annual space-conditioned electricity consumption of buildings with TARC roofs can decrease by 8.9 kWh/m$^2$ in Changchun, 9.1 kWh/m$^2$ in Beijing, 4.0 kWh/m$^2$ in Kunming, 2.2 kWh/m$^2$ in Chongqing, and 0.6 kWh/m$^2$ in Guangzhou compared to that of buildings with PDRC roofs.

**Table 3.** Annual electricity savings for buildings with different roofs.

| City | Compared to AS Roof (kWh/m$^2$) | Compared to PDRC Roof (kWh/m$^2$) |
|---|---|---|
| Changchun | 8.6 | 8.9 |
| Beijing | 12.0 | 9.1 |
| Kunming | 5.7 | 4.0 |
| Chongqing | 5.0 | 2.2 |
| Guangzhou | 6.6 | 0.6 |

Accordingly, approximately 17.8–43.0 MJ/m$^2$ and 2.0–32.6 MJ/m$^2$ of energy consumption can be reduced for buildings with TARC roofs compared to those with AS roofs and PDRC roofs, respectively. Furthermore, TARC roofs are more suitable for buildings located in zones with high heating demands, while more building energy consumption can be achieved using PDRC roofs in buildings located in zones with high cooling demands.

### *3.2. Monthly Energy Consumption of Buildings with TARC Roofs*

To determine the energy savings potential of applying TARC roofs in detail, the monthly energy consumption of buildings with different roofs is illustrated in Figure 8. As shown in Figure 8, the average monthly energy consumption of applying TARC roofs is generally lower than that of AS roofs. Compared to those of the buildings with AS roofs, the average monthly cooling energy reductions are 0.2–4.8 MJ/m$^2$ in Changchun, 0.4–7.5 MJ/m$^2$ in Beijing, 0.2–4.2 MJ/m$^2$ in Kunming, 0.0055.5 MJ/m$^2$ in Chongqing, and 0.03–4.5 MJ/m$^2$ in Guangzhou. However, the cooling benefits of applying TARC are generally lower than those of PDRC roofs. Especially for the Jun. to Aug. in Changchun,

May to Sept. in Beijing, Jun. to Aug. in Kunming, Jun. to Sept. in Chongqing, and Apr. to Nov. in Guangzhou, in which the average monthly cooling energy consumptions increase by 5.6–6.8 $MJ/m^2$, 4.2–9.2 $MJ/m^2$, 6.7–7.3 $MJ/m^2$, 6.5–10.2 $MJ/m^2$, and 1.2–5.4 $MJ/m^2$, respectively. This is mainly because the mid-infrared emissivity of TARC roofs in cooling mode is reduced to balance the cooling and heating capacity in these modes. Although the application of TARC roofs deteriorates the cooling benefits, heating penalties are more effectively mitigated than they are for PDRC roofs.

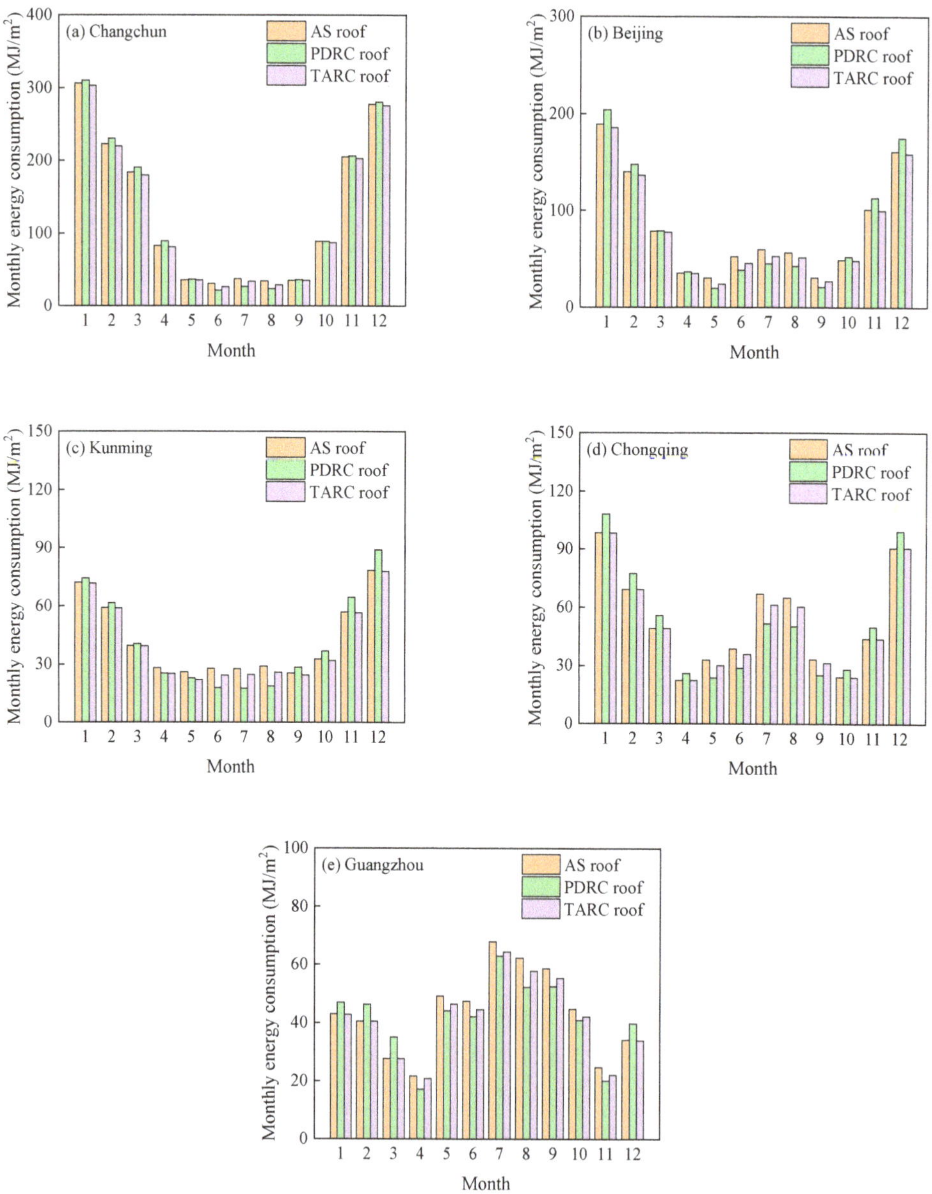


**Figure 8.** Monthly energy consumption for buildings with different roofs.

In addition, a comparison of the energy reduction for buildings with different roofs is also summarized in Table 4. As shown in Table 4, the energy savings of applying TARC roofs during the cooling season were 12.6/−17.3 MJ/m$^2$ in Changchun, 29.8/−34.1 MJ/m$^2$ in Beijing, 13.9/−20.1 MJ/m$^2$ in Kunming, 17.3/−40.0 MJ/m$^2$ in Chongqing, and 23.2/−21.3 MJ/m$^2$ in Guangzhou compared to those of AS roofs/PDRC roofs. However, the energy savings compared to those of AS roofs/PDRC roofs in the heating season are 18.0/49.1 MJ/m$^2$ in Changchun, 13.2/66.8 MJ/m$^2$ in Beijing, 6.5/34.5 MJ/m$^2$ in Kunming, 0.5/47.8 MJ/m$^2$ in Chongqing, and 0.5/23.3 MJ/m$^2$ in Guangzhou compared to those of AS roofs/PDRC roofs. It should be noted that heating energy consumption is almost unchanged for the application of TARC roofs in buildings located in Guangzhou and Chongqing compared to that for AS roofs. This is mainly because the two cities are in hot summer and warm winter and hot summer and cold winter climatic zones, respectively, which have low heating demand throughout the year.

**Table 4.** Comparison of energy reduction for buildings with different roofs.

| City | Energy Reduction (MJ/m$^2$) | | | |
|---|---|---|---|---|
| | Compared to AS Roof | | Compared to PDRC Roof | |
| | Cooling Season | Heating Season | Cooling Season | Heating Season |
| Changchun | 12.6 | 18.0 | −17.3 | 49.1 |
| Beijing | 29.8 | 13.2 | −34.1 | 66.8 |
| Kunming | 13.9 | 6.5 | −20.1 | 34.5 |
| Chongqing | 17.3 | 0.5 | −40.0 | 47.8 |
| Guangzhou | 23.2 | 0.5 | −21.3 | 23.3 |

### 3.3. Building Carbon Reduction of Applying TARC

The objective of reducing building energy consumption is to achieve carbon neutrality, and, thus, carbon emissions are also quantified in this study to further clarify the benefit of the application of TARC roofs in buildings. The reduction in carbon emissions can be derived from the following [39]:

$$E_{carbon} = \eta \times E_{ele} \tag{8}$$

where $E_{carbon}$ (kgCO$_2$/m$^2$) is the space-conditioned carbon emissions; $E_{ele}$ (kWh/m$^2$) is the space-conditioned energy; and $\eta$ is the average emission coefficient accounting for the power grid of the city, which is presented in Table 5 [40].

**Table 5.** Average emission coefficients for energy consumption in the selected cities.

| City | Emission Factors (kg CO$_2$/kWh) |
|---|---|
| Changchun | 0.7769 |
| Beijing | 0.8843 |
| Kunming | 0.5271 |
| Chongqing | 0.5257 |
| Guangzhou | 0.5271 |

As shown in Figure 9, the annual carbon emissions reductions caused by applying the TARC roof are 23.8 kgCO$_2$/m$^2$ in Changchun, 38.0 kgCO$_2$/m$^2$ in Beijing, 10.7 kgCO$_2$/m$^2$ in Kunming, 9.4 kgCO$_2$/m$^2$ in Chongqing, and 12.5 kgCO$_2$/m$^2$ in Guangzhou compared to those caused by applying the AS roof. Furthermore, the annual carbon emissions of applying TARC roofs decreased by 24.7 kgCO$_2$/m$^2$ in Changchun, 28.9 kgCO$_2$/m$^2$ in Beijing, 7.6 kgCO$_2$/m$^2$ in Kunming, 4.1 kgCO$_2$/m$^2$ in Chongqing, and 1.0 kgCO$_2$/m$^2$ in Guangzhou in this model.

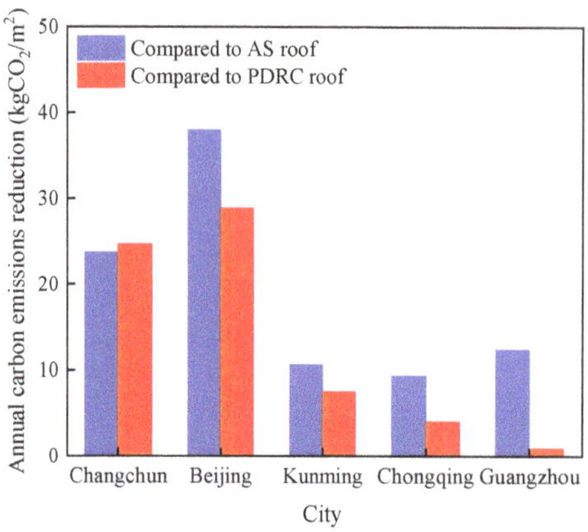


**Figure 9.** Carbon emissions reduction of applying the TARC roof.

To further indicate the differences in carbon emissions for buildings with different roofs, annual space-conditioned carbon emissions are also given in Table 6. As shown in Table 6, the carbon emissions of the buildings with TARC roofs are significantly lower than those with AS roofs and PDRC roofs. However, although the application of PDRC roofs can reduce carbon emissions from cooling, it also leads to an increase in carbon emissions from heating compared to that of the AS roof. Especially for Guangzhou, the annual carbon emissions of the AS roof and PDRC are identical, but the initial cost of applying the PDRC roof is greater than that of applying the AS roof, which limits its application in hot summer and cold winter zones. Compared with those of the AS roof, the carbon emissions from applying the TARC roof for both cooling and heating are lower, which demonstrates the significant carbon emissions reduction potential of the TARC roof. However, it should also be noted that although the annual carbon emissions of applying TARC roofs are lower than those of PDRC roofs, they are mainly due to the mitigation of heating penalties. However, the carbon emissions from cooling still increase.

**Table 6.** Annual space-conditioned carbon emissions.

| City | Annual Carbon Emission ($kgCO_2/m^2$) | | | | | | | | |
|---|---|---|---|---|---|---|---|---|---|
| | AS Roof | | | PDRC Roof | | | TARC Roof | | |
| | Cooling | Heating | Total | Cooling | Heating | Total | Cooling | Heating | Total |
| Changchun | 79.0 | 1117.8 | 1196.8 | 55.8 | 1141.9 | 1197.7 | 69.3 | 1103.8 | 1173.0 |
| Beijing | 204.0 | 666.7 | 870.7 | 147.5 | 714.0 | 861.6 | 177.7 | 655.0 | 832.7 |
| Kunming | 58.3 | 206.8 | 265.0 | 40.4 | 221.5 | 261.9 | 51.0 | 203.4 | 254.3 |
| Chongqing | 124.5 | 209.0 | 333.6 | 94.4 | 233.9 | 328.3 | 115.4 | 208.8 | 324.2 |
| Guangzhou | 198.5 | 76.5 | 275.1 | 175.0 | 88.6 | 263.6 | 262.6 | 76.3 | 262.6 |

## 4. Conclusions and Outlook

This paper proposes a thermochromic powder-based TARC coating, and a typical two-story residential building is modeled to determine the energy efficiency of the proposed TARC coating. Furthermore, the building energy efficiency enhancement of applying TARC roofs under different climates in China is discussed via comparisons with AS roofs and PDRC roofs. Compared to those for buildings with AS roofs and PDRC roofs in China, the main conclusions are as follows: (1) the energy consumptions for buildings with TARC roofs can be reduced by approximately 17.8–43.0 $MJ/m^2$ and 2.0–32.6 $MJ/m^2$, respectively; (2) the annual carbon emissions reductions are 9.4–38.0 $kgCO_2/m^2$ and 1.0–28.9 $kgCO_2/m^2$, respectively.

This study demonstrates the energy efficiency of applying TRAC roofs in buildings. However, TARC roofs are more suitable for buildings located in zones with high heating demands, while more building energy consumption can be achieved using PDRC roofs in buildings located in zones with high cooling demands. Furthermore, the building envelope has a great impact on the energy consumption of TARC roof-based buildings. This paper only investigates the material's energy performance based on the code in China, and the building model should be modified according to the code for countries where the buildings are located.

**Author Contributions:** Conceptualization, K.Z.; methodology, K.Z.; software, G.S.; validation, G.S. and K.Z.; formal analysis, G.S.; investigation, G.S. and K.Z.; writing—original draft preparation, G.S.; writing—review and editing, K.Z., F.X., Z.Z., S.J. and Y.G.; supervision, K.Z.; project administration, K.Z.; funding acquisition, K.Z. All authors have read and agreed to the published version of the manuscript.

**Funding:** This research was funded by the National Natural Science Foundation of China, grant number 51878342.

**Data Availability Statement:** The data presented in this study are available on request from the corresponding author. The data are not publicly available due to privacy.

**Conflicts of Interest:** The authors declare no conflicts of interest.

## Abbreviations

| | | | |
|---|---|---|---|
| AS | Original roof | PDMS | Polydimethylsiloxane |
| PDRC | Passive daytime radiative cooling | PE | Polyethylene |
| TARC | Temperature-adaptive radiative cooling | TPX | Ploymethylpentene |

## References

1. Li, R.; You, K.R.; Cai, W.G.; Wang, J.B.; Liu, Y.; Yu, Y.H. Will the Southward Center of Gravity Migration of Population, Floor Area, and Building Energy Consumption Facilitate Building Carbon Emission Reduction in China. *Build. Environ.* **2023**, *242*, 110576. [CrossRef]
2. Kwok, K.Y.G.; Kim, J.; Chong, W.K.O.; Ariaratnam, S.T. Structuring a Comprehensive Carbon-Emission Framework for the Whole Lifecycle of Building, Operation, and Construction. *J. Archit. Eng.* **2016**, *22*, 04016006. [CrossRef]
3. Nutakki, T.U.K.; Kazim, K.W.U.; Alamara, K.; Salameh, T.; Abdelkareem, M.A. Experimental Investigation on Aging and Energy Savings Evaluation of High Solar Reflective Index (Sri) Paints: A Case Study on Residential Households in the GCC Region. *Buildings* **2023**, *13*, 419. [CrossRef]
4. Li, H.R.; Zhang, K.; Shi, Z.J.; Jiang, K.Y.; Wu, B.Y.; Ye, P.L. Cooling Benefit of Implementing Radiative Cooling on a City-Scale. *Renew. Energy* **2023**, *212*, 372–381. [CrossRef]
5. Jiang, K.Y.; Zhang, K.; Shi, Z.J.; Li, H.R.; Wu, B.Y.; Mahian, O.; Zhu, Y.T. Experimental and Numerical Study on the Potential of a New Radiative Cooling Paint Boosted by $SiO_2$ Microparticles for Energy Saving. *Energy* **2023**, *283*, 128473. [CrossRef]
6. Ma, M.Q.; Zhang, K.; Chen, L.F.; Tang, S.H. Analysis of the Impact of a Novel Cool Roof on Cooling Performance for a Low-Rise Prefabricated Building in China. *Build. Serv. Eng. Res. Technol.* **2020**, *42*, 26–44. [CrossRef]
7. Zhuang, Z.Y.; Yang, X.B.; Xie, K.; Tang, M.Y.; Xu, Y.B.; Ben, X.Y. The Mathematical Modeling and Performance of Sky Radiative Coolers. *Buildings* **2023**, *13*, 2972. [CrossRef]
8. Xiang, M.L.; Liao, Y.X.; Jia, Y.H.; Zhang, W.T.; Long, E.S. Summer Thermal Challenges in Emergency Tents: Insights into Thermal Characteristics of Tents with Air Conditioning. *Buildings* **2024**, *14*, 710. [CrossRef]
9. Lin, K.X.; Du, Y.W.; Chen, S.R.; Chao, L.K.; Lee, H.H.; Ho, C.T.; Zhu, Y.H.; Zeng, Y.J.; Pan, A.Q.; Yan, T.C. Nanoparticle-Polymer Hybrid Dual-Layer Coating with Broadband Solar Reflection for High-Performance Daytime Passive Radiative Cooling. *Energy Build.* **2022**, *276*, 112507. [CrossRef]
10. Liu, Y.M.; Bu, X.H.; Liu, R.Q.; Feng, M.X.; Zhang, Z.W.; He, M.; Huang, M.J.; Zhou, Y.M. Construction of Robust Silica-Hybridized Cellulose Aerogels Integrating Passive Radiative Cooling and Thermal Insulation for Year-Round Building Energy Saving. *Chem. Eng. J.* **2024**, *481*, 148780. [CrossRef]
11. Chen, J.H.; Lu, L.; Gong, Q.; Wang, B.X.; Jin, S.G.; Wang, M. Development of a New Spectral Selectivity-Based Passive Radiative Roof Cooling Model and Its Application in Hot and Humid Region. *J. Clean. Prod.* **2021**, *307*, 127170. [CrossRef]
12. Ning, Y.; Xuan, Q.D.; Fu, Y.; Ma, X.; Lei, D.Y.; Niu, J.L.; Dai, J.G. Phosphate Activated Geopolymer-Based Coating with High Temperature Resistance for Sub-Ambient Radiative Cooling. *Sustain. Cities Soc.* **2024**, *100*, 104992.

13. Cai, C.Y.; Chen, W.B.; Wei, Z.C.; Ding, C.X.; Sun, B.J.; Gerhard, C.; Fu, Y.; Zhang, K. Bioinspired "Aerogel Grating Metasurfaces Durable Daytime Radiat. Cool. Year-Round Energy Savings". *Nano Energy* **2023**, *114*, 108625. [CrossRef]
14. Yang, S.S.; Lei, S.; Wang, F.J.; Long, H.B.; Ou, J.F.; Amirfazli, A.; Baldelli, A. A Comprehensive Investigation of Zeolite/Polyurea Cooling Coating on Concrete for Building Energy Conservation. *Prog. Org. Coat.* **2024**, *188*, 108265. [CrossRef]
15. Wang, N.S.; Lv, Y.Y.; Zhao, D.L.; Zhao, W.B.; Xu, J.T.; Yang, R.G. Performance Evaluation of Radiative Cooling for Commercial-Scale Warehouse. *Mater. Today Energy* **2022**, *24*, 100927. [CrossRef]
16. Giulia, U.; Gianluca, R.; Kwok, W.S.; Jie, F.; Mattheos, S. On the Energy Modulation of Daytime Radiative Coolers: A Review on Infrared Emissivity Dynamic Switch Against Overcooling. *Sol. Energy* **2020**, *209*, 278–301.
17. Michał, M.; Lech, L. The Impact of a Mobile Shading System and a Phase-Change Heat Store on the Thermal Functioning of a Transparent Building Partition. *Materials* **2021**, *14*, 14102512.
18. Zhang, K.; Chen, L.F.; Song, G.; Niu, X.F.; Li, F. Passive Cooling/Heating Double-Effect Material. International Patent WO2021120706A1, 24 June 2021.
19. Tang, K.; Dong, K.C.; Li, J.C.; Gordon, M.P.; Reichertz, F.G.; Kim, H.; Rho, Y.; Wang, Q.J.; Lin, C.Y.; Grigoropoulos, C.P.; et al. Temperature-Adaptive Radiative Coating for All-Season Household Thermal Regulation. *Science* **2021**, *374*, 1504–1509. [CrossRef] [PubMed]
20. Wang, S.C.; Jiang, T.Y.; Meng, Y.; Yang, R.G.; Tan, G.; Long, Y. Scalable Thermochromic Smart Windows with Passive Radiative Cooling Regulation. *Science* **2021**, *374*, 1501–1504. [CrossRef] [PubMed]
21. Zhang, R.; Li, R.Z.; Xu, P.; Zhong, W.H.; Zhang, Y.; Luo, F.; Xiang, B. Thermochromic Smart Window Utilizing Passive Radiative Cooling for Self-Adaptive Thermoregulation. *Chem. Eng. J.* **2023**, *471*, 144527. [CrossRef]
22. Huang, J.C.; Zhang, X.-k.; Yu, X.Y.; Tang, G.H.; Wang, X.Y.; Du, M. Scalable Self-Adaptive Radiative Cooling Film through $VO_2$-Based Switchable Core–Shell Particles. *Renew. Energy* **2024**, *224*, 120208. [CrossRef]
23. Wang, J.Y.; Li, G.; Zhao, D.L. Multi-Objective Optimization of an Anti-Reflection AIN/$VO_2$/AIN Thermochromic Window for Building Energy Saving. *Energy* **2024**, *288*, 129798. [CrossRef]
24. Min, X.Z.; Wang, X.Y.; Li, J.L.; Xu, N.; Du, X.R.; Zeng, M.Y.; Li, W.; Zhu, B.; Zhu, J. A Smart Thermal-Gated Bilayer Membrane for Temperature-Adaptive Radiative Cooling and Solar Heating. *Sci. Bull.* **2023**, *18*, 2054–2062. [CrossRef] [PubMed]
25. Guo, N.; Liu, S.Q.; Chen, C.X.; Song, C.X.; Mo, S.H.; Yan, H.J.; Chen, M.J. Outdoor Adaptive Temperature Control Based on a Thermochromic Hydrogel by Regulating Solar Heating. *Sol. Energy* **2024**, *270*, 112405. [CrossRef]
26. Jiang, N.; Chen, S.M.; Wang, J.T.; He, C.Y.; Fang, K.; Yin, H.L.; Liu, Y.T.; Li, Y.; Yu, D. Smart Thermally Responsive Perovskite Materials: Thermo-Chromic Application and Density Function Theory Calculation. *Heliyon* **2023**, *9*, e12845. [CrossRef] [PubMed]
27. Du, Y.W.; Liu, S.; Zhou, Z.W.; Lee, H.H.; Ho, T.C.; Feng, S.-P.; Tso, C.Y. Study on the Halide Effect of MA4PbX6·$2H_2O$ Hybrid Perovskites–from Thermochromic Properties to Practical Deployment for Smart Windows. *Mater. Today Phys.* **2022**, *23*, 100624. [CrossRef]
28. Yan, C.Q.; Li, A.K.; Wu, H.L.; Tong, Z.P.; Qu, J.H.; Sun, W.; Yang, Z.W. Scalable and All-Season Passive Thermal Modulation Enabled by Radiative Cooling, Selective Solar Absorption, and Thermal Retention. *Appl. Therm. Eng.* **2023**, *221*, 119707. [CrossRef]
29. Makarevich, A.M.; Sobol, A.G.; Sadykov, L.; Sharovarov, D.I.; Amelichev, V.A.; Tsymbarenko, D.M.; Boytsova, O.V.; Kaul, A.R. Delicate Tuning of Epitaxial $VO_2$ Films for Ultra-Sharp Electrical and Intense IR Optical Switching Properties. *J. Alloys Compd.* **2021**, *853*, 157214. [CrossRef]
30. Wong, R.Y.M.; Tso, C.Y.; Chao, C.Y.H.; Huang, B.; Wan, M.P. Ultra-Broadband Asymmetric Transmission Metallic Gratings for Subtropical Passive Daytime Radiative Cooling. *Sol. Energy Mater. Sol. Cells* **2018**, *186*, 330–339. [CrossRef]
31. Tang, S.H.; Akkurt, N.; Zhang, K.; Chen, L.F.; Ma, M.Q. Effect of Roof and Ceiling Configuration on Energy Performance of a Metamaterial-based Cool Roof for Low-rise Office Building in China. *Indoor Built Environ.* **2021**, *30*, 1739–1750. [CrossRef]
32. Weather Date. Available online: https://energyplus.net/weather-region (accessed on 10 April 2024).
33. *GB 55015-2021*; Code for Building Energy Efficiency and Renewable Energy Utilization. Ministry of Housing: Beijing, China, 2021.
34. Yang, Y.; Long, L.S.; Meng, S.; Denisuk, N.; Chen, G.Z.; Zhu, L.P. Bulk Material Based Selective Infrared Emitter for Sub-Ambient Daytime Radiative Cooling. *Sol. Energy Mater. Sol. Cells* **2020**, *211*, 110548. [CrossRef]
35. Huang, J.; Fan, D. Core-Shell Microspheres Hybridized Membrane for Light Emitting and Radiative Cooling. *J. Alloys Compd.* **2022**, *924*, 166480. [CrossRef]
36. Liu, J.W.; Zhang, D.B.; Jiao, S.F.; Zhou, Z.H.; Zhang, Z.F.; Gao, F. Preliminary Study of Radiative Cooling in Cooling Season of the Humid Coastal Area. *Sol. Energy Mater. Sol. Cells* **2020**, *208*, 110412. [CrossRef]
37. Kontoleon, K.J.; Eumorfopoulou, E.A. The Influence of Wall Orientation and Exterior Surface Solar Absorptivity on Time Lag and Decrement Factor in the Greek Region. *Renew. Energy* **2008**, *7*, 1652–1664. [CrossRef]
38. Wu, B.Y.; Zhang, K.; Ye, P.L.; Niu, Z.Y.; Song, G. Effect of Electronic and Phonon Properties on Polar Dielectric Embedded Polymer-Based Radiative Cooling Materials. *Sol. Energy Mater. Sol. Cells* **2018**, *224*, 371–381. [CrossRef]

39. Myint, N.N.; Shafique, M. Embodied Carbon Emissions of Buildings: Taking a Step Towards Net Zero Buildings. *Case Stud. Constr. Mater.* **2023**, *260*, 112473. [CrossRef]
40. Chen, Z.Y.; Dong, M.Y.; Wang, C.H. Passive Interfacial Photothermal Evaporation and Sky Radiative Cooling Assisted All-Day Freshwater Harvesting: System Design, Experiment Study, and Performance Evaluation. *Appl. Energy* **2023**, *475*, 146431. [CrossRef]

**Disclaimer/Publisher's Note:** The statements, opinions and data contained in all publications are solely those of the individual author(s) and contributor(s) and not of MDPI and/or the editor(s). MDPI and/or the editor(s) disclaim responsibility for any injury to people or property resulting from any ideas, methods, instructions or products referred to in the content.

buildings

MDPI

Article

# Comparative Evaluation of Cool Roofs and Photovoltaic Roofs in Sustainable Buildings Within the Scope of the 3-E Static Payback Period Framework

**Shanguo Zhao [1,2,*], Xiaosong Zhang [2] and Xing Jin [3]**

1 Electrization and Intelligence, College of Marine Engineering, Jiangsu Maritime Institute, Nanjing 211199, China
2 School of Energy and Environment, Southeast University, Nanjing 211189, China; rachpe@126.com
3 School of Architecture and Urban Planning, Nanjing University, Nanjing 210093, China; jxining@163.com
* Correspondence: 20221094@jmi.edu.cn

**Abstract**

Building envelopes play a pivotal role in influencing building energy consumption. Among its components, the roof, as a critical element, directly absorbs solar radiation and serves as a primary medium for external heat exchange. Its thermal performance significantly impacts the overall energy consumption of buildings. This study focuses on cool roofs as the research subject to investigate their thermal performance and its effects on building energy consumption. Drawing on the principles of life cycle assessment (LCA), a novel concept of environmental payback period is introduced. By comparing cool roofs with photovoltaic roofs, this research employs energy consumption simulation and life cycle assessment to evaluate their performance across three dimensions: economic, energy, and environmental impacts. A comprehensive 3-E (Economic, Energy, Environmental) static payback period theoretical framework based on LCA is established. Within this framework, the concepts of economic static payback period, energy static payback period, and environmental static payback period are explicitly defined, and corresponding calculation formulas are provided. A case study in Nanjing is conducted to validate the proposed framework. The results indicate that the economic payback periods for cool roofs and photovoltaic roofs are 1.75 years and 10.90 years, respectively; the energy payback periods are 13.6 years and 43.7 years, respectively; and the environmental payback periods are 2.2 years and 7.6 years, respectively. In terms of energy savings, photovoltaic roofs outperform cool roofs significantly, with an annual energy saving of 139 kWh/m$^2$ for photovoltaic roofs compared to 6.5 kWh/m$^2$ for cool roofs. However, cool roofs demonstrate clear advantages in the comparison of payback periods.

**Keywords:** cool roof; photovoltaic; life cycle assessment; energy consumption simulation; static payback period

## 1. Introduction

The increasing demand for sustainable building solutions has intensified research into energy-efficient roofing technologies, particularly cool roofs and photovoltaic (PV) systems. These technologies offer significant potential to reduce building energy consumption and mitigate urban heat island effects, yet their comprehensive environmental and economic viability remains underexplored. While prior studies have evaluated cool roofs and PV roofs separately using life cycle assessment (LCA) [1] or cost–benefit analysis [2], a unified framework that integrates economic, energy, and environmental payback periods has not been sufficiently developed.

Cool roofs, designed to reflect more sunlight and absorb less heat than conventional roofs, have been widely studied for their energy-saving benefits in cooling-dominated climates [3,4]. Similarly, PV roofs contribute to renewable energy generation, reducing reliance on fossil fuels [5,6]. However, existing evaluations often focus on single dimensions—either energy savings, economic feasibility, or environmental impact—without considering their interdependencies. For instance, while a cool roof may reduce cooling loads, its manufacturing process may involve materials with high embodied carbon [4]. Conversely, PV roofs require significant upfront investments but offer long-term energy and emissions savings [7]. A holistic assessment is therefore necessary to guide decision-making in building design and policy.

The evaluation of sustainable roofing technologies has undergone significant advancements in recent years, with increasing emphasis on life cycle assessment (LCA) methodologies to quantify environmental impacts [8,9]. Previous studies have predominantly focused on comparing conventional roofing systems with emerging alternatives, such as green roofs, cool roofs, and photovoltaic (PV) integrated systems. However, these assessments often employ fragmented methodologies, addressing energy efficiency, economic feasibility, or environmental benefits in isolation, rather than adopting a holistic, integrated approach [10,11]. This segmented focus limits the ability to comprehensively evaluate the multi-dimensional performance of sustainable roofing technologies, highlighting the need for more unified analytical frameworks that simultaneously consider energy, economic, and environmental dimensions.

The adoption of cool roofs has emerged as a promising strategy to mitigate urban heat island effects and reduce building energy consumption [4]. Recent studies have extensively evaluated the environmental and economic performance of cool roofs through life cycle assessment (LCA) methodologies. For instance, Scolaro and Ghisi [8] conducted a comprehensive review of green roof systems, highlighting the critical role of material selection in optimizing energy efficiency and environmental impacts. While their focus was on green roofs, their findings underscore the importance of material innovation for sustainable roofing technologies, including cool roofs.

Building on this, Pirvaram et al. [10] systematically analyzed radiative cooling technologies for buildings, integrating techno-enviro-economic assessments with LCA frameworks. Their research demonstrated that cool roofs significantly reduce cooling energy demand, achieving reductions of 20–40%, while simultaneously lowering greenhouse gas emissions over their lifecycle. However, they emphasized that performance is highly contingent on climatic conditions—a finding further substantiated by the work of Pique et al. This variability underscores the importance of contextual factors in assessing the effectiveness of cool roofs, highlighting the need for region-specific evaluations to optimize their implementation and benefits [12], who compared the global warming potential (GWP) of green and conventional roofs in cold climates. Although their results indicated limited benefits of green roofs in such regions, cool roofs exhibited superior adaptability due to their simpler design and lower maintenance requirements.

From an economic perspective, Katebi et al. [1] evaluated multiple roof systems and found that cool roofs achieved the highest cost–benefit ratio among sustainable options, reducing annual energy costs by 15–30% while maintaining minimal environmental footprints. This aligns with the technical analysis by Ma et al. [13], who tested a novel cool roof prototype on low-rise prefabricated buildings in China. Their field measurements confirmed a 25% reduction in peak cooling loads, emphasizing the technology's scalability for rapidly urbanizing regions.

LCA has become a standard tool for assessing the environmental footprint of building materials and systems across their entire life cycle, from raw material extraction to disposal [14,15]. Several studies have applied LCA to compare traditional gravel-ballasted roofs with green roofs and reflective cool roofs. For instance, Rasul et al. demonstrated that extensive green roofs exhibit lower global warming potential (GWP) over their lifespan compared to conventional roofs, primarily due to reduced urban heat island effects and stormwater management benefits [16]. Similarly, Koroxenidis et al. highlighted that intensive green roofs, while offering greater ecological benefits, often incur higher initial embodied energy due to complex substrate layers and irrigation requirements [17].

Cool roofs have been primarily evaluated for their operational energy savings in cooling-dominated climates. Research by Aggarwal demonstrated that reflective roofing membranes can reduce cooling energy demand by up to 20%; however, their manufacturing processes often involve polymer-based materials with significant carbon footprints [18]. In contrast, Cubi's findings revealed that in cold climates, the heating penalty associated with cool roofs may offset their cooling benefits, highlighting the necessity for region-specific adjustments in life cycle assessment (LCA) methodologies [19]. These contrasting outcomes emphasize the importance of tailoring cool roof evaluations to specific climatic conditions to ensure their effectiveness and sustainability across diverse geographic contexts.

Photovoltaic roofs represent another sustainable alternative, with studies like [20] quantifying their energy payback periods (typically 2–4 years) and carbon offset potential. However, the integration of PV with roofing systems introduces additional complexity in LCA due to material compatibility, maintenance requirements, and end-of-life recycling challenges.

Beyond environmental impacts, the economic feasibility of roofing technologies has been assessed through life cycle costing (LCC) and static/dynamic payback period analyses. The conventional static payback period, which calculates the time required to recover initial investments through energy savings, has been widely used for its simplicity. For example, Ref. [21] compared radiative cooling roofs and PV systems, showing that radiative cooling achieves faster economic payback in hot climates due to lower installation costs.

Energy payback period (EPP) analysis extends this concept by evaluating the time needed for a system to generate energy equivalent to its embodied energy. Studies such as [12] have applied EPP to green roofs, revealing that their long lifespan (40+ years) compensates for higher initial energy inputs. However, these analyses often neglect the interplay between energy savings and environmental impacts, leading to incomplete sustainability assessments.

To address this gap, we propose a novel 3-E static payback period framework that evaluates roofing technologies across three dimensions: Economic, Energy, and Environmental payback periods. This approach extends the conventional static payback period [22] by incorporating environmental impact metrics derived from Life Cycle Assessment (LCA). Specifically, the economic payback period assesses the time required to recover initial costs, the energy payback period quantifies the duration needed to offset energy inputs, and the environmental payback period measures the time required to neutralize lifecycle environmental impacts. By integrating these metrics, our framework enables a more holistic and balanced comparison of roofing alternatives, accounting for both short-term financial considerations and long-term sustainability benefits.

The novelty of this work lies in its multi-dimensional payback analysis, which synthesizes LCA results with financial and energy performance metrics into a unified evaluation framework. Unlike prior research that often examines these dimensions in isolation, this paper offers a consolidated approach designed to serve the diverse needs of stakeholders, including architects, policymakers, and building owners. Furthermore, we apply this framework to a comparative case study of cool roofs and PV roofs, investigating how material choices, regional climate conditions, and energy mixes influence payback timelines. Our findings reveal critical trade-offs between initial investments and long-term gains, providing actionable insights for advancing sustainable building practices and informing decision-making processes.

## 2. Target Selection and Scope Definition

From the perspective of energy utilization, roof energy-saving technologies can be divided into two categories: active and passive. Active roof technologies, such as photovoltaic roofs and solar thermal systems, convert solar energy into electricity or heat for human use. The generated electricity can power lighting, sockets, and air conditioning systems, while the collected thermal energy can be used for cooling, heating, and domestic hot water, thereby reducing building energy consumption. Passive roof technologies, such as green roofs, cool reflective roofs, and ventilated roofs, modify the building envelope to reduce heat transfer from the external environment into the interior, thereby lowering air conditioning energy demand and conserving building energy [23].

These two approaches differ fundamentally: one directly reduces the building's cooling load at the source, while the other compensates for the existing load by utilizing solar energy on the demand side. The former achieves direct energy savings at the source, whereas the latter relies on additional energy compensation. To determine which technology holds greater promise in energy efficiency, a quantitative analysis is necessary.

From a societal perspective, energy savings during the operational phase of any system or technology may not represent true energy conservation. In reality, operational-phase savings often merely shift energy consumption from the usage stage to the production stage. Thus, from a holistic societal viewpoint, energy is still being consumed. This necessitates the introduction of a life cycle assessment (LCA) framework. Additionally, the question of whether the energy consumed in manufacturing solar photovoltaic panels and thermal collectors outweighs the energy they generate during operation must also be analyzed from a full life cycle perspective.

Given the uncertainties and knowledge gaps in assessing the life cycle impacts of green vegetation, this study focuses on comparing cool reflective roofs and photovoltaic roofs. A comprehensive life cycle assessment is conducted from economic, energy, and environmental (3-E) perspectives, establishing a 3-E payback period framework for evaluating roof energy-saving technologies.

A complete life cycle consists of five stages: raw material extraction, equipment manufacturing, system installation, operation and maintenance, and end-of-life disposal [24]. In this study, due to the impracticality of recycling roof coatings and the challenges in recycling photovoltaic components, the final waste disposal stage is omitted.

For this experiment, a pure titanium dioxide-epoxy resin coating was chosen as the cool roof reflective coating, with the mass ratio of water-epoxy resin-titanium dioxide approximately set at 2:2:1. The life cycle assessment (LCA) input and output parameters for both epoxy resin and titanium dioxide were sourced from Simapro (version 9.5), a widely used LCA software tool. The energy input for the manufacturing process was electricity, which was accounted for in the LCA to evaluate the environmental impacts associated with the production and application of the coating. This approach ensures a comprehensive assessment of the coating's lifecycle, from raw material extraction to its functional use as a cool roof solution.

## 3. Energy Consumption Analysis

To compare the energy-saving performance between photovoltaic roofs and cool reflective roofs, it is necessary to simulate the energy consumption of both roof types. To simplify the model, a 10 m × 10 m building model was constructed in EnergyPlus (version 9.4.0). For the cool roof, the reflectance was set at 0.85 and emissivity at 0.9, covering the entire 100 $m^2$ roof surface. For the photovoltaic roof, photovoltaic panels with a total area of 73.8 $m^2$ were installed on the roof of the aforementioned model. Considering efficient utilization of electricity to avoid waste, a 10 kW grid-tied inverter was incorporated—drawing power from the grid when photovoltaic production is insufficient and feeding excess power back to the grid when production exceeds demand.

The energy performance of the building, including the reflective effect of the BIPV roof, was modeled using EnergyPlus. Key material properties of the PV panels, such as reflectivity, absorptivity, and emissivity, were input into the simulation to accurately capture their impact on the building's thermal performance. The PV panels were treated as an integral part of the roof, ensuring that their reflective effect on solar radiation was fully accounted for in the energy calculations. The building thermal performance and operating parameters can be seen in Table 1.

**Table 1.** Building thermal performance and operating parameters.

| | Specific Parameters |
|---|---|
| Air Conditioning System | VAV + gas boiler |
| Lighting power | 6 $W/m^2$ |
| plug load power | 20 $W/m^2$ |
| Refrigeration unit COP | 4.5 |
| boiler efficiency | 0.9 |
| Air tightness | 7.5 $m^3/(m^2\ h)$ |
| ventilation rate | 3 $m^3/(h\ person)$ |

The climate profile data for Jiangsu Province were obtained from the EnergyPlus website, specifically the EPW (EnergyPlus Weather) file for Nanjing, which is representative of the region. This file includes hourly weather data for temperature, solar radiation, humidity, and wind speed, essential for accurate energy modeling.

Annual building energy consumption simulations were conducted, with the comparative energy performance between the cool roof and conventional roof illustrated in Figure 1.

As shown in Figure 2, the implementation of the cool roof resulted in an annual electricity savings of 6.5 $kWh/m^2$, representing an 8.6% reduction in total energy consumption. Specifically, heating energy consumption increased by 6.3%, while cooling energy consumption decreased by 15.0%.

The power generation/supply profile of the photovoltaic panels on the PV roof is presented in Figure 3.

The data reveals that the photovoltaic (PV) panels generated a total annual electricity output of 10,283 kWh. Of this amount, 3516 kWh (34.2% of total generation) was exported to the grid during periods of low building demand, while the remaining 6767 kWh (65.8%) was consumed on-site to meet the building's electrical load. The building's total annual electricity consumption reached 15,806 kWh, with 9038 kWh (57.2% of total consumption) sourced from grid purchases and the balance supplied by the PV system's self-generation.

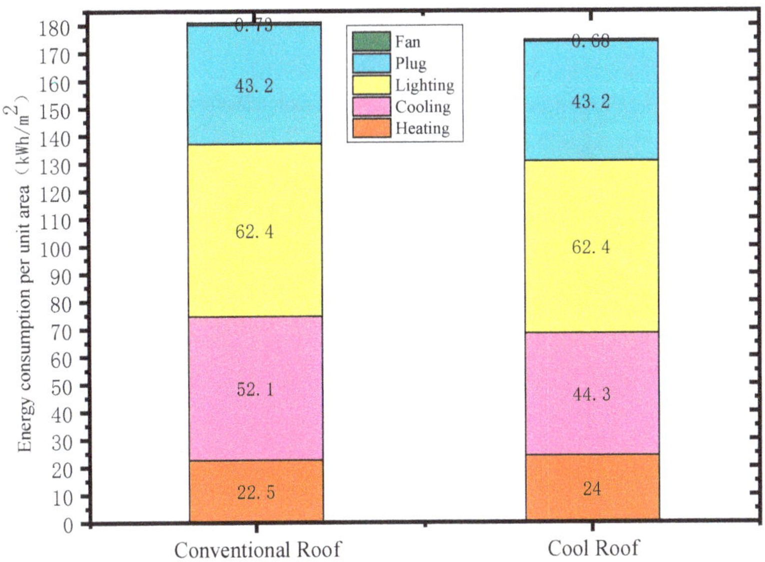


**Figure 1.** Bar chart of energy consumption for different roofs.

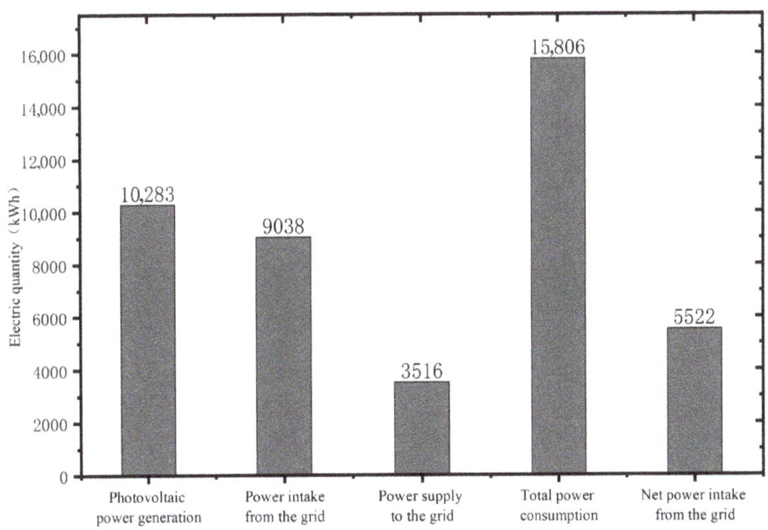


**Figure 2.** Power generation/power extraction status of the photovoltaic panels on the photovoltaic roof.

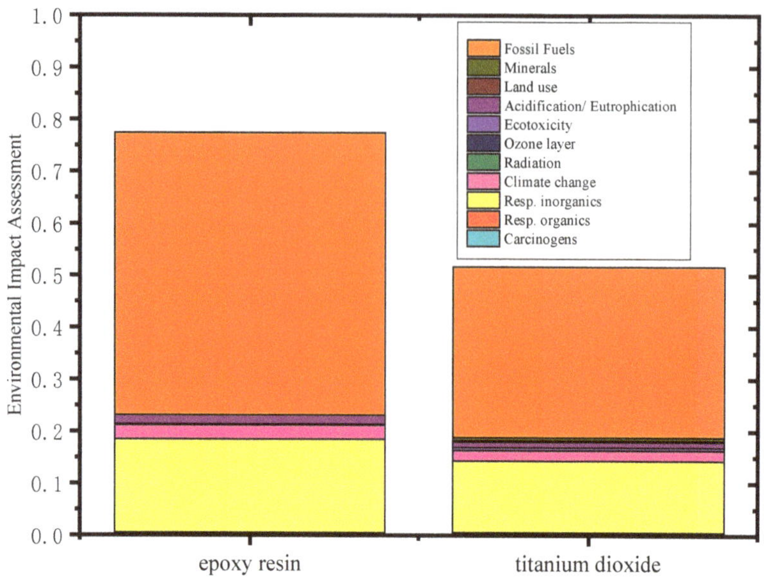


**Figure 3.** Environmental impact assessment of titanium dioxide and epoxy resin.

## 4. Life Cycle Assessment of Roof Energy Efficiency

From the above analysis, it can be observed that for annual building energy consumption, the cool roof achieves an annual energy saving of 6.5 kWh/m$^2$, while a 73.8 m$^2$ photovoltaic roof generates 10,283 kWh annually. This translates to an energy saving of 10,283/73.8 = 139 kWh/m$^2$ for the photovoltaic roof. Therefore, in terms of annual energy savings alone, photovoltaic roofs are significantly superior to cool roofs.

For the heat-reflective cool roof, actual tests show that 20 kg of coating can cover 40 m$^2$ of roof area, meaning the required coating mass per square meter is 0.5 kg. Based on the approximate ratio of 2:2:1 for water–epoxy resin–titanium dioxide in the titanium dioxide–epoxy resin coating, it can be deduced that 1 m$^2$ of roof area requires about 0.2 kg of water, 0.2 kg of epoxy resin, and 0.1 kg of $TiO_2$ crystals. The energy consumption during coating production mainly comes from the mechanical work consumed during the grinding and mixing of solids and the electricity consumed during the high-temperature calcination stage. Calculations show that producing 1 kg of coating consumes approximately 1.6 kWh of electricity, based on the average energy consumption observed during the preparation process in our laboratory.

For the photovoltaic roof, its main components include photovoltaic panels, inverters, connectors, and support structures. On average, 1 m$^2$ of polycrystalline silicon photovoltaic panels requires about 3.35 kg of aluminum and 3.21 kg of steel. The input–output parameters for titanium dioxide (sulfate method), epoxy resin, and photovoltaic panels can be found in the Simapro database and imported into the life cycle assessment model for calculation.

The environmental factors in this study are quantified through a series of normalization and weighting operations performed using SimaPro software. While the detailed LCA calculation process is not the focus of this paper, the results are incorporated into the 3-E framework to assess environmental payback

The data presented in Figures 4–7 were calculated using Simapro software, which incorporates an extensive built-in database for life cycle assessment. The database is an integral component of the software and provides the necessary datasets for modeling the environmental impacts of cool roofs and photovoltaic roofs. The calculations were conducted based on the specific input parameters and assumptions detailed in the study.

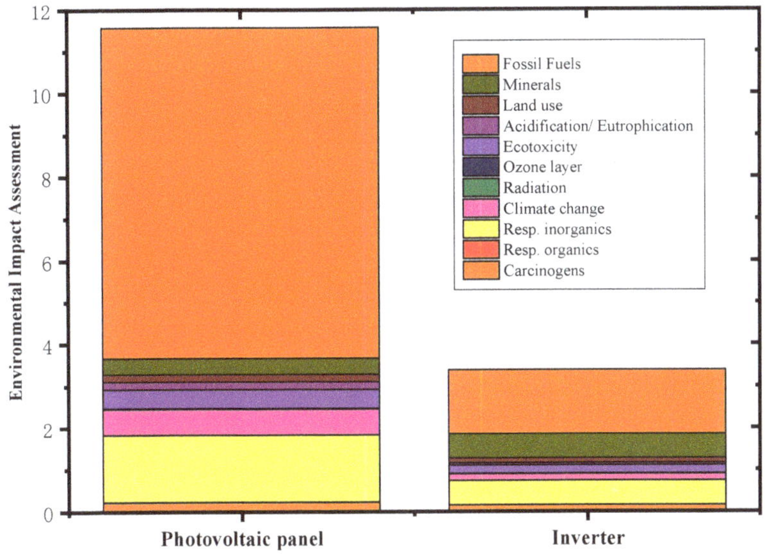


**Figure 4.** Environmental impact assessment of polycrystalline silicon photovoltaic panels and inverters.

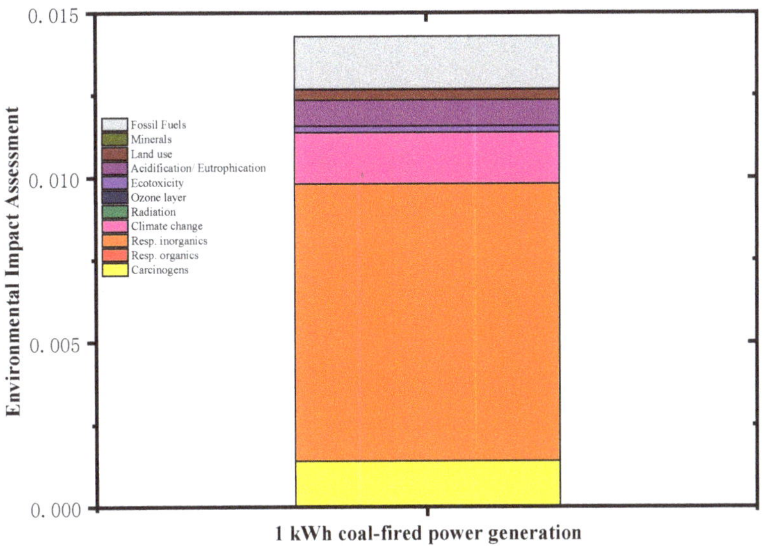


**Figure 5.** Environmental impact indicators of 1 kWh coal-fired power generation.

First, the life cycle assessment models for titanium dioxide & epoxy resin and solar photovoltaic panels & inverters were calculated in Simapro. Based on the input-output parameters in Simapro, the full life cycle impact indicators for 1 $m^2$ of polycrystalline silicon solar photovoltaic panels and one 500 W inverter are presented in Figure 4. Pt represents "points," a dimensionless unit commonly used in SimaPro software for life cycle assessment (LCA). This unit quantifies the relative contributions of different environmental impact categories (e.g., global warming potential, acidification potential, eutrophication potential) and allows for easier comparison and aggregation of overall environmental impacts.

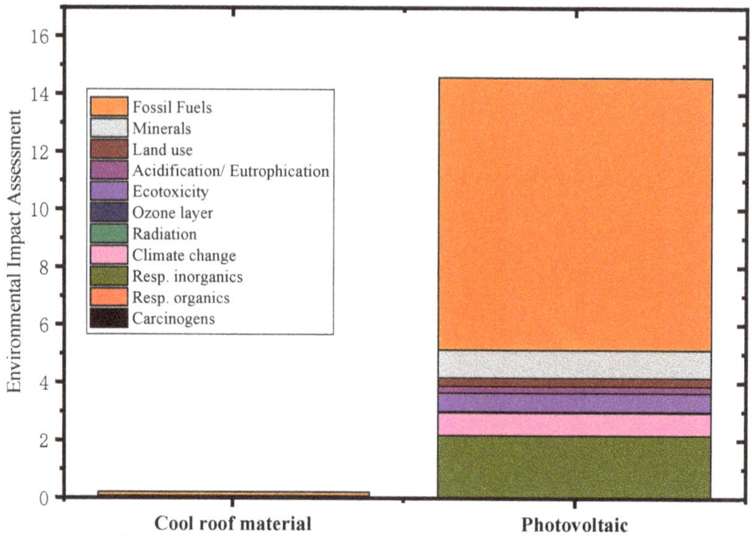


**Figure 6.** Environmental impact indicators of 1 $m^2$ of cool roof materials and 1 $m^2$ of photovoltaic roof components.

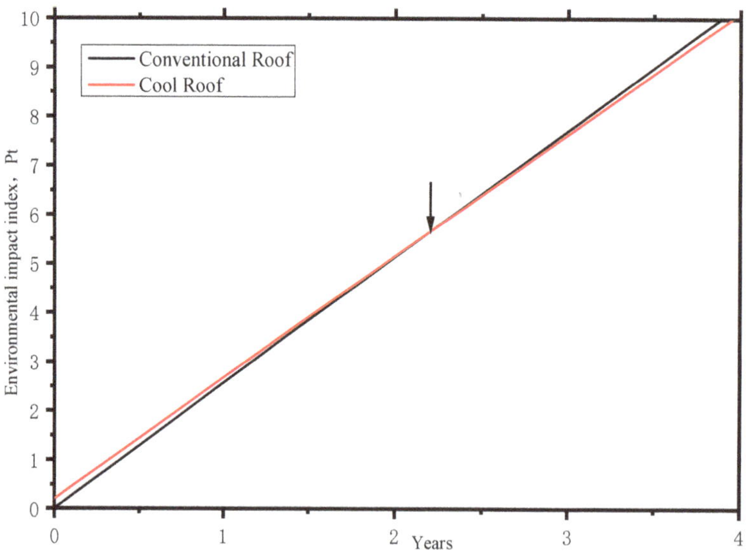


**Figure 7.** Schematic diagram of the environmental impact index of the cool roof vs. the number of years.

The development of raw materials for products, as well as the production, assembly, and transportation during the manufacturing phase, all require electricity input. At the same time, the operation of photovoltaic roofs also generates and exports electricity. Currently, coal-fired power plants remain the primary source of electricity generation in China. In SimaPro, the full life cycle environmental impact indicators for 1 kWh of coal-fired power generation were calculated based on the corresponding input-output parameters, as shown in Figure 5.

The above data indicate that titanium dioxide and epoxy resin contribute most significantly to fossil fuel consumption and inhalable inorganic particulate matter emissions. This is primarily due to the substantial energy demands and pollutant releases during raw material extraction processes.

For photovoltaic (PV) panels, fossil fuel consumption constitutes the dominant environmental impact category. This stems from the intensive energy requirements for silicon purification and wafer production in the PV manufacturing chain.

Regarding coal-fired power generation, inhalable inorganic particulates represent the most severe impact, followed by climate change effects. This pattern occurs because both coal mining operations and combustion processes in thermal power plants emit substantial quantities of hazardous substances, while the released $CO_2$ and other greenhouse gases contribute to global warming and climate change, as shown in Figure 6.

The data above demonstrates that titanium dioxide and epoxy resin have the most significant impact on fossil fuel consumption and inhalable inorganic substances in the air, primarily due to the substantial energy requirements and emissions of harmful substances during raw material extraction. For photovoltaic panels, fossil fuel consumption accounts for the vast majority of the impact, as the silicon purification and silicon wafer production processes in the photovoltaic industry consume enormous amounts of energy. In coal-fired power generation, inhalable inorganic substances in the air have the greatest impact, followed by climate change, because both coal mining and the combustion process in thermal power plants emit large quantities of harmful substances, while $CO_2$ and other gases contribute to the greenhouse effect and global climate change.

Based on the above data, the corresponding full life cycle environmental impact indicators for 1 $m^2$ of cool roofing material and 1 $m^2$ of photovoltaic roofing components were calculated in Simapro, as shown in Figure 7.

The above data shows that although the annual energy savings per square meter of photovoltaic roofs are greater than those of cool roofs, their energy consumption and environmental impact over the entire life cycle far exceed those of cool roofing materials. Therefore, a thorough analysis and quantitative calculation are still needed to determine which roofing system—cool or photovoltaic—consumes more energy and which has a better or worse environmental impact across its full life cycle.

From a product-centric perspective, economic viability and energy efficiency may be prioritized by producers, while energy savings and environmental friendliness may appeal to consumers. However, when examined through the lens of societal and planetary well-being, the issues of energy efficiency and environmental sustainability are far more complex. This complexity arises because both producers and consumers are confined to specific stages of the product's life cycle, and the benefits they perceive are inherently limited to their respective phases. Typically, neither party considers the upstream (e.g., raw material extraction, manufacturing) or downstream (e.g., disposal, recycling) impacts of the product. Whether at the individual, national, or regional level, a comprehensive understanding of the entire life cycle impact of a product is essential to make meaningful contributions to energy conservation and environmental protection on a societal scale. Only by adopting a holistic life cycle assessment (LCA) approach can we address these interconnected challenges effectively and ensure sustainable outcomes for the planet as a whole.

To fully assess the role of cool roofs and photovoltaic roofs in building energy efficiency within society, this study introduces the concept of "payback period" from economics. Based on life cycle assessment (LCA) theory, a 3-E (Energy, Economy, Environment) static payback period theoretical framework for roofing energy-saving technologies is established. Using the cool roof and photovoltaic roof cases discussed in this paper as examples, an empirical analysis is conducted to reveal whether these technologies are genuinely energy-efficient and environmentally sustainable over their entire life cycles.

## 5. Results and Discussion

The 3-E static payback period refers to the Economic Static Payback Time (ECSPT), Energy Static Payback Time (ENSPT), and Environmental Static Payback Time (EVSPT) of roofing systems.

### 5.1. Economic Static Payback Period

The concept of the payback period originates from economics, referring to the time required for the cumulative cash inflows generated by an investment to equal the initial investment amount. It represents the number of years needed to recoup the investment. A shorter payback period indicates a more favorable project. The payback period includes both static and dynamic payback periods, and this study focuses exclusively on the static payback period.

The static payback period does not account for the time value of money. Instead, it directly uses the time taken for the cumulative net future cash flows to equal the original investment amount as the payback period. Generally, investors prefer to recover their investments as quickly as possible, meaning the shorter the payback period, the better.

For roof energy-saving technologies, the economic static payback period can be defined as: the total cost of roof construction/retrofitting technology divided by the annual electricity cost savings achieved through the roof's energy efficiency.

$$ECSPT = \frac{C_{ini}}{E_{anual} \ * \ p_e} \tag{1}$$

where

$ECSPT$ represents the Economic Static Payback Period (years);

$C_{ini}$ denotes the initial investment cost of the roof technology (RMB);

$E_{anual}$ indicates the annual electricity savings of the roof technology (kWh/year);

$p_e$ is the equivalent electricity price (RMB/kWh).

For the cool roof discussed in this paper a cost of approximately RMB 240 for 20 kg can cover 40 m$^2$, resulting in a unit cost of RMB 6 per square meter. The annual electricity savings amount to 6.5 kWh/m$^2$. Taking Nanjing as an example, where the first-tier electricity price is 0.528 RMB/kWh, the economic static payback period of the cool roof in Nanjing can be calculated as follows:

$$ECSPT_{Cool} = \frac{6}{6.5 * 0.528} = 1.75 \tag{2}$$

For photovoltaic roofs, specifically flat roofs, the price per square meter of photovoltaic modules is approximately RMB 800, based on current market research and average pricing in China. With an average annual electricity savings of 139 kWh/m$^2$, the static payback period for photovoltaic roofs in the Nanjing region can be calculated as follows:

$$ECSPT_{PV} = \frac{800}{139 * 0.528} = 10.9 \tag{3}$$

It can be observed that in Nanjing, the static payback period for cool roofs is 1.75 years, while that for photovoltaic roofs is 10.9 years. In terms of economic payback period, cool roofs hold a clear advantage.

### 5.2. Energy Static Payback Period

The energy payback period refers to the total energy consumed over the entire life cycle of a photovoltaic (PV) power generation system divided by its average annual energy output, with the unit being years. In other words, it indicates how many years a PV system takes to recover the energy consumed during its life cycle. Clearly, a shorter payback period is preferable. The energy payback period is one of the indicators used to evaluate renewable energy sources. It can be said that the concept of the energy payback period emerged alongside PV technology, but the full life cycle energy payback period differs from the conventional energy payback period.

In the conventional concept of the energy payback period, the energy consumption of a PV power station is calculated primarily from three aspects: equipment, component auxiliary materials, and power station auxiliary materials. Taking silicon-based components as an example, the conventional energy payback period only accounts for the energy consumption in the aforementioned processes, i.e., the energy consumption directly involved by the producer. However, processes prior to silicon material production, such as the exploration, mining, grinding, beneficiation, and transportation of silicon ore, are not considered by the producer. Similarly, for auxiliary materials like aluminum in components, only the energy consumption during aluminum smelting is counted, while the energy consumed in earlier processes such as aluminum ore mining and transportation is excluded. Therefore, the energy input in the conventional energy payback period is inaccurate and does not represent a true full life cycle assessment.

This paper redefines the concept of the energy static payback period based on life cycle assessment (LCA) theory. In LCA, the indicator for resource consumption is based on the depletion of fossil fuels and mineral resources. That is, all types of energy and natural resource consumption throughout the entire life cycle are converted into equivalent impacts on fossil fuel consumption and mineral resource depletion using corresponding impact factors. Among the 11 impact indicators in the LCA system, fossil fuel consumption reflects the total energy consumption across the entire life cycle, including extraction, transportation, manufacturing, assembly, usage, and recycling. Based on this theory, the energy static payback period is defined as follows:

$$ENSPT = \frac{INDEX_{FF.ini}}{E_{anual} * INDEX_{FF.e}} \tag{4}$$

where

$ENSPT$ is the Energy Static Payback Period (years);

$INDEX_{FF.ini}$ is the Fossil Fuels (Fossil Fuels) impact score in the initial environmental impact assessment of the rooftop technology (Pt);

$INDEX_{FF.e}$ is the Fossil Fuels (Fossil Fuels) impact score per kilowatt-hour (kWh) of electricity generated, derived from previous data as $1.6 \times 10^{-3}$ Pt.

For the cool roof discussed in this paper, based on the calculations presented earlier, it can be determined that in the environmental impact assessment of 1 m$^2$ of heat-reflective coating, the fossil fuel resource consumption indicator scores 0.141 Pt, with an annual electricity savings of 6.5 kWh/m$^2$. Therefore, the energy static payback period for cool roofs in Nanjing can be calculated as:

$$ENSPT_{Cool} = \frac{0.141}{1.6 * 10^{-3} * 6.5} = 13.6 \tag{5}$$

For the photovoltaic roof discussed in this paper, according to the previous calculations, it can be determined that in the environmental impact assessment of 1 m$^2$ of photovoltaic modules, the fossil fuel resource consumption indicator scores 9.43 Pt, with an annual electricity savings of 139 kWh/m$^2$. Therefore, the energy static payback period for photovoltaic roofs in Nanjing can be calculated as:

$$ENSPT_{PV} = \frac{9.43}{1.6 * 10^{-3} * 135} = 43.7 \quad (6)$$

It can be observed that in Nanjing, the energy static payback period for cool roofs is 13.6 years, while that for photovoltaic roofs is 43.7 years, which essentially exceeds the entire service life of photovoltaic systems.

### 5.3. Environmental Static Payback Period

The concept of Environmental Static Payback Period is newly proposed in this paper. It refers to the ratio of the negative environmental impact indicators caused by all resource consumption throughout a product's life cycle to the positive environmental impact indicators converted from its annual energy savings. This concept can be understood similarly to the energy payback period.

The development and deployment of energy-efficient products necessitate an initial energy investment, which is irrecoverable once expended. However, the integration of energy-saving technologies can significantly reduce annual energy consumption. While energy consumption persists in absolute terms, the adoption of such technologies effectively lowers the annual consumption rate. In the absence of these technologies, energy consumption would escalate more rapidly each year. Consequently, the reduction in energy consumption facilitated by energy-efficient products can be conceptualized as a form of "energy recovery." The time required to achieve this recovery, often referred to as the energy payback period, can be quantified by dividing the initial energy investment by the annual energy savings.

The environmental payback period follows a similar logic. For an energy-efficient product, the initial resource consumption (from processes such as production, transportation, and installation) inevitably has adverse environmental effects. The implementation of energy-saving technologies (e.g., roof insulation) can reduce annual energy consumption (e.g., building air conditioning loads). These energy savings can be viewed as having a positive environmental impact in two ways:

Avoidance of Additional Environmental Damage: Without energy-saving measures, more energy would be consumed, leading to greater environmental harm. The adoption of energy-efficient technology prevents this additional damage, analogous to how economic payback avoids extra costs and energy payback avoids additional energy consumption.

Hypothetical "Environmental Purifier": One can imagine an "environmental purifier" that continuously generates positive environmental effects as long as energy is supplied. The energy savings from efficient technologies can be conceptualized as input to this purifier, thereby producing beneficial environmental outcomes. This perspective helps clarify why energy savings can be considered environmentally "favorable."

This dual interpretation provides a comprehensive framework for understanding the environmental payback period and its significance in evaluating sustainable technologies.

Based on the above perspectives, the calculation of environmental static payback period is defined as:

$$EVSPT = \frac{INDEX_{whole.ini}}{E_{annual} * INDEX_{whole.e}} \quad (7)$$

where

$EVSPT$ represents the environmental payback period (years);

$INDEX_{whole.ini}$ denotes the total environmental impact indicator of the initial investment under life cycle assessment (Pt);

$NDEX_{whole.e}$ signifies the total environmental impact indicator per kilowatt-hour under life cycle assessment (Pt). From previous calculations, $E$ = 0.0142 Pt.

For the cool roof discussed in this paper, based on the previous calculations, the total environmental impact score of 1 $m^2$ of heat-reflective coating under life cycle assessment is 0.206 Pt, with an annual electricity savings of 6.5 kWh/$m^2$. Thus, the environmental static payback period of the cool roof in Nanjing can be calculated as:

$$EVSPT_{Cool} = \frac{0.206}{0.0143 * 6.5} = 2.2 \tag{8}$$

Figure 8 illustrates the environmental payback period diagram for cool roofs. For conventional roofs, the annual electricity consumption per square meter is 181 kWh. Thus, on the environmental impact index versus years diagram, it is represented by a straight line passing through the origin with a slope corresponding to the environmental impact index of 2.57 Pt (associated with the annual consumption of 181 kWh). For cool roofs, the annual electricity consumption per square meter is 174.5 kWh. Due to the initial investment in cool roof materials, the initial environmental impact index is 0.206 Pt per square meter (as derived earlier). Consequently, its representation on the diagram is a straight line with an intercept of 0.206 Pt and a slope corresponding to the environmental impact index of 2.478 Pt (associated with the annual consumption of 174.5 kWh). The intersection point of these two lines indicates the environmental payback period for cool roofs, which is calculated to be 2.2 years.

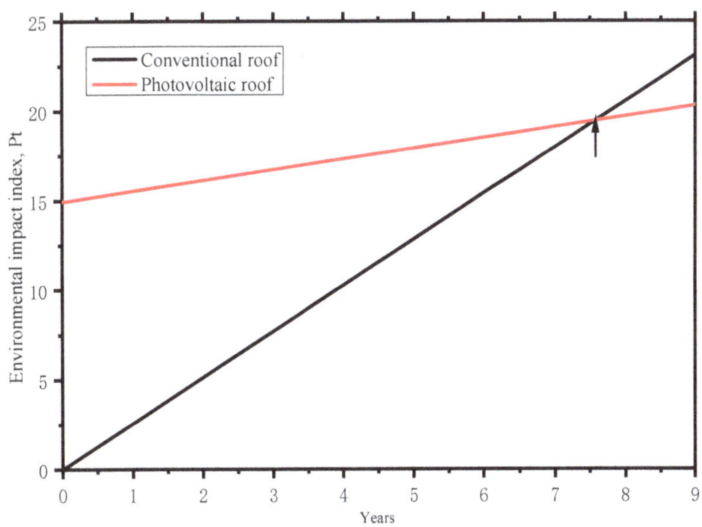


**Figure 8.** Schematic diagram of the environmental impact index of photovoltaic roofs versus the number of years.

For the photovoltaic roof discussed in this paper, based on the previous calculations, we know that the total environmental impact score for 1 $m^2$ of PV modules over their entire life cycle is 14.950 Pt, with an annual electricity savings of 139 kWh/$m^2$. Therefore, the static environmental payback period for the cool roof in Nanjing can be calculated as:

$$ENSPT_{PV} = \frac{14.950}{0.0142 * 139} = 7.6 \tag{9}$$

Figure 9 illustrates the environmental payback period diagram for photovoltaic roofs. For conventional roofs, the annual electricity consumption per square meter is 181 kWh, resulting in a straight line on the environmental impact index versus years diagram that passes through the origin with a slope corresponding to the environmental impact index of 2.57 Pt (associated with the annual consumption of 181 kWh). For photovoltaic roofs, the annual electricity consumption per square meter is 42 Wh. Due to the initial investment in cool roof materials, the corresponding initial environmental impact index is 14.95 Pt per square meter (as derived earlier). Thus, its representation on the diagram is a straight line with an intercept of 14.95 Pt and a slope corresponding to the environmental impact index of 0.596 Pt (associated with the annual consumption of 42 kWh). The intersection point of these two lines represents the environmental payback period of the cool roof, calculated to be 7.6 years.

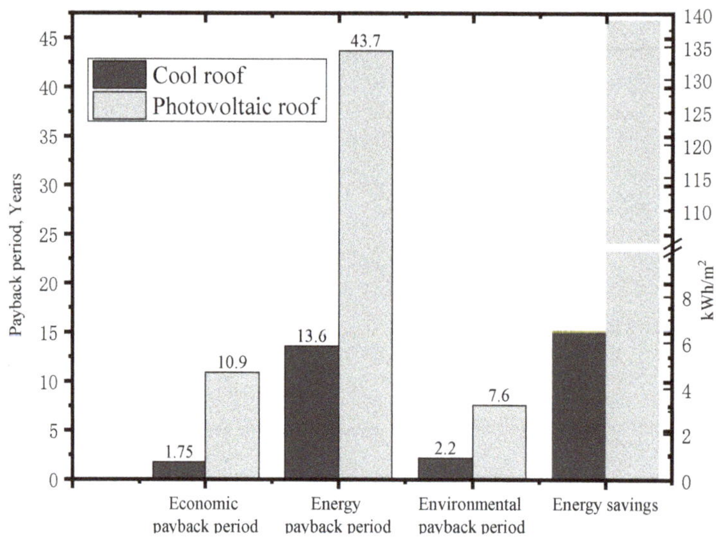


**Figure 9.** Comparison of payback periods and energy savings between cool roofs and photovoltaic roofs.

From the *roof environmental impact index versus years* diagram, the environmental payback period is defined as the intersection point where the cumulative environmental impact of the energy-saving technology—comprising its initial environmental impact and its operational impact over the years—equals the environmental impact of conventional roofs over the same duration. Beyond this payback point, the environmental impact index of conventional roofs exceeds the total environmental impact (including the initial investment) of energy-saving roofs over an equivalent time frame. For society as a whole, if the operational duration of the energy-saving technology surpasses the environmental payback period, adopting such measures—compared to conventional roofs—can significantly reduce environmental impact. This constitutes the practical significance of the environmental payback period.

Figure 9 presents a comparative analysis of the payback periods (economic, energy, and environmental) and annual energy savings between cool roofs and photovoltaic roofs. The data clearly indicate that cool roofs outperform photovoltaic roofs across all three payback metrics. However, it is important to note that photovoltaic roofs exhibit far superior energy-saving and power-generation intensities compared to cool roofs. The 3-E (economic, energy, and environmental) static payback period model proposed in this study is not limited to roof applications but can also be extended to other energy-saving technologies. The introduction of this theoretical framework provides valuable guidance for the implementation of various energy efficiency initiatives, offering a robust analytical tool for evaluating the long-term sustainability and feasibility of such technologies.

It can be seen that, whether it is the economic payback period, energy payback period or environmental payback period, the performance of cool roofs is better than that of photovoltaic roofs. However, the energy-saving and energy-production intensity of photovoltaic roofs cannot be matched by cool roofs. The 3-E static payback period model proposed in this paper can be applied not only to roofs but also to other energy-saving technologies. The proposal of this theory can play a guiding role in the implementation of many activities.

## 6. Conclusions

This paper analyzes and elaborates on the fundamental concepts and operational procedures of Life Cycle Assessment (LCA). Using cool roofs and photovoltaic roofs as comparative subjects, the study conducts a comprehensive comparison from economic, energy, and environmental perspectives through energy consumption simulation and LCA software. The conclusions are as follows:

(1) The energy savings per square meter of cool roofs are 6.5 kWh, representing an annual energy reduction of 8.6%. Photovoltaic roofs generate 10,283 kWh annually, with 65.8% used to meet their own energy demands and the remainder supplied to the grid. In terms of energy savings, photovoltaic roofs significantly outperform cool roofs.

(2) The life cycle environmental impact indicators for 1 $m^2$ of cool roof materials and 1 $m^2$ of photovoltaic roof components are 0.206 Pt and 14.950 Pt, respectively. Therefore, from an environmental perspective, photovoltaic roofs cause substantially greater environmental damage compared to cool roofs.

(3) To comprehensively compare the energy-saving effects of cool roofs and photovoltaic roofs, a 3-E static payback period theoretical framework based on life cycle assessment was established. This framework includes calculation formulas for economic, energy, and environmental static payback periods. The calculated results show that the economic payback periods for cool roofs and photovoltaic roofs are 1.75 years and 10.90 years, respectively; the energy payback periods are 13.6 years and 43.7 years, respectively; and the environmental payback periods are 2.2 years and 7.6 years, respectively. In terms of payback periods, cool roofs demonstrate significant advantages.

Beyond technical performance, the study highlights broader implications for policy and practice. The divergence between economic and environmental payback periods suggests that market forces alone may not always drive optimal sustainable choices, particularly in heating-dominated climates where cool roofs incur heating penalties. Targeted incentives, such as carbon pricing or reflectance-based building codes, could accelerate adoption where energy savings alone are insufficient. Additionally, the framework's adaptability to hybrid systems and emerging materials, such as thermochromic coatings, opens avenues for future innovation in building envelope design.

While the current model focuses on static assessments, its foundational principles can be extended to dynamic analyses, incorporating evolving grid decarbonization and material efficiency trends. Future work should also explore urban-scale microclimate effects and human-centric metrics, such as thermal comfort, to further refine the framework's applicability. By bridging technical performance with real-world feasibility, this research contributes to the growing emphasis on whole-building sustainability, offering a transparent and scalable tool for architects, policymakers, and building owners. Ultimately, the

3-E framework advances the prioritization of energy-efficient solutions in urban planning, supporting the transition toward climate-responsive building strategies.

The proposed 3-E static payback period framework provides a robust foundation for evaluating roofing technologies, yet several avenues remain for refinement and expansion. Addressing these opportunities will enhance the framework's applicability across diverse building contexts and emerging technologies.

(1) Dynamic Life Cycle Assessment Integration. The current model employs static emission factors and energy prices, which may not accurately reflect future decarbonization trends or market fluctuations. Incorporating dynamic life cycle assessment (DLCA) methodologies would enable time-dependent adjustments to grid carbon intensity, material efficiency improvements, and policy-driven cost variations. For instance, as renewable energy penetration increases, the environmental payback period of PV roofs could shorten significantly, altering their comparative advantage over cool roofs. Similarly, integrating probabilistic forecasting for electricity prices would improve the economic payback period's reliability in volatile energy markets.

(2) Urban-Scale Microclimate Modeling. While the framework evaluates roofing systems at the building level, their aggregate impact on urban heat islands (UHI) remains unquantified. Future work should expand the system boundary to include neighborhood-scale effects, such as reduced ambient temperatures from widespread cool roof adoption. Coupling the 3-E framework with computational fluid dynamics (CFD) or urban canopy models could quantify secondary energy savings for surrounding buildings, potentially revising payback periods upward in densely populated areas. This expansion would align the analysis with municipal sustainability goals, where UHI mitigation often takes precedence over individual building performance.

**Author Contributions:** Conceptualization, X.Z.; experiment and data curation, S.Z.; original draft preparation, S.Z., X.J.; review and editing, X.Z. All authors have read and agreed to the published version of the manuscript.

**Funding:** This work is supported by the Jiangsu Maritime Institute Doctoral Research Initiation Fund (2023BSKY04), and the Jiangsu Province Young Scientific and Technological Talent Support Project (JSTJ-2024-JS016).

**Data Availability Statement:** The data presented in this study are available on request from the corresponding author.

**Conflicts of Interest:** The authors declare no conflicts of interest.

## References

1. Katebi, A.; Tushmanlo, H.S.; Asadollahfardi, G. Environmental life cycle assessment and economic comparison of different roof systems. *J. Build. Eng.* **2023**, *76*, 107316. [CrossRef]
2. Hekrle, M.; Liberalesso, T.; Macháč, J.; Silva, C.M. The economic value of green roofs: A case study using different cost–benefit analysis approaches. *J. Clean. Prod.* **2023**, *413*, 137531. [CrossRef]
3. Alhazmi, M.; Sailor, D.J.; Levinson, R. A review of challenges, barriers, and opportunities for large-scale deployment of cool surfaces. *Energy Policy* **2023**, *180*, 113657. [CrossRef]
4. Tian, D.; Zhang, J.; Gao, Z. The advancement of research in cool roof: Super cool roof, temperature-adaptive roof and crucial issues of application in cities. *Energy Build.* **2023**, *291*, 113131. [CrossRef]
5. Houchmand, L.J.; Martí, M.M.; Gassó-Domingo, S. Photovoltaics and green roofs: Holistic analysis in built environments. *Renew. Sustain. Energy Rev.* **2025**, *207*, 114987. [CrossRef]
6. Yao, H.; Zhou, Q. Research status and application of rooftop photovoltaic Generation Systems. *Clean. Energy Syst.* **2023**, *5*, 100065. [CrossRef]
7. Agdas, D.; Barooah, P. On the economics of rooftop solar PV adoption. *Energy Policy* **2023**, *178*, 113611. [CrossRef]
8. Scolaro, T.P.; Ghisi, E. Life cycle assessment of green roofs: A literature review of layers materials and purposes. *Sci. Total. Environ.* **2022**, *829*, 154650. [CrossRef] [PubMed]

9. Li, Z.; Zhang, W.; He, B.; Xie, L.; Chen, M.; Li, J.; Zhao, O.; Wu, X. A comprehensive life cycle assessment study of innovative bifacial photovoltaic applied on building. *Energy* **2022**, *245*, 123212. [CrossRef]
10. Pirvaram, A.; Talebzadeh, N.; Leung, S.N.; O'BRien, P.G. Radiative cooling for buildings: A review of techno-enviro-economics and life-cycle assessment methods. *Renew. Sustain. Energy Rev.* **2022**, *162*, 112415. [CrossRef]
11. Le, A.B.D.; Whyte, A.; Biswas, W.K. Carbon footprint and embodied energy assessment of roof-covering materials. *Clean Technol. Environ. Policy* **2019**, *21*, 1913–1923. [CrossRef]
12. Pique, L.; Blanchet, P.; Breton, C. Global warming potential comparison between green and conventional roofs in cold climate using life cycle assessment. *J. Clean. Prod.* **2023**, *420*, 138314. [CrossRef]
13. Ma, M.; Zhang, K.; Chen, L.; Tang, S. Analysis of the impact of a novel cool roof on cooling performance for a low-rise prefabricated building in China. *Build. Serv. Eng. Res. Technol.* **2021**, *42*, 26–44. [CrossRef]
14. Ben-Alon, L.; Loftness, V.; Harries, K.; Hameen, E.C. Life cycle assessment (LCA) of natural vs conventional building assemblies. *Renew. Sustain. Energy Rev.* **2021**, *144*, 110951. [CrossRef]
15. Füchsl, S.; Rheude, F.; Röder, H. Life cycle assessment (LCA) of thermal insulation materials: A critical review. *Clean. Mater.* **2022**, *5*, 100119. [CrossRef]
16. Rasul, M.G.; Arutla, L.K.R. Environmental impact assessment of green roofs using life cycle assessment. *Energy Rep.* **2020**, *6*, 503–508. [CrossRef]
17. Koroxenidis, E.; Theodosiou, T. Comparative environmental and economic evaluation of green roofs under Mediterranean climate conditions–Extensive green roofs a potentially preferable solution. *J. Clean. Prod.* **2021**, *311*, 127563. [CrossRef]
18. Contarini, A.; Meijer, A. LCA comparison of roofing materials for flat roofs. *Smart Sustain. Built Environ.* **2015**, *4*, 97–109. [CrossRef]
19. Cubi, E.; Zibin, N.F.; Thompson, S.J.; Bergerson, J. Sustainability of rooftop technologies in cold climates: Comparative life cycle assessment of white roofs, green roofs, and photovoltaic panels. *J. Ind. Ecol.* **2016**, *20*, 249–262. [CrossRef]
20. Herrando, M.; Elduque, D.; Javierre, C.; Fueyo, N. Life Cycle Assessment of solar energy systems for the provision of heating, cooling and electricity in buildings: A comparative analysis. *Energy Convers. Manag.* **2022**, *257*, 115402. [CrossRef]
21. Li, H.; Zhang, J.; Liu, X.; Zhang, T. Comparative investigation of energy-saving potential and technical economy of rooftop radiative cooling and photovoltaic systems. *Appl. Energy* **2022**, *328*, 120181. [CrossRef]
22. Van de Moortel, E.; Allacker, K.; De Troyer, F.; Schoofs, E.; Stijnen, L. Dynamic versus static life cycle assessment of energy renovation for residential buildings. *Sustainability* **2022**, *14*, 6838. [CrossRef]
23. Ling, H.M.; Yew, M.C.; Saw, L.H. Analyzing recent active and passive cool roofing technology in buildings, including challenges and optimization approaches. *J. Build. Eng.* **2024**, *89*, 109326. [CrossRef]
24. Fnais, A.; Rezgui, Y.; Petri, I.; Beach, T.; Yeung, J.; Ghoroghi, A.; Kubicki, S. The application of life cycle assessment in buildings: Challenges, and directions for future research. *Int. J. Life Cycle Assess.* **2022**, *27*, 627–654. [CrossRef]

**Disclaimer/Publisher's Note:** The statements, opinions and data contained in all publications are solely those of the individual author(s) and contributor(s) and not of MDPI and/or the editor(s). MDPI and/or the editor(s) disclaim responsibility for any injury to people or property resulting from any ideas, methods, instructions or products referred to in the content.

buildings

MDPI

Article

# An Analysis of the Influence of Cool Roof Thermal Parameters on Building Energy Consumption Based on Orthogonal Design

**Shanguo Zhao [1,*], Guangmei Hai [1] and Xiaosong Zhang [2]**

[1] Jiangsu Maritime Institute, College of Marine Engineering, Electrization and Intelligence, Nanjing 211199, China
[2] School of Energy and Environment, Southeast University, Nanjing 211189, China
* Correspondence: brozhao@126.com

**Abstract:** An analytical hierarchy model of the impact of solar reflectance, thermal emittance, heat transfer coefficient, and heat storage coefficient on building energy consumption was established through the implementation of orthogonal design experiments. The EnergyPlus software (v9.0.1) was utilized to simulate building energy consumption across diverse climatic regions in China, providing essential benchmarks for the orthogonal design. The results of the range analysis consistently indicate that, barring regions characterized by extremely cold climates, solar reflectance emerges as the predominant factor exerting an influence on building energy consumption. As geographical latitude increases, the impact of the heat transfer coefficient becomes progressively larger, while the weight of thermal reflectance concurrently diminishes. Drawing upon the principles rooted in the gradient refractive rate theory and the concept of atmospheric window radiation, a range of high-reflectance and high-emittance cool roof coatings in various colors were meticulously developed. A spectrophotometer was employed to precisely quantify their reflectance properties, and simulations were subsequently conducted to scrutinize their energy-saving characteristics. The results demonstrate that the cool roof coatings that were developed using the methodology described in this paper exhibit substantial enhancements in reflectance, with increases of 0.24, 0.25, 0.37, and 0.35 for the yellow, red, blue, and green cool roofing materials, respectively, in comparison to conventional colored coatings. Under typical summer conditions, these enhancements translate to significant reductions in roof temperatures, ranging from 9.4 °C to 14.0 °C. Moreover, the simulations exploring the cooling loads for the roofs of differing colors consistently revealed remarkable energy savings. These savings were quantified to be 4.1%, 3.9%, 5.5%, and 5.4%, respectively, when compared to conventional coatings of the corresponding colors. These findings offer valuable insights into strategies for optimizing the energy efficiency of buildings through the application of high-reflectance cool roofing materials.

**Keywords:** cool roof; orthogonal design; range analysis; reflectance; energy consumption simulation

## 1. Introduction

The three predominant sectors contributing to society's overall energy consumption are industrial energy, transportation energy, and building energy [1]. As economic and societal progress marches forward, the expansion of built environments becomes more prominent, leading to a steady rise in the share of energy attributed to buildings [2]. Of this, the energy required for heating, ventilation, and air conditioning (HVAC) commonly stands out as the most significant component, often constituting between 40% and 50% of a building's total energy footprint [3].

Given the magnitude of HVAC energy utilization, it is paramount to prioritize strategies that enhance building energy efficiency, specifically focusing on HVAC systems [4]. In the area of curbing HVAC energy consumption, there are typically two primary strategies. The first entails enhancing the efficiency of HVAC systems, making them more proficient in delivering the desired output with reduced energy consumption. The second strategy is centered on reducing the energy demands imposed on these systems by optimizing building design and utilizing appropriate materials [5,6].

One crucial aspect that significantly dictates a building's heating and cooling needs is the thermal performance of its envelope, which includes elements like windows, walls, floors, and notably, roofs [7]. The heating and cooling requirements of buildings are significantly impacted by solar radiation. Roofs, in particular, are exposed to prolonged periods of sunlight, resulting in the absorption of a substantial amount of heat, which subsequently affects the indoor temperature of buildings [8].

Recognizing the pivotal role that roofs play, numerous innovative techniques are being integrated into building design. These techniques aim to mitigate the amount of heat that roofs retain and include strategies such as enhanced insulation, improved ventilation cooling, innovative water storage cooling, planting vegetation, and applying high-reflectance coatings [9–13]. Each of these methods offers its unique advantages, paving the way for more energy-efficient buildings in the future.

The pioneering concept of cool roofs was first unveiled by Parker [14]. In contrast to conventional reflective roofs, the uniqueness of cool roofs lies in their dual-pronged approach. These roofs not only aim to reduce the absorption of solar radiation by enhancing solar reflectance but also endeavor to increase the emissivity of roofing materials for long-wave atmospheric radiation. This dual strategy primarily focuses on minimizing absorption while maximizing emission, ultimately resulting in a significant reduction in the roof's surface temperature. Consequently, it diminishes the direct heat transfer from the roof into the indoor spaces of the building.

Through meticulous testing carried out on nine representative buildings located in Florida, USA, Parker was able to substantiate the effectiveness of employing cool roofing materials. The findings revealed a notable energy-saving trend, registering an impressive 19% reduction in energy consumption for residential buildings, and an even higher reduction of 25% for commercial structures.

The success showcased in Parker's investigation underscores the substantial energy-saving potential inherent in the adoption of cool roof technologies. By tackling both solar absorption and material emissivity, cool roofs present a robust solution to one of the primary challenges in building energy efficiency. Furthermore, the empirical evidence obtained from real-world application in both residential and commercial settings in Florida provides a compelling case for the broader adoption of cool roof methodologies. Through continued exploration and implementation of such innovative roofing solutions, it is conceivable that significant strides can be made toward achieving enhanced energy efficiency in buildings, which is a critical step in navigating the broader challenges of societal energy consumption [15].

Levinson et al. [16,17] conducted a robust and thorough exploration of cool roofs. This exploration was multidimensional, combining both theoretical analyses with hands-on empirical studies. Central to this examination was an in-depth evaluation of a plethora of materials and colors typically used in crafting cool roofs.

Levinson focused on their respective capabilities in terms of absorbing solar radiation and their radiation properties across the solar spectrum. This meticulous scrutiny gave rise to a comprehensive database that detailed the distinct material properties that define cool roofs.

To transition from the theoretical to the practical, Levinson's research leveraged a broad array of engineering field tests to provide empirical substantiation. A salient revelation from these tests was the definitive energy-saving potential of cool roofs, particularly in the context of air conditioning. In the warm Californian summer months, buildings fitted with cool roofs were shown to achieve energy savings in air-conditioning cooling energy consumption, with savings rates ranging impressively from 7% to 21.5%.

Beyond the direct energy implications, Levinson's study also delved into the broader environmental benefits of cool roofs. A noteworthy contribution was the assessment of cool roofs' role in counteracting the urban heat island effect—a phenomenon where urban regions experience heightened temperatures due to human activities. Levinson's findings highlighted the potency of cool roofs in substantially alleviating this effect.

Guo et al. [18] conducted a comprehensive study that involved both experimental work during the summer and transition seasons and annual simulations. Their aim was to assess the thermal performance, energy savings, and enhancement of thermal comfort achieved by integrating a cool roof with night ventilation. Their research conclusively demonstrated that roof albedo is the most influential parameter affecting both building energy performance and indoor thermal comfort. Furthermore, they identified an optimal reduction of 28% in cooling energy consumption.

Rawat and Singh [19] provided a comprehensive summary of cool roof thermal performance involving various surface coatings in distinct climate zones. Their work not only presents the advantages of cool roofs but also acknowledges their limitations while offering valuable recommendations for future research in this area. As per Rawat and Singh's findings, the potential energy-saving impact of cool roofs ranges from 15% to 35.7% across diverse climate zones. Moreover, their research suggests that implementing cool roof technology can lead to an average reduction in roof surface temperatures ranging from 1.4 °C to as much as 4.7 °C.

In the contemporary academic landscape, there has been a marked surge in interest toward the study of radiative cooling mechanisms [20–22]. This area of study seeks to harness the vast and unobstructed expanse of the sky as a reservoir for heat dissipation. Scholars from around the world, as well as those conducting region-specific research, have plunged into rigorous theoretical and empirical analyses. Their studies are aimed at gaining a deeper understanding of the foundational concepts, functional approaches, and intricate material engineering details inherent to radiative cooling.

In terms of the preparation of cool roofing materials, $TiO_2$ has conventionally served as the primary component for reflective "white" pigments or coatings due to its effective light-scattering characteristics [23]. Consequently, $TiO_2$ was most commonly employed in the early stages for fabricating heat-reflective materials. However, recent studies in the literature have emphasized the exploration and advancement of alternative near-infrared (NIR) pigments to titanium dioxide, such as zinc oxide [24], bismuth vanadate [25], and inorganic salts [26]. Rosati performed a comprehensive review of the state of the art of synthesis methods for pigments employed for cool roofs. However, the impact of reflectance and emittance on roof energy efficiency is intricate and conflicting between the cooling and heating seasons. The overall effect across the entire year is contingent on local climate factors, geographical considerations, the proportion of cooling to heating loads, and the prevailing energy usage patterns.

The impact of cool roofs extends beyond direct effects on building energy consumption, holding significant importance in ameliorating urban climate, particularly the urban heat island (UHI) effect [27]. Prior research on urban climate models like ENVI-met, PALM, SOLWEIG, PALM-4U, RayMan, and TEB has underscored the significance of thermal performance parameters—such as absorptance, reflectance, emissivity, and heat capacity of urban structural elements, including rooftops—in influencing urban climate [28]. Roof areas comprise over 25% of the urban surface and directly receive intense solar radiation, making their performance pivotal in the context of the urban heat island effect. Santamouris' comprehensive research on the UHI effect, spanning its causes, energy and environmental impacts, modeling, and mitigation measures, particularly emphasizes rooftop performance, thereby providing a robust theoretical foundation for UHI studies [29]. Zhao proposed a coupling effect of cool roofs under the UHI effect on building energy consumption, establishing a triple-model for assessing energy-saving effects. This study confirmed the significant role of cool roofs in enhancing urban climate conditions under UHI effects [30]. The combined findings from these studies corroborate the direct and indirect effects of diverse cool roof parameters on building energy usage, underscoring reflectance and emissivity as the paramount factors in this regard [31].

The aforementioned research has largely investigated the inherent characteristics of roofing materials and their energy-saving properties during cooling. However, there is a lack of research concerning the distinct performance of materials during both cooling and heating processes, particularly within the specific climatic regions of China. The specific details of material behavior in the unique climate zones of China have not yet been extensively studied.

Together, these dual axes of research—coating reflectivity and infrared emission—represent the cutting edge in the field of radiative cooling. The advancements in this domain not only hold promise for enhancing building efficiency but also offer potential solutions for a range of heat management challenges in various sectors.

A review of the existing literature highlights a predominant focus in domestic research on cool roofs in two key areas. Firstly, researchers have focused on roofing materials, with a particular emphasis on thermal reflective materials. Secondly, there has been a substantial body of work analyzing the impact of specific cool roof applications on building energy consumption, primarily through simulations or experimental methods. However, there exists a significant research gap concerning the intrinsic thermal characteristics of cool roofs, especially with regard to a detailed investigation of the individual parameters' influence on building energy consumption.

This study aims to address this research gap by deconstructing a heat transfer model of cool roofs and conducting an exhaustive analysis of the diverse factors that influence their performance. To facilitate this exploration, we delved into the operational mechanisms of cool roofing materials. In this context, we prepared five distinct types of high-reflectance roof coatings based on their underlying principles. These coatings were subjected to rigorous testing, including a measurement of their near-infrared thermal reflectance.

Furthermore, to ascertain their real-world energy-saving potential, a comprehensive comparative analysis against conventional colored coatings was conducted. Subsequently, the energy-saving rates achieved following the application of these high-reflectance coatings to roofing surfaces were simulated and quantified in different climate zones in China. In conclusion, this study undertook a meticulous comparative analysis, juxtaposing its findings with the existing literature. This comparative assessment was detailed, focusing on two pivotal aspects: rooftop surface temperature and the effectiveness of energy-saving measures. Additionally, a thorough and insightful elucidation was presented to expound upon the underlying reasons for any divergences noted between the outcomes of this research and the insights documented in the literature. This in-depth investigation not only broadens current understanding of cool roof performance but also offers practical insights into the implementation of these materials in building energy efficiency strategies.

## 2. Methodology

The critical thermal parameters impacting the indoor cooling load of buildings during the summer include solar reflectance, thermal emittance, heat transfer coefficient, and heat storage coefficient. To assess the relative influence of each parameter on building energy consumption, we employed an orthogonal design approach. The findings indicate that, across the majority of climate regions in China, solar reflectance exerts the most significant impact on building energy consumption. Therefore, enhancing rooftop reflectance emerges as a pivotal strategy for achieving energy efficiency.

In an effort to balance esthetic appeal with energy efficiency requirements, our study introduced a novel, colored cool roofing material developed based on the gradient refractive index theory. We measured its near-infrared reflectance and assessed its energy-saving performance on typical meteorological days using an energy simulation software, as illustrated in Figure 1. This innovative approach aims to contribute to both visually appealing building designs and energy-efficient building practices.

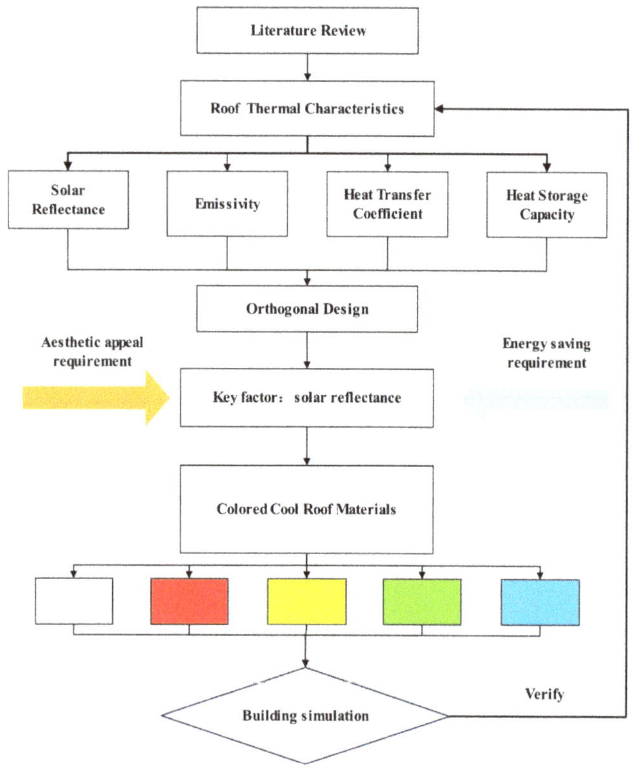


**Figure 1.** Flowchart of the research method and logic. Different colored rectangular boxes represent materials of different colors.

### 2.1. Orthogonal Design

During the architectural design phase, it is imperative to tailor the building envelope designs to accommodate the distinct load characteristics present in various climate zones. This approach ensures a harmonious balance between energy efficiency and cost-effectiveness. The roof, being a pivotal component of the building enclosure, plays a particularly critical role in the overall energy-efficient design of low-rise buildings. To align with China's climatic zoning criteria for building thermal design, this study chose Harbin, Beijing, Nanjing, Guangzhou, and Kunming as representative cities, each representing a specific climate zone, for detailed design analysis.

To analyze the impact of reflectance, emissivity, thermal conductivity, and heat storage coefficient on a building's annual energy consumption, simulations of the building's yearly heating and cooling energy consumption were conducted using EnergyPlus. The annual average energy consumption per unit area serves as the evaluation criterion. An orthogonal design experiment was employed to analyze the weight of each influencing parameter.

The China Academy of Building Research's generic building energy consumption simulation model was chosen for this study [32]. The building, measuring 50 m × 30 m, serves as a two-story office, as shown in Figure 2. Parameters like air conditioning system type, set temperatures for cooling and heating, operational strategy, personnel distribution, building lighting indicators, air tightness, efficiency in heating and cooling, and roof thermal settings were all based on the stipulations in GB50189-2015 [33]. Some indicators are detailed in Table 1. For the walls, floor, and roof, a 50-mm-thick XPS insulation board structure was used, consisting of the following layers from the innermost to the outermost layers: the inner protective coating, the XPS insulation layer (thermal conductivity of

0.028 W/m·K), the fiberglass reinforcement layer, and the outer protective layer. The window-to-wall ratio was set to be 0.3.

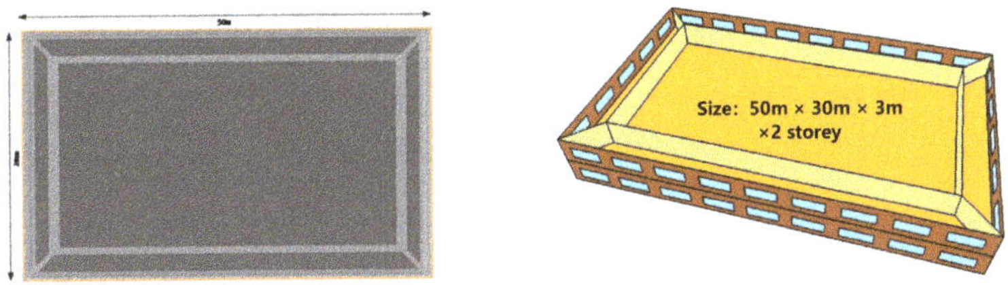


(**a**) (**b**)

**Figure 2.** General building energy consumption simulation building model plan structure diagram. (**a**) Plan view; (**b**) Section view.

**Table 1.** Building thermal performance and operating parameters.

| | Specific Parameters |
|---|---|
| Shape | 50 m × 30 m |
| Air Conditioning System | VAV + gas boiler |
| Lighting power | 6 W/m$^2$ |
| Plug load power | 20 W/m$^2$ |
| Refrigeration unit COP | 4.5 |
| Boiler efficiency | 0.9 |
| Air tightness | 7.5 m$^3$/(m$^2$ h) |
| Ventilation rate | 3 m$^3$/(h person) |

If a particular result has multiple influencing parameters, due to practical limitations that prevent conducting numerous experiments to discern the primary and secondary factors, orthogonal design can be employed to analyze the influence weight.

Orthogonal design is an experimental methodology utilized to discern factors impacting experimental outcomes and their interactions. By systematically altering and combining various factor levels, it minimizes noise and errors, thereby facilitating a clearer understanding and analysis of the results. Within orthogonal design, factors are distributed across different experimental combinations, ensuring each factor level interacts evenly and equitably with every other factor level. This process effectively identifies the primary influencing factors while reducing the number of experiments required. Through orthogonal design, researchers can efficiently pinpoint key factors influencing the outcomes and comprehend their interactions, minimizing the need for extensive experimentation. This approach has found widespread application in engineering, scientific research, and experimentation, enabling the optimization of product design, process improvements, and efficiency enhancements.

A typical layout of roof construction can be seen in Figure 3. Taking the roof's solar emissivity, thermal emissivity, thermal conductivity, and heat storage capability as the four variables, we adopted the L9(34)-type orthogonal table (Table 2) to conduct a comprehensive comparative analysis of the aforementioned influencing factors. This helps identify the primary and secondary relationships of these major factors on building air-conditioning energy consumption. Specific values are shown in the table. Here, the roof's overall thermal conductivity coefficient K is chosen to express the thermal conductivity capability, and the specific heat of the roofing material is selected to represent the heat storage capability. The heat transfer coefficient of the roof represents the insulation capacity of the walls, indicating the rate of heat transfer from the temperature difference between the interior and exterior surfaces. In the simulation process, the thermal conductivity was altered by varying the thickness of the XPS material layer.

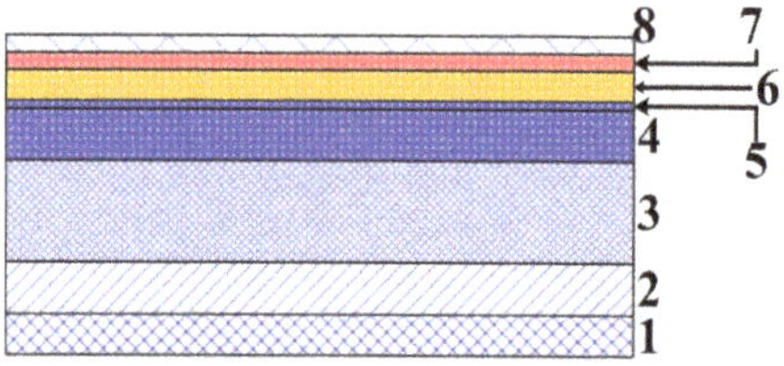


**Figure 3.** Typical layout of roof construction: 1—cement plaster, 2—thermal insulation, 3—reinforced concrete, 4—foam concrete, 5—membrane, 6—sand fill, 7—mortar bed, 8—tiles/coatings.

**Table 2.** Orthogonal table parameter selection.

| Level | Factor | | | |
|---|---|---|---|---|
| | Solar Reflectance (A) | Emissivity (B) | Heat-Transfer Coefficient (W/m²·K) (C) | Heat Storage Capacity Index kJ/(m²·K) (D) |
| 1 | 0.65 | 0.9 | 0.4 | 122 |
| 2 | 0.75 | 0.85 | 0.45 | 143 |
| 3 | 0.85 | 0.8 | 0.5 | 163 |

Regarding the heat storage capacity, the heat storage capacity index ($K$) is utilized as the metric. Its definition and calculation are presented in the following equation:

$$K = \sum \frac{\rho_i V_i c_i}{A_r} = \sum \rho_i d_i c_i \quad (1)$$

where $i$ represents the $i$-th layer of the roof, $\rho$ is the density of the material layer in kg/m$^3$, $V$ is the volume of the material layer in m$^3$, $c$ is the specific heat capacity of the material layer in J/(kg·K), $c_i$ is the thickness of the layer in m, $A_r$ is the roof area in m$^2$, and $K$ is the heat storage capacity index in J/(m$^2$·K).

In this simulation, roof thicknesses of 0.06 m, 0.08 m, and 0.1 m were chosen, and their respective $K$ values were calculated.

In accordance with a 4-factor, 3-level orthogonal design table, the parameters of solar reflectivity, thermal emissivity, thermal conductivity coefficient, and specific heat were systematically assigned to their respective levels, thereby generating a total of 9 unique test scenarios, as shown in Table 3. These scenarios were subsequently employed in EnergyPlus, a comprehensive building energy simulation software, to yield annual energy consumption data. This dataset encompassed various components, including energy utilization for the air-conditioning and heating systems, indoor lighting, electrical equipment, and the ventilation system.

**Table 3.** L9(34)-type orthogonal table.

| Test Number | Factor | | | |
|---|---|---|---|---|
| | Solar Reflectance (A) | Emissivity (B) | Heat-Transfer Coefficient (W/m²·K) (C) | Heat Storage Capacity Index kJ/(m²·K) (D) |
| 1 | 0.7 | 0.9 | 0.4 | 122 |
| 2 | 0.7 | 0.85 | 0.45 | 143 |
| 3 | 0.7 | 0.8 | 0.5 | 163 |
| 4 | 0.8 | 0.9 | 0.45 | 163 |
| 5 | 0.8 | 0.85 | 0.5 | 122 |
| 6 | 0.8 | 0.8 | 0.4 | 143 |
| 7 | 0.9 | 0.9 | 0.5 | 143 |
| 8 | 0.9 | 0.85 | 0.4 | 163 |
| 9 | 0.9 | 0.8 | 0.45 | 122 |

followed by the heat radiation rate, while the heat conductivity coefficient and heat storage capacity exert the least impact on energy consumption.

For extremely cold regions, the heat transfer coefficient shows the most significant variance, indicating its predominant influence on building energy consumption in these areas, and relegating solar reflectance to the second place. This observation can be primarily attributed to two reasons:

(1) In extremely cold regions, where lower temperatures prevail, a building's heating load significantly outweighs the cooling load in its annual energy consumption. While high solar reflectance offers advantages during the summer months by reducing the roof's surface temperature and minimizing indoor heat gain through the reflectance of solar radiation, it poses challenges during the winter heating periods. Solar reflectance inhibits the building's ability to absorb heat from solar radiation at these times. In these regions, which are characterized by substantial heating demands, the adverse impact of solar reflectance is amplified, thereby diminishing its overall annual effectiveness.

(2) In extremely cold regions, particularly during the winter heating season when there is a substantial temperature difference between the indoor and outdoor environments, heat dissipation from the building interior to the exterior through the roof becomes more pronounced. This heat transfer is primarily driven by roof conduction, emphasizing the crucial role of the heat transfer coefficient in influencing annual energy consumption.

On a nationwide scale, as latitude increases, the influence of solar reflectance and heat radiation rate gradually diminishes, while the impact of the heat transfer capability steadily intensifies. This trend is primarily due to the variations in environmental temperatures and the proportional shift in winter/summer heating and cooling loads as latitude rises. The influence of specific heat capacity (i.e., heat storage capability) on the annual building energy consumption remains relatively stable. This is because conventional roofing materials typically have a lower specific heat, and their capacity to store or absorb heat/cold is comparatively minimal against the overall heating and cooling loads, making their effect less pronounced. However, this dynamic might significantly shift if phase-change materials are integrated into roofing systems, potentially enhancing their thermal performance.

### 3.2. Material Indicator Testing

We methodically applied the coatings, each differing in color, onto white panels to ensure uniform coverage. These coatings were allowed to completely dry before progressing to the subsequent phase, which involved the rigorous process of reflectance testing. During this stage, the solar reflectance (SR) values were meticulously measured using a highly specialized spectrophotometer, specifically the Japanese Shimadzu UV3600 model. The use of this sophisticated instrumentation, illustrated in Figure 5, ensured the precision and accuracy of our measurements.

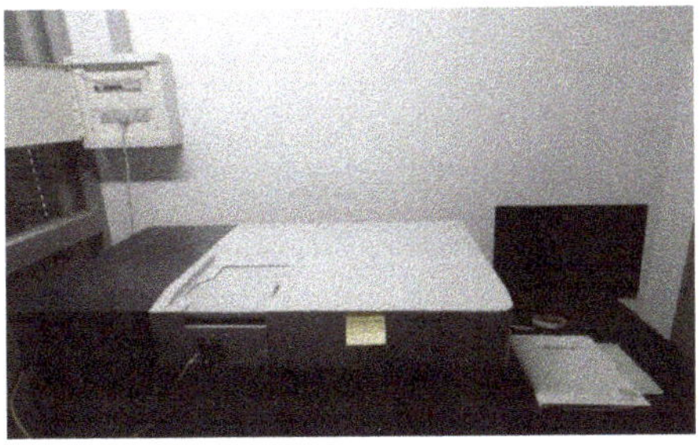

**Figure 5.** UV3600 spectrophotometer.

The SR values, acquired through rigorous testing and measurement, serve as crucial indicators of the materials' solar reflectance capabilities. The results obtained from this comprehensive procedure are presented graphically in Figure 6, offering a clear visual representation of how different coating materials, with their distinct color variations, perform in terms of solar reflectance. Emittance measurements for the coatings were also conducted. However, the results reveal that the emittance values for the coatings of different colors are quite consistent, hovering around 0.85.

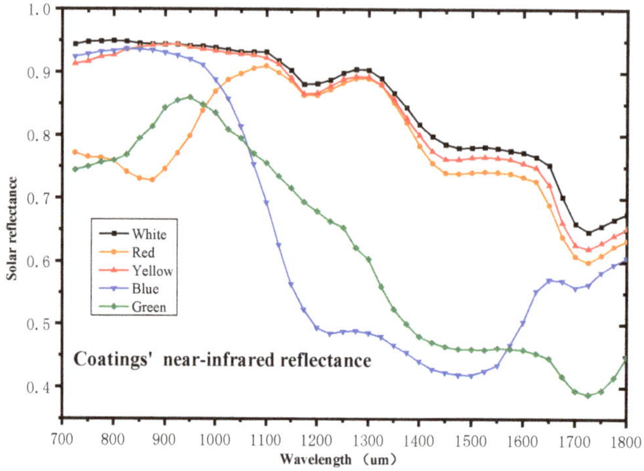


**Figure 6.** Coating near-infrared reflectance.

For pure white pigment, i.e., 100% titanium dioxide + 5 μm hollow $SiO_2$ glass beads + water-based epoxy resin coating, the calculated reflectance is 0.85. For yellow pigment, i.e., 50% titanium dioxide + 50% Nickel Titanium Yellow + 5 μm hollow $SiO_2$ glass beads + water-based epoxy resin coating, the calculated reflectance is 0.75. For red pigment, i.e., 50% titanium dioxide + 50% Silicon Iron Red + 5 μm hollow $SiO_2$ glass beads + water-based epoxy resin coating, the calculated reflectance is 0.68. For blue pigment, i.e., 50% titanium dioxide + 50% Cobalt Blue + 5 μm hollow $SiO_2$ glass beads + water-based epoxy resin coating, the calculated reflectance is 0.62. For green pigment, i.e., 50% titanium dioxide + 50% Cobalt Green + 5 μm hollow $SiO_2$ glass beads + water-based epoxy resin coating, the calculated reflectance is 0.56.

The results are summarized below and compared with the conventional pigments in Table 7.

**Table 7.** Reflectivity of heat-reflective coatings of different colors.

| Color | (white) | (yellow) | (red) | (blue) | (green) |
|---|---|---|---|---|---|
| Cool roof pigments | 0.85 | 0.75 | 0.68 | 0.62 | 0.56 |
| Conventional pigments | | 0.51 | 0.43 | 0.25 | 0.21 |

## 3.3. *Energy Saving Analysis*

For the energy saving analysis of the colored high-reflectance coatings that were prepared as mentioned earlier and to study their energy-saving characteristics compared to conventional pigments, this study employed EnergyPlus to compare the roof surface temperatures and energy consumption between cool roofs (CRs), conventional paint roofs (CPRs), and standard asphalt roofs (SARs) in a typical summer for a two-story office building, measuring 50 m × 30 m, located in a region with hot summers and cold winters.

Based on the GB50189-2015 various parameters were set for the building's thermal performance based on the enclosure structure, airtightness, and efficiency of the heating and cooling systems [33]. This included considerations of the cooling and heating loads, personnel distribution, lighting, and electrical use.

Figure 7 displays the external surface temperatures and hourly heat loads for the yellow-pigmented roofs during the summer. From the graph, it is evident that, compared to conventional yellow pigments, the titanium nickel yellow cool material exhibits an increase of 0.24 in reflectance. Consequently, the maximum external roof surface temperature during the summer decreases by 9.4 °C, resulting in a maximum instantaneous energy savings of 4.1%. Compared to a black roof, this translates to energy savings of 9.5%.

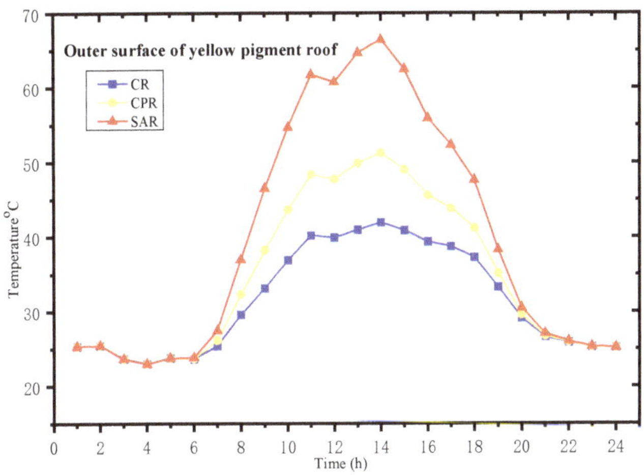


(a)

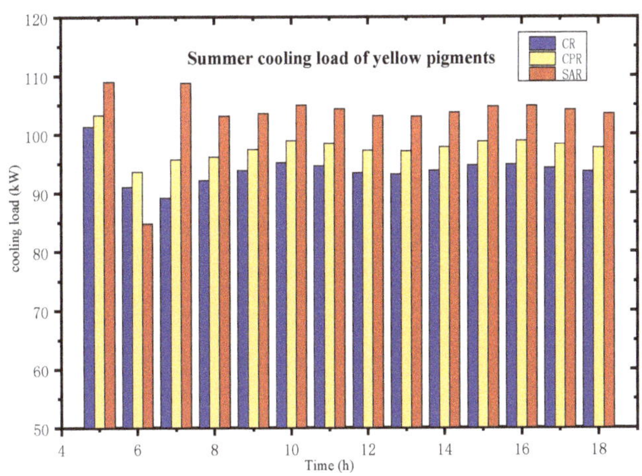


(b)

**Figure 7.** (**a**) Comparison of temperature on the outer surface of yellow pigment roofs. (**b**) Comparison of summer cooling load of yellow pigments.

Other colored materials were simulated using the same methodology to derive their roof surface temperatures and energy consumption, as illustrated in Table 8.

**Table 8.** Energy saving performance results.

| Roof Color | Reflectance Enhancement | Surface Temperature Reduction (°C) | Energy-Saving Rate Compared to Conventional Roofs | Energy Efficiency Compared to Typical Color |
|---|---|---|---|---|
| white | / | / | / | / |
| yellow | 0.24 | 9.4 | 4.1% | 9.5% |
| red | 0.25 | 9.2 | 3.9% | 8.2% |
| blue | 0.37 | 14.0 | 5.5% | 7.3% |
| green | 0.35 | 13.9 | 5.4% | 6.7% |

The results depicted in Table 8 indicate temperature reductions ranging from 9.4 °C to 13.9 °C on the roof surface for materials of different colors under typical meteorological conditions. Additionally, they showcase total energy savings ranging between 3.9% and 5.5%.

*3.4. Discussion*

The outcomes of energy saving performance, as detailed in Table 9, illuminate a significant and consistent trend across all cool roof coatings—they consistently manifest higher reflectance values in comparison to their conventional counterparts. This pattern underscores their inherent capacity to efficiently redirect a more substantial portion of incident solar radiation. Consequently, this redirection leads to a noteworthy reduction in heat absorption and, consequently, a substantial decrease in rooftop temperatures. A positive correlation becomes evident, indicating that an escalation in reflectance levels corresponds proportionally to a reduction in roof temperatures. This temperature decline directly translates into immediate and greater energy savings. This observation underscores the pivotal role played by enhanced reflectance in bolstering building energy efficiency. Moreover, this correlation emphasizes the potential for substantial real-world benefits associated with the adoption of cool roof coatings. Beyond the immediate energy savings, the diminished rooftop temperatures can contribute to a more comfortable indoor environment, potentially reducing the reliance on air-conditioning systems and further enhancing the overall energy efficiency of buildings.

**Table 9.** Annual energy consumption comparison across different climatic regions.

| Annual Energy-Saving Rate | Severe Cold Areas | Cold Areas | Hot Summer and Cold Winter Areas | Hot Summer and Warm Winter Area |
|---|---|---|---|---|
| white | / | / | / | / |
| yellow | −1.49% | 1.94% | 3.32% | 5.06% |
| red | −1.23% | 1.70% | 3.18% | 4.73% |
| blue | −2.10% | 3.00% | 4.39% | 5.80% |
| green | −2.03% | 3.06% | 4.31% | 5.75% |

The results of Table 9 were compared with findings from other studies. A review on cool roof performance showed that the energy-saving impact of cool roofs varies from 15% to 35.7% across different climatic regions, and the reduction in rooftop surface temperature ranges from 1.4 °C to 4.7 °C [19]. In terms of energy efficiency, the energy savings achieved in this study are approximately 5%, notably lower than values reported in the literature. A comprehensive analysis reveals two key factors contributing to this difference.

Firstly, previous studies did not consider aspects of building energy consumption such as lighting and equipment use, resulting in a high proportion of solar radiative heat being included in the total cooling load. Cool roofs primarily derive energy savings from the reduction in solar radiative heat. Therefore, the energy-saving effect of cool roofs is particularly pronounced when these heat loads are considered.

Secondly, the simulations conducted in this study employed a double-layer structure. It is important to note that the energy-saving benefits of cool roofs are significantly higher for the second layer compared to the first layer, to the extent that cool roofs may have negligible energy-saving benefits for the first layer. As a result, the more floors a building has, the smaller the proportion of energy savings attributable to cool roofs in the total building cooling load, leading to less pronounced energy efficiency gains. Hence, the application of cool roofs is more impactful in single-story buildings compared to multi-story or multi-level structures.

Considering the reduction in rooftop temperatures, the materials utilized in this study have the capacity to lower rooftop surface temperatures by approximately 10 °C. This performance surpasses the data available in the literature. This outcome can be attributed to the fact that rooftop temperature has a weak correlation with building floors and indoor loads, but it has a direct positive correlation with solar radiation and rooftop materials. As such, the materials developed in this study exhibit more pronounced benefits in terms of reducing rooftop temperatures.

By delving into these factors, this study provides a more comprehensive understanding of the variations in energy-saving outcomes and the specific conditions under which cool roofs can offer substantial energy efficiency improvements.

Considering the varied effects of cool roofs during the summer and winter, simulation was conducted to determine the applicability of different colored materials across diverse climatic regions in China, assessing their year-round energy saving performance. The outcomes of this simulation are presented in Table 9.

Table 9 reveals that, concerning the annual energy consumption, in cold regions, hot-summer/cold-winter regions, and hot-summer/warm-winter regions, increasing heat reflectance significantly enhances energy savings. As the dimension of a building decreases and the cooling load increases, the energy-saving effect becomes more pronounced. However, for cold regions, due to the higher proportion of heating load, cool roofing materials could paradoxically increase building energy consumption. This phenomenon is influenced by the distinct operational characteristics of cool roofs in the summer and winter, aligning with the results obtained from the orthogonal analysis discussed earlier.

The limitations of this study are mainly demonstrated in the following aspects: (1) as observed from the analysis, the energy-saving effectiveness of cool roofs is more pronounced in low-rise buildings, while in high-rise constructions, the energy-saving efficiency diminishes due to a reduced proportion of heat absorption through the roof; (2) the actual effectiveness of cool roofs varies significantly across different climatic regions, necessitating a consideration of regional factors when deploying cool roofs; and (3) the annual energy consumption in this study is primarily derived from the simulation data. Given the challenge of finding buildings with identical thermal parameters that can sustain consistent loads over extended periods, it is difficult to calibrate these findings through experiments. Future research will focus on verifying the simulated outcomes through methods like small-sample analysis or small-scale models. Additionally, a multidimensional study on energy efficiency and economics will be conducted, comparing cool roofs with other roof types such as green roofs, ventilated roofs, and photovoltaic roofs.

## 4. Conclusions and Outlook

By establishing and analyzing a heat transfer model for cool roofs, we identified four primary parameters that significantly influence the thermal performance of building roofs. These parameters include solar reflectance, thermal emissivity, thermal conductivity, and heat capacity. By utilizing EnergyPlus to conduct annual building energy consumption simulations across diverse climatic conditions and implementing an orthogonal experimental design, we assessed the relative importance of these parameters in shaping annual building energy consumption.

Based on this analysis, we developed high-reflectance and high-emissivity coatings in various colors, which were subsequently tested for their reflectance capacity using a spectrophotometer. Through comprehensive simulations, we investigated the energy-saving attributes of these coatings, and the results are as follows:

(1) Except for severe cold regions, in the other three climate zones with air-conditioning and heating loads, solar reflectance dominates the impact on building energy consumption. As latitude increases, roof surface temperature decreases, the proportion of cooling and heating loads changes, and the influence of solar reflectance gradually decreases, while the influence of thermal conductivity increases.
(2) In severe cold regions, thermal conductivity plays a dominant role in annual building energy consumption. During the building design phase, it is advisable to pay more attention to the material's thermal conductivity and insulation performance. Roof heat capacity has a relatively stable influence on building materials, and future research can explore the thermal properties of phase-change materials combined with cool roofing.
(3) The cool roof coatings that were developed using the methods described in this paper showed significant reductions in roof surface temperatures under typical summer conditions when compared to conventional coatings of the same color. Yellow, red, blue, and green cool roofing materials achieved temperature reductions of 9.4 °C, 9.2 °C, 14.0 °C, and 13.9 °C, respectively, in the summer. This translated to energy savings of 4.1%, 3.9%, 5.5%, and 5.4%, respectively.

Overall, this study demonstrates the effectiveness of cool roofing materials in reducing building energy consumption, particularly in regions with air-conditioning and heating needs. These findings provide valuable insights for energy-efficient building design and can contribute to more sustainable and environmentally friendly construction practices. Future research can explore the integration of phase-change materials into cool roofing systems for further improvements in thermal performance.

**Author Contributions:** Conceptualization, X.Z.; experiment and data curation, S.Z.; original draft preparation, S.Z. and G.H.; review and editing, X.Z. All authors have read and agreed to the published version of the manuscript.

**Funding:** This work is supported by the Jiangsu Maritime Institute Doctoral Research Initiation Fund (2023BSKY04), the Natural Science Foundation of the Jiangsu Higher Education Institutions of China (23KJB470005) and High-end Research Project of Academic Leaders in Jiangsu Higher Vocational Colleges.

**Data Availability Statement:** The data presented in this study are available on request from the corresponding author. The data are not publicly available due to privacy.

**Conflicts of Interest:** The authors declare no conflict of interest.

## Nomenclature

| | | | |
|---|---|---|---|
| $A$ | roof area, $m^2$ | Subscript | |
| $\rho$ | density, $kg/m^3$ | $c$ | layer number |
| $V$ | volume, $m^3$ | $r$ | roof |
| $c$ | specific heat capacity, $J/(kg{\cdot}K)$ | | |
| $t$ | temperature, °C | Acronyms | |
| K | Heat storage capacity index, $J/(m^2{\cdot}K)$ | HVAC | Heating Ventilation and Air Conditioning |
| $d$ | thickness, m | CR | cool roof |
| $R$ | reflection coefficient of light | CPR | conventional paint roof |
| $n$ | refractive indices | SAR | standard asphalt roof |
| $SR$ | Solar Reflectance | | |

## References

1. Abbasabadi, N.; Ashayeri, M. Urban energy use modeling methods and tools: A review and an outlook. *Build. Environ.* **2019**, *161*, 106270. [CrossRef]
2. Lanau, M.; Liu, G.; Kral, U.; Wiedenhofer, D.; Keijzer, E.; Yu, C.; Ehlert, C. Taking stock of built environment stock studies: Progress and prospects. *Environ. Sci. Technol.* **2019**, *53*, 8499–8515. [CrossRef] [PubMed]

3. Huang, S.; Yu, H.; Zhang, M.; Qu, H.; Wang, L.; Zhang, C.; Yuan, Y.; Zhang, X. Advances, challenges and outlooks in frost-free air-source heat pumps: A comprehensive review from materials, components to systems. *Appl. Therm. Eng.* **2023**, *234*, 121163. [CrossRef]
4. Amanowicz, Ł.; Ratajczak, K.; Dudkiewicz, E. Recent Advancements in Ventilation Systems Used to Decrease Energy Consumption in Buildings—Literature Review. *Energies* **2023**, *16*, 1853. [CrossRef]
5. Ahmed, A.; Ge, T.; Peng, J.; Yan, W.-C.; Tee, B.T.; You, S. Assessment of the renewable energy generation towards net-zero energy buildings: A review. *Energy Build.* **2022**, *256*, 111755. [CrossRef]
6. Li, W.; Koo, C.; Hong, T.; Oh, J.; Cha, S.H.; Wang, S. A novel operation approach for the energy efficiency improvement of the HVAC system in office spaces through real-time big data analytics. *Renew. Sustain. Energy Rev.* **2020**, *127*, 109885. [CrossRef]
7. Kamel, E.; Memari, A.M. Residential building envelope energy retrofit methods, simulation tools, and example projects: A review of the literature. *Buildings* **2022**, *12*, 954. [CrossRef]
8. Vassiliades, C.; Agathokleous, R.; Barone, G.; Forzano, C.; Giuzio, G.F.; Palombo, A.; Buonomano, A.; Kalogirou, S. Building integration of active solar energy systems: A review of geometrical and architectural characteristics. *Renew. Sustain. Energy Rev.* **2022**, *164*, 112482. [CrossRef]
9. Sharma, V.; Rai, A.C. Performance assessment of residential building envelopes enhanced with phase change materials. *Energy Build.* **2020**, *208*, 109664. [CrossRef]
10. Veloso, R.C.; Souza, A.; Maia, J.; Ramos, N.M.M.; Ventura, J. Nanomaterials with high solar reflectance as an emerging path towards energy-efficient envelope systems: A review. *J. Mater. Sci.* **2021**, *56*, 19791–19839. [CrossRef]
11. Shafique, M.; Xue, X.; Luo, X. An overview of carbon sequestration of green roofs in urban areas. *Urban For. Urban Green.* **2020**, *47*, 126515. [CrossRef]
12. Heidari, A.; Roshandel, R.; Vakiloroaya, V. An innovative solar assisted desiccant-based evaporative cooling system for co-production of water and cooling in hot and humid climates. *Energy Convers. Manag.* **2019**, *185*, 396–409. [CrossRef]
13. Zhang, C.; Pomianowski, M.; Heiselberg, P.K.; Yu, T. A review of integrated radiant heating/cooling with ventilation systems-Thermal comfort and indoor air quality. *Energy Build.* **2020**, *223*, 110094. [CrossRef]
14. Parker, D.S.; Barkaszi, S.F. Roof solar reflectance and cooling energy use: Field research results from Florida. *Energy Build.* **1997**, *25*, 105–115. [CrossRef]
15. Florida Solar Energy Center; Parker, D. *Demonstration of Cooling Savings of Light Colored Roof Surfacing in Florida Commercial Buildings: Retail Strip Mall*; FSEC Energy Research Center: Cocoa, FL, USA, 1997.
16. Levinson, R.; Berdahl, P.; Akbari, H. Solar spectral optical properties of pigments—Part II: Survey of common colorants. *Sol. Energy Mater. Sol. Cells* **2005**, *89*, 351–389. [CrossRef]
17. Levinson, R.; Berdahl, P.; Akbari, H. Solar spectral optical properties of pigments—Part I: Model for deriving scattering and absorption coefficients from transmittance and reflectance measurements. *Sol. Energy Mater. Sol. Cells* **2005**, *89*, 319–349. [CrossRef]
18. Guo, R.; Gao, Y.; Zhuang, C.; Heiselberg, P.; Levinson, R.; Zhao, X.; Shi, D. Optimization of cool roof and night ventilation in office buildings: A case study in Xiamen, China. *Renew. Energy* **2020**, *147*, 2279–2294. [CrossRef]
19. Rawat, M.; Singh, R.N. A study on the comparative review of cool roof thermal performance in various regions. *Energy Built Environ.* **2022**, *3*, 327–347. [CrossRef]
20. Zhang, K.; Zhao, D.; Yin, X.; Yang, R.; Tan, G. Energy saving and economic analysis of a new hybrid radiative cooling system for single-family houses in the USA. *Appl. Energy* **2018**, *224*, 371–381. [CrossRef]
21. Zhao, B.; Hu, M.; Ao, X.; Chen, N.; Pei, G. Radiative cooling: A review of fundamentals, materials, applications, and prospects. *Appl. Energy* **2019**, *236*, 489–513. [CrossRef]
22. Yu, X.; Chan, J.; Chen, C. Review of radiative cooling materials: Performance evaluation and design approaches. *Nano Energy* **2021**, *88*, 106259. [CrossRef]
23. Jose, S.; Joshy, D.; Narendranath, S.B.; Periyat, P. Recent advances in infrared reflective inorganic pigments. *Sol. Energy Mater. Sol. Cells* **2019**, *194*, 7–27. [CrossRef]
24. Kiomarsipour, N.; Razavi, R.S.; Ghani, K.; Kioumarsipour, M. Evaluation of shape and size effects on optical properties of ZnO pigment. *Appl. Surf. Sci.* **2013**, *270*, 33–38. [CrossRef]
25. Kumari, L.S.; Rao, P.P.; Radhakrishnan, A.N.P.; James, V.; Sameera, S.; Koshy, P. Brilliant yellow color and enhanced NIR reflectance of monoclinic BiVO4 through distortion in $VO_4^{3-}$ tetrahedra. *Sol. Energy Mater. Sol. Cells* **2013**, *112*, 134–143. [CrossRef]
26. Ding, C.; Han, A.; Ye, M.; Zhang, Y.; Yao, L.; Yang, J. Synthesis and characterization of a series of new green solar heat-reflective pigments: Cr-doped BiPO4 and its effect on the aging resistance of PMMA (Poly(methyl methacrylate)). *Sol. Energy Mater. Sol. Cells* **2019**, *191*, 427–436. [CrossRef]
27. Sinsel, T.; Simon, H.; Broadbent, A.M.; Bruse, M.; Heusinger, J. Modeling impacts of super cool roofs on air temperature at pedestrian level in mesoscale and microscale climate models. *Urban Clim.* **2021**, *40*, 101001. [CrossRef]
28. Vieira Zezzo, L.; Pereira Coltri, P.; Dubreuil, V. Microscale models and urban heat island studies: A systematic review. *Environ. Monit Assess* **2023**, *195*, 1284. [CrossRef]

29. Santamouris, M. Recent progress on urban overheating and heat island research. Integrated assessment of the energy, environmental, vulnerability and health impact. Synergies with the global climate change. *Energy Build.* **2020**, *207*, 109482. [CrossRef]
30. Zhao, S.; Zhang, X. Energy consumption and heat island effect mitigation analysis of different roofs considering superposition coupling. *Front. Energy Res.* **2023**, *10*, 1047614. Available online: https://www.frontiersin.org/articles/10.3389/fenrg.2022.1047614 (accessed on 11 December 2023). [CrossRef]
31. Wang, C.; Wang, Z.-H.; Kaloush, K.E.; Shacat, J. Cool pavements for urban heat island mitigation: A synthetic review. *Renew. Sustain. Energy Rev.* **2021**, *146*, 111171. [CrossRef]
32. An, J.; Wu, Y.; Gui, C.; Yan, D. Chinese Prototype Building Models for Simulating the Energy Performance of the Nationwide Building Stock. In *Building Simulation*; Tsinghua University Press: Beijing, China, 2023; Volume 16, pp. 1559–1582.
33. *GB 50189-2015*; Energy Efficiency Design Standards for Public Buildings. China Construction Industry Press: Beijing, China, 2015. (In Chinese)
34. Gao, Q.; Wu, X.; Shi, F. Novel superhydrophobic NIR reflective coatings based on Montmorillonite/SiO2 composites for Energy-saving building. *Constr. Build. Mater.* **2022**, *326*, 126998. [CrossRef]
35. Zhu, J.; Chew, D.A.S.; Lv, S.; Wu, W. Optimization method for building envelope design to minimize carbon emissions of building operational energy consumption using orthogonal experimental design (OED). *Habitat Int.* **2013**, *37*, 148–154. [CrossRef]


**Disclaimer/Publisher's Note:** The statements, opinions and data contained in all publications are solely those of the individual author(s) and contributor(s) and not of MDPI and/or the editor(s). MDPI and/or the editor(s) disclaim responsibility for any injury to people or property resulting from any ideas, methods, instructions or products referred to in the content.

buildings

MDPI

Article

# Effect of Air Parameters on LiCl-$H_2O$ Film Flow Behavior in Liquid Desiccant Systems

Yue Lyu [1,2,*], Yonggao Yin [3] and Jingjing Wang [1,2]

[1] School of Environmental & Municipal Engineering, Lanzhou Jiaotong University, Lanzhou 730070, China; wangjingjing@mail.lzjtu.cn
[2] Key Laboratory of Yellow River Water Environment in Gansu Province, Lanzhou Jiaotong University, Lanzhou 730070, China
[3] School of Energy and Environment, Southeast University, Nanjing 210096, China; y.yin@seu.edu.cn
* Correspondence: lvyuelzjt@163.com; Tel.: +86-19993090805

**Abstract:** The wettability and stability of a solution's film on the filler surface are the key factors determining heat and mass transfer efficiency in liquid desiccant air conditioning systems. Therefore, this study investigates the effects of different air parameters on the flow behavior of a lithium chloride solution's film. The effects of air velocity, air flow pattern, and pressure on the wettability and critical amount of spray are discussed. The results show that the main mechanism by which the air velocity affects the wettability is that the shear stress generated by the direction of the air velocity disperses the direction of the surface tension and weakens its effect on the liquid film distribution. In addition, in the counter flow pattern, the air flow blocks the liquid film from spreading longitudinally and destroys the stability of the liquid film at the liquid outlet, which increases the critical amount of spray. The pressure distribution is similar under different operating pressures when the flow is stable; thus, pressure has little effect on wettability. The simulation results under 8 atm are compared with the experimental results. It is found that the sudden increase in the amount of moisture removal when the amount of spray changes from 0.05 to 0.1 $m^3/(m{\cdot}h)$ in the experiment is caused by the change in the liquid film flow state. In addition, the results show that within the range of air flow parameters for the liquid desiccant air conditioning system, air flow shear force is not the main factor affecting the stability of the solution's film, and there is no secondary breakage of the solution's film during the falling-film flow process.

**Keywords:** liquid desiccant; compressed air; lithium chloride solution; wettability; stability

## 1. Introduction

Compressed air is widely used in various industries as both a power source as well as a process gas source for various applications such as drying and pneumatics [1,2]. However, compressed air can be harmful to production processes when its humidity is too high, leading to problems such as reducing electrical insulation and causing valve corrosion [3]. As the use of liquid desiccants is advantageous in lowering energy costs [4–6], a new method of compressed air drying using liquid desiccant has been proposed. Studies have shown that this system could effectively use the waste heat of air compressors [7,8]. However, because the liquid desiccant systems are still in the development stage, there are still some shortcomings that make them difficult to be applied. First of all, poor heat and mass transfer will make the device huge [9]. Secondly, studies have shown that droplet entraining will occur when there is gas–liquid contact, resulting in the corrosion of indoor surfaces [10]. Therefore, controlling droplet entrainment effectively while improving heat and mass exchange efficiency is the key to optimizing liquid desiccant systems. The current research on optimizing the performance of this new method has only focused on system optimization and control strategies [11,12]. The solution flow behavior on the surface of the filler is an important part in the dehumidification and regeneration processes. The wettability of the solution's film determines the heat and mass transfer efficiency [13–15], and the stability of the solution's film determines the droplet entrainment. Therefore, it is necessary to systematically study the solution flow behavior during dehumidification/regeneration processes.

In terms of liquid film wettability, due to the numerous variables of a solution'sfalling-film process in the liquid desiccant air conditioning system, much work is required for its theoretical analysis and experimental research. With the development of computer technology, computational fluid dynamics (CFD) has become an effective method for studying such problems. Sakhnov et al. [16] researched the effect of the contact angle on the spreading of a refrigerant mixture over a vertical cylinder using CFD. They established mathematical models of R21 and R114 refrigerant mixtures flowing over a vertical cylinder, which were in good agreement with the experimental results [17]. Tan et al. [18] studied the effect of liquid viscosity and surface tension on surface wettability. They simulated the flow behavior of several fluids (e.g., ethylene glycol, and acetone) using a volume of fluid (VOF) model and proposed a formula for predicting the effective interface area ratio. These studies indicate that the VOF model is suitable for capturing the gas–liquid two-phase flow interface in simulations. However, in these studies, the gas-phase parameters are ignored due to the small flow rate. Wen et al. [19] proposed a three-dimensional model to describe the shrinkage of the liquid film during falling-film flow on a plate and verified the accuracy of the model through experiments. The results showed that the poor wettability and mass transfer resistance of the air affected the mass transfer capacity, but they did not elaborate on the effect of the air parameters on wettability. At present, it is necessary to maintain a large amount of solution spray to ensure sufficient wetting. However, a larger amount of spray will consume more pump power and may cause droplet carrying problems. Zhang et al. [20] argued that there must be a reasonable value for the solution flow rate, but their research also did not determine the reasonable amount of spray to ensure adequate wetting. From the above studies, it can be seen that the CFD method can be used to conveniently and accurately study a solution's film wettability on the surface of a filler, but the study of the influence of air parameters on a solution's film wettability is still scarce.

In terms of the stability of the liquid film, previous studies have shown that the liquid film may break due to the shear force of the airflow [21]. Wang [22] believes that a wavy liquid film drawn into the airflow is mainly controlled by shear force and surface tension. When the shear force causes the extreme deformation of the interface, it will cause a certain section of the wave to break. In addition, studies have shown that droplets can splash and form microdroplets when they collide with the liquid film. Dai et al. [23] used the coupled level set and VOF method to simulate the morphological evolution and splashing process of liquid droplets impacting an inclined liquid film. They summarized the main water splashing mechanism caused by droplet impact on a liquid film (with water in the liquid phase) as follows: the water splashes in the first half of the liquid are mainly caused by the neck jet resulting from pressure difference, whereas the water splashes in the second half are formed by the change in the direction of the flow of the impacted liquid film. From the above studies, the fragmentation of liquid films is related to the shear force of the air flow, and the impact velocity of droplets in the secondary fragmentation of liquid films is also affected by the incoming air flow velocity. However, most of these studies on the stability of liquid films focused on water, and the research on salt solutions is still scarce.

Based on the discussion above, previous studies of surface wettability have focused on solution properties and filler properties. However, gas-phase parameters also affect the wettability in the gas–liquid falling-film flow for liquid desiccant processes. Especially for compressed air drying systems using liquid desiccants, whether high pressure air will have an effect on the wettability needs to be further discussed. In addition, studies show that air parameters are one of the key factors affecting the stability of the liquid film. However, in these studies, the working fluid used is mostly water, the physical properties of which differ greatly from those of the solution. And the air conditions are also not the same. Therefore, studying the influence of air parameters on the stability of a solution's film according to the working condition and working fluid of liquid desiccant air conditioning systems is the key to further control droplet entrainment and optimize the systems. At present, halogen salt solutions such as lithium chloride, lithium bromide, and calcium chloride solutions are often used as liquid desiccants in liquid desiccant air conditioning systems. Among them, a lithium chloride solution ($LiCl-H_2O$) has the lowest surface partial pressure of water vapor and a low crystallization risk, so it has the highest system efficiency as a liquid desiccant. Therefore, this study takes $LiCl-H_2O$ as the working fluid to explore the air parameters affecting the solution distribution in a liquid desiccant. The effect of the air velocity, the gas–liquid flow pattern, and the gas pressure on the wettability of the solution's falling film is summarized, and the corresponding critical amount of spray is discussed, which provides a theoretical reference for choosing the amount of spray and the optimal air parameters. The causes of liquid film breakage during the flow process are analyzed, which provides a theoretical basis for further controlling droplet entrainment.

## 2. Materials and Methods

### 2.1. Physical Model

Figure 1 shows the physical model of the air–solution flow channel. The corresponding dimensions and structural vertices are marked in the figure. The dimensions of the falling-film plates remained the same as those in the previous study [24]. The size of the channel is $50 \times 60 \times 7$ mm$^3$, and the filler surface is at an angle of 60° relative to the horizontal. The size of the solution inlet is $0.4 \times 50$ mm$^2$, which is calculated using the Nusselt film thickness theory. The position of the air inlet and outlet are opposite in the cases of parallel flow and counter flow, with the corresponding boundary conditions for the two flow patterns are shown in Tables 1 and 2, respectively. For parallel flow, the size of the air inlet is $6.6 \times 50$ mm, which is the channel size minus the solution inlet size. For counter flow, as the liquid film thickness at the bottom becomes thicker with the development of the flow, to ensure that the air inlet does not coincide with the liquid film (causing flooding in the counter flow case), the size of the air inlet is set to half the channel size, which is $3.5 \times 50$ mm. Therefore, the size of the solution outlet for the counter flow case is $3.5 \times 50$ mm, which is the channel size minus the air inlet size. As the size of the solution inlet for the counter flow case is the same as that for the parallel flow case, the size of the air outlet is $6.6 \times 50$ mm.

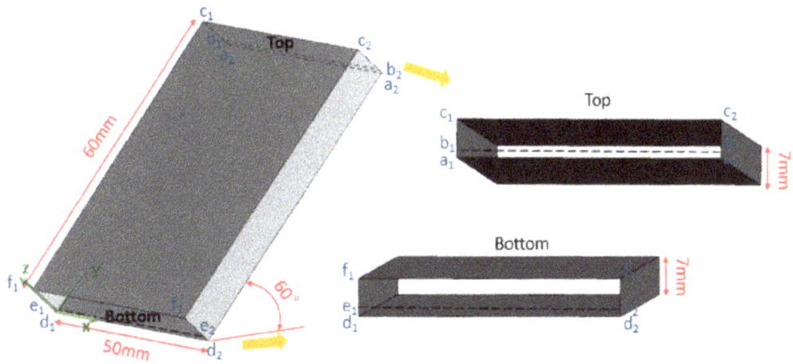


**Figure 1.** A physical model of the air–solution flow channel.

**Table 1.** The boundary condition settings for the parallel flow pattern.

| Face | Boundary Condition |
|---|---|
| $a_1b_1b_2a_2$, $b_1c_1c_2b_2$ | Velocity inlet |
| $f_1e_1e_2f_2$, $e_1d_1d_2e_2$ | Pressure outlet |
| $c_1f_1f_2c_2$, $c_2f_2d_2a_2$, $a_1d_1f_1c_1$, $a_1d_1d_2a_2$ | No-slip wall |

**Table 2.** The boundary condition settings for the counter flow pattern.

| Face | Boundary Condition |
|---|---|
| $a_1b_1b_2a_2$, $f_1e_1e_2f_2$ | Velocity inlet |
| $b_1c_1c_2b_2$, $e_1d_1d_2e_2$ | Pressure outlet |
| $c_1f_1f_2c_2$, $c_2f_2d_2a_2$, $a_1d_1f_1c_1$, $a_1d_1d_2a_2$ | No-slip wall |

### 2.2. Mathematical Modeling

In this paper, Fluent (holding by Ansys in the Canonsburg, PA, United States) is used for the simulation, in which it is important to simulate the change in the interface of the air–solution flow. Compared with other interface capturing methods [25,26], the convergence and mass conservation of the VOF method is superior [27,28]; therefore, it is used in this study to simulate the falling-film flow. The VOF method defines a volume fraction in a cell as $\alpha$, which follows Equation (1):

$$\alpha_s + \alpha_a = 1, \tag{1}$$

where $\alpha_s$ is the volume fraction of the solution, and $\alpha_a$ is the volume fraction of the air.

The momentum is defined as follows:

$$\frac{\partial(\rho v)}{\partial t} + \nabla(\rho v v) = -\nabla p + \nabla \cdot \left[\mu\left(\nabla v + \nabla v^T\right)\right] + \rho g + F_{vol}, \tag{2}$$

where $\rho$ is the density, $\mu$ is the viscosity, and $F_{vol}$ is the volume force, which reflects the effect of surface tension.

The continuum surface force model [29] is used to simulate the surface tension. In this model, $F_{vol}$ is defined as follows:

$$F_{vol} = \sum_{i<j} \sigma_{ij} \frac{\alpha_i \rho_j \kappa_j \nabla \alpha_j + \alpha_j \rho_j \kappa_i \nabla \alpha_i}{\frac{1}{2}(\rho_i + \rho_j)}, \tag{3}$$

where $\sigma$ is the surface tension coefficient, and $\kappa$ is the curvature.

Because the air flow is turbulent in this study, the viscous model is chosen as the standard k-epsilon model. To solve the coupling of pressure and velocity, the SIMPLE method is used.

The surface wettability of the filler surface is expressed by the wetting ratio ($WR$), which is defined as follows:

$$WR = \frac{A_w}{A_b}, \tag{4}$$

where $A_w$ is the projection area of the solution's film on the filler surface, and $A_b$ is the area of the filler surface.

A flat falling film is usually realized by overflow, corresponding to the liquid phase inlet of the model in this study, which has a line shape. Therefore, when analyzing the solution flow behavior, the amount of spray of the solution is expressed by the volumetric flow rate per unit of length per unit of time. The amount of spray ($SP$) is defined as follows:

$$SP = \frac{Q_s}{W} \times 3600 \tag{5}$$

where $Q_s$ is the volumetric flow rate of the solution, and $W$ is the inlet width.

The ranges of the gas-phase parameters are selected according to the liquid desiccant evaporative cooling air conditioning system, the liquid desiccant system driven by a heat pump [24], and the compressed air drying system using a liquid desiccant [7]. The conditions of the LiCl solution in this study are the most unfavorable conditions, as shown in Table 3. To ensure the accuracy and speed of the simulation results, the grid is divided into three different densities and we do the same simulation, with grid numbers of 168,000,

228,000, and 288,000. After a grid independence verification, the mesh with 228,000 grids is selected. The accuracy of the mathematical model in this simulation has been verified in previous studies.

**Table 3.** The LiCl solution's conditions.

| Process | Concentration (%) | Temperature (°C) |
|---|---|---|
| Dehumidification | 35 | 20 |
| Regeneration | 30 | 60 |

### 2.3. Experimental Methods

A schematic and physical diagrams of the flat plate falling-film experimental device are shown in Figures 2 and 3, respectively. The names of each component are marked in the figures. The device comprises an arched organic glass plate forming two symmetrical falling-film surfaces. The falling-film surfaces and the shell form two gas–liquid two-phase flow channels. The channels are rectangular channels with a width of 15 mm. The flat parts of the arched glass plate in the channels are treated with surface modification and used as filler surfaces. Therefore, the filler surfaces inside the channels are flat surfaces, corresponding to the physical model in the simulation.

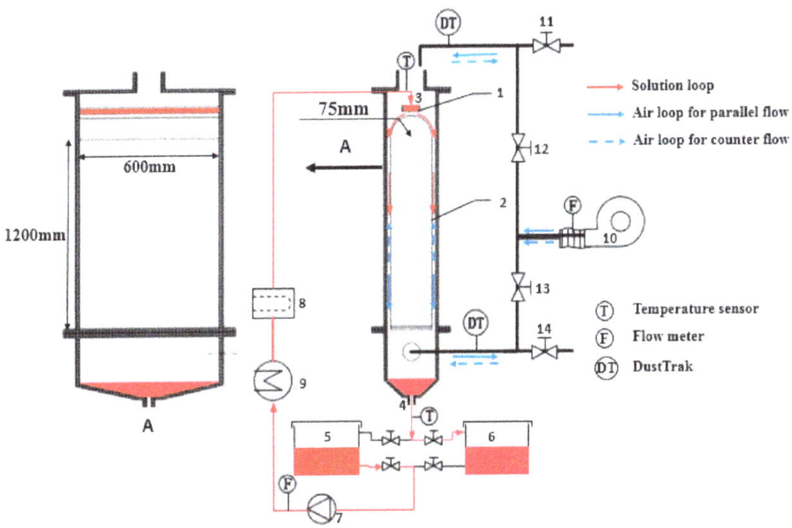


**Figure 2.** A schematic of the flat falling-film experimental device: 1, the liquid distributor; 2, the falling-film plate; 3, the solution inlet; 4, the solution outlet; 5, tank A; 6, tank B; 7, the solution pump; 8, the cooler; 9, the electric heater; 10, the fan; 11, valve A; 12, valve B; 13, valve C; 14, valve D.

According to the liquid desiccant evaporative cooling air conditioning system, the liquid desiccant system driven by a heat pump and a compressed air drying system using liquid desiccant, the experimental conditions are shown in Table 4.

In order to ensure a high measurement accuracy and a lower cost, the measuring range of the measuring equipment is usually about 1.5–2 times the required measuring range. Considering the range of the solution temperature, the solution flow rate and air flow rate to be measured in the experiment (as shown in Table 4), and the relevant parameters of the actual products, the measuring equipment adopted in this study are shown in Table 5. The table also presents the specific parameters of the measurement equipment. The labels T and F in Figure 2 represent K-type thermocouples, and micro flow turbine sensors and gas vortex flow meters are used to measure the temperature and the flow rate of the solution and air flow rate, respectively.

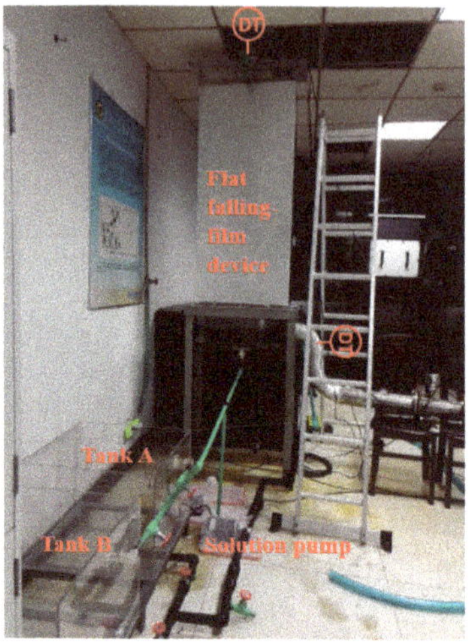


**Figure 3.** A photograph of the flat falling-film experimental device.

**Table 4.** The experimental conditions.

| | |
|---|---|
| LiCl solution concentration | 35% |
| Solution temperature during dehumidification process | 20 °C |
| Solution temperature during regeneration process | 60 °C |
| Maximum solution flow rate | 2 L/min |
| Maximum air flow rate | 129.6 $m^3$/h |

**Table 5.** The equipment measurement parameters.

| Parameter | Equipment | Accuracy | Range |
|---|---|---|---|
| Air flow rate | Gas turbine flowmeter | ±1% | 12–360 $m^3$/h |
| Solution flow rate | Micro flow turbine sensorDigmesa 939–1525 | ±2% | 0.15–3.74 L/min |
| Solution temperature | K-type thermocouples | ±0.2 °C | −10 to 120 °C |
| Solution density | Density meter | ±0.5 kg/$m^3$ | 1200–400 kg/$m^3$ |

Aerosol refers to a gaseous dispersion system consisting of solid or liquid particles suspended in a gaseous medium. The microparticles contained in the air in this study may include solid particles contained in the treated air or solution microdroplets. According to the definition of an aerosol, it can be judged that the microparticles contained in the treated air are aerosols. In addition, the gas–liquid two-phase flow rate is low in the liquid desiccant air conditioning system, so that the microparticles can settle in the air duct. Studies have shown that particles larger than 10 μm can be rapidly precipitated by their own gravity [30]. Therefore, when considering the problem of indoor air quality, the measurement of microparticles in the air mainly focuses on microparticles with a particle size-range below 10 μm. The TSI DustTrak II 8530 desktop aerosol detector (providing by TSI Corporation in the Shoreview, MN, United States) can measure PM1, PM2.5, PM4, and PM10 in real time. Therefore, to analyze the effect of the heat and mass exchange between the air and the solution on the contained microparticles, we measure the mass concentration of the microparticles with different particle size ranges in the air at the air inlet and outlet of the device, which is labelled as DT in Figures 1 and 2, using the TSI DustTrak II 8530 desktop aerosol detector. The accuracy of the device is 0.001 mg/m$^3$ and the zero drift rate of the device is $\pm$0.002 mg/m$^3$. The mass concentration increment is defined as the $\Delta C$, which represents the difference between the mass concentration of particles at the air outlet and that at the air inlet.

## 3. Results and Discussion

### *3.1. Effect of Air Velocity on Wettability*

In this section, the effect of air velocity on the *WR* in a parallel flow during the dehumidification and regeneration processes is discussed. The air velocity in the system is mainly below 1.5 m/s. The results are shown in Figures 4 and 5. To analyze the reasonable amount of spray, the critical amount of sprays for different air velocities are shown in Table 6. The critical amount of spray is defined as the minimum amount of spray to ensure a full-film flow, as denoted by the dashed lines in Figures 4 and 5.

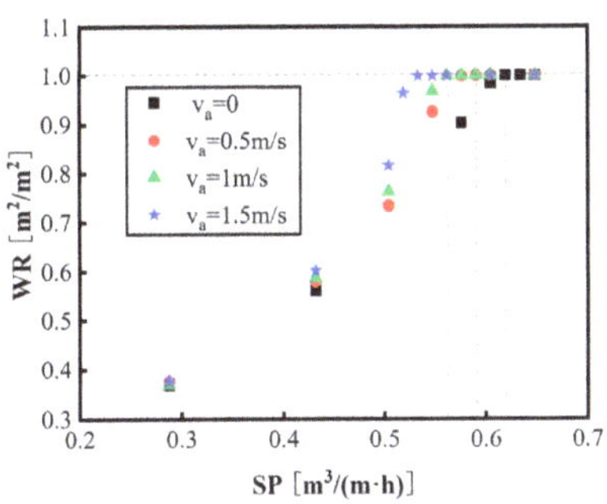


**Figure 4.** The effect of air velocity on the wetting ratio (*WR*) during dehumidification.

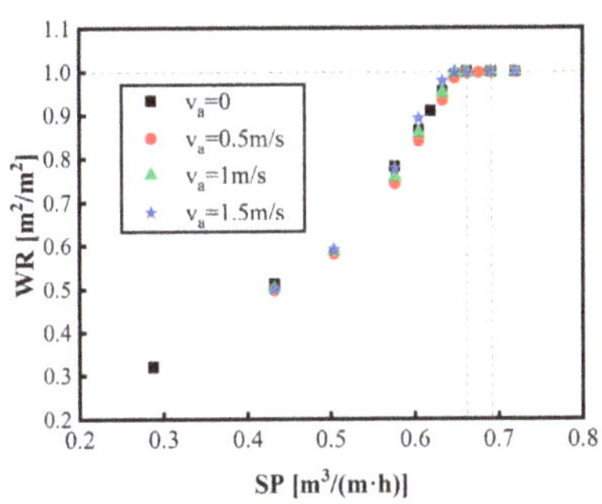


**Figure 5.** The effect of air velocity on the *WR* during regeneration.

**Table 6.** The critical amount of spray for different air velocities.

| Air Velocity (m/s) | Critical Amount of Spray (m$^3$/(m·h)) | |
|---|---|---|
| | Dehumidification | Regeneration |
| 0 | 0.62 | 0.66 |
| 0.5 | 0.59 | 0.69 |
| 1 | 0.59 | 0.69 |
| 1.5 | 0.56 | 0.69 |

For the dehumidification process, Figure 4 shows that before a full-film flow is reached, the *WR* increases with increasing air velocity under the same *SP*. As seen in Table 6, the larger the air velocity, the smaller the critical spray amount, that is, the better the wettability. This is because the effect of surface tension causes the solution's film to converge toward the center during the flow. When the *SP* is small, the effect of surface tension causes the liquid film to form a stream. The shear stress generated by the direction of the air velocity weakens the component of the surface tension that points toward the center, making the liquid film easier to spread. For the regeneration process, Figure 5 shows that before a full-film flow is reached, increasing the air velocity leads to a slight increase in the WR of the solution under the same *SP*. However, the critical amount of spray under a non-air flow is smaller than that where the air velocity is not zero. When the air velocity increases from 0.5 m/s to 1.5 m/s, the critical spray amount remains the same. In general, the air velocity has little effect on the *WR* for the regeneration process.

The difference between the dehumidification and regeneration processes is mainly the temperature. The physical properties of the solution (i.e., its density, viscosity, and surface tension) change with its temperature. Surface tension causes the liquid film to converge toward the center, which reduces the wetted area. At the same time, the surface tension also keeps the liquid film stable and not easily broken, which is conducive to the formation of a full-film flow. Viscosity is described as the resistance of the fluid to flow, which prevents the liquid film from converging toward the center during the flow. Therefore, the higher the viscosity, the better the wettability. According to the previous research [24], during dehumidification, surface tension is the major factor that limits wettability. Therefore, the effect of air velocity on the improvement of wettability is apparent. In contrast, during regeneration, viscosity is the major factor that limits wettability. Thus, the air velocity has little effect on the wettability. At the same time, the stability of the liquid film becomes worse when the air velocity is not zero, which will make the critical amount of spray increase.

### 3.2. The Effect of Air Flow Patterns on Wettability

The gas–liquid flow in the flat falling-film liquid desiccant system is mainly divided into parallel flow and counter flow. The simulation results show that the effect of air velocity on the wettability during the dehumidification process for counter flow is the same as that for parallel flow, that is, the greater the air velocity, the better the wettability and the smaller the critical amount of spray. However, the improvement of wettability for counter flow is not as great as that in parallel flow. Figure 6 compares the effect of different flow patterns on the wettability when the flow velocity is 0.5 m/s and 1.5 m/s during the dehumidification process. The dashed lines indicate the critical amount of spray corresponding to the liquid film just reaching a full-film flow, the values of which are given in Table 7.

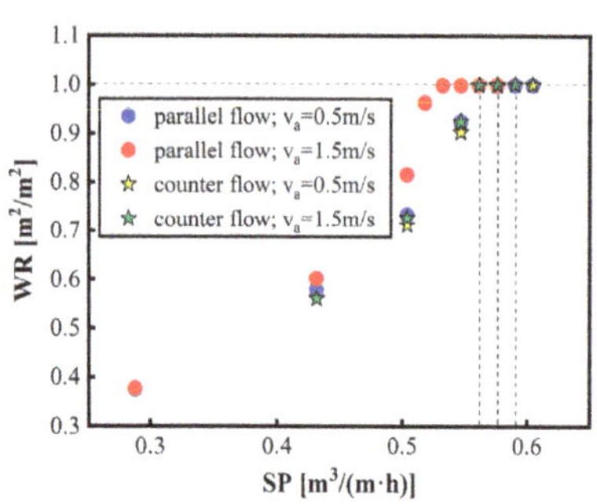


**Figure 6.** A comparison of the flow patterns at different air velocities under 1 atm.

**Table 7.** The critical amount of spray for different flow patterns.

| Flow Pattern | Air Velocity (m/s) | Critical Amount of Spray ($m^3/(m \cdot h)$) |
|---|---|---|
| Parallel flow | 0.5 | 0.59 |
| Parallel flow | 1.5 | 0.56 |
| Counter flow | 0.5 | 0.59 |
| Counter flow | 1.5 | 0.58 |

As seen in Figure 6, for dehumidification, when a full-film flow is not reached and the amount of spray is the same, the *WR* increases with increasing air velocity for both flow patterns. However, the wettability of parallel flow (circle point) is markedly improved, whereas the wettability of counter flow (star point) is only slightly improved. Similarly, as seen in Table 7, the critical amount of spray decreases more for the parallel flow when the air velocity increases, whereas the air velocity has little effect on the critical amount of spray for the counter flow. This is because, in the counter flow, a nonzero air velocity hinders the longitudinal spreading of the solution's film and affects its stability. Therefore, the improvement of wettability at high air velocities is much smaller for counter flow than for parallel flow.

The stability of the liquid film at the liquid outlet becomes worse due to the influence of the reverse air velocity, which is evident during the regeneration process. Table 8 compares gas–liquid phase diagrams near a full-film flow during regeneration under a non-air flow, a parallel flow, and a counter flow, where the air velocity is 1.5 m/s. The corresponding *WR* values are shown under each phase diagram. The *WR* that first reaches 1 is marked in red font, the corresponding *SP* of which is the critical amount of spray.

**Table 8.** A diagram of a comparison of the phases of different flow patterns near a full-film flow.

| *SP* ($m^3/(m \cdot h)$) | **0.63** | **0.66** | **0.69** | **0.71** | **0.72** |
|---|---|---|---|---|---|
| $v_a = 0$ |  |  |  |  |  |
| *WR* | 0.953 | 1 | 1 | 1 | 1 |
| Parallel flow $v_a = 1.5$ m/s |  |  |  |  |  |
| *WR* | 0.979 | 0.999 | 1 | 1 | 1 |
| Counter flow $v_a = 1.5$ m/s |  |  |  |  |  |
| *WR* | 0.936 | 0.998 | 0.999 | 0.999 | 1 |

As seen in Table 8, the liquid film immediately reaches a full-film flow with an increasing *SP* in the case where the air velocity is zero. In the case of the parallel flow, although the decrease in surface tension can improve the wettability during the wetting process ($SP$ = 0.63 ($m^3$/(m·h)) in Table 6), near full-film flow conditions, the decreased stability will make it difficult to reach a full-film flow, which will increase the critical amount of spray. This situation is even more pronounced in the case of the counter flow. As seen in Table 8, when the liquid phase is close to a full-film flow, the $WR$ approaches infinitely close to 1 but does not reach a full-film flow as the $SP$ continues to increase. Therefore, although the air velocity weakens the surface tension to improve the wettability, it also reduces the stability. The adverse effect on stability is especially pronounced at the liquid outlet and air inlet in the case of the counter flow, which increases the critical amount of spray. Furthermore, during the regeneration process, as the air velocity has little effect on the improvement of wettability, the disadvantage of the weakened stability becomes apparent. Therefore, from the viewpoint of wettability, parallel flow is better than counter flow.

### 3.3. Effect of Pressure on Wettability

The research shows that the use of a liquid desiccant in dry compressed air has great energy saving potential, but the current research on solution flow is mainly under atmospheric pressure. The influence of air pressure on the wettability of a solution's flow is still unclear, which brings hidden dangers to the application of compressed air drying using a liquid desiccant. To compare the effect of pressure on wettability more clearly, this section simulates the effect of different pressures on wettability when the liquid film is in a stream-flow state ($SP$ = 0.432 $m^3$/m·h). The simulation results are shown in Figure 7, which indicate that an increase in pressure has little effect on wettability, even that under different pressures has no effect on the degree of improvement. As for the effect on the critical amount spray, the changes in the $WR$ with $SP$ are compared separately under air velocities of 0.5 m/s and 1.5 m/s, and pressures of 1 atm and 8 atm, as shown Figure 8. The corresponding critical spray amounts under different conditions are shown in Table 9.

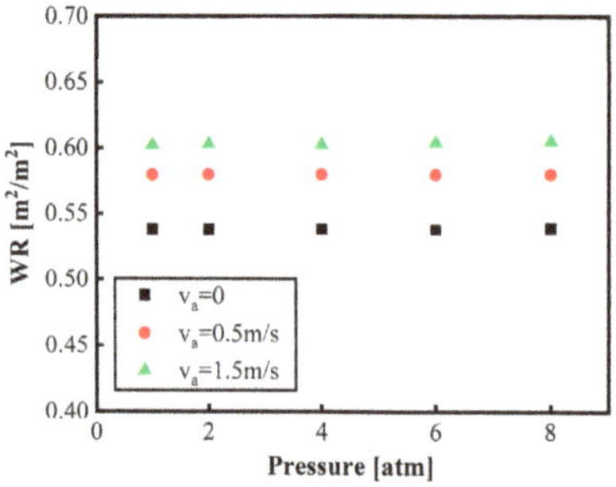


**Figure 7.** The effect of pressure on wettability under a non-full-film flow.

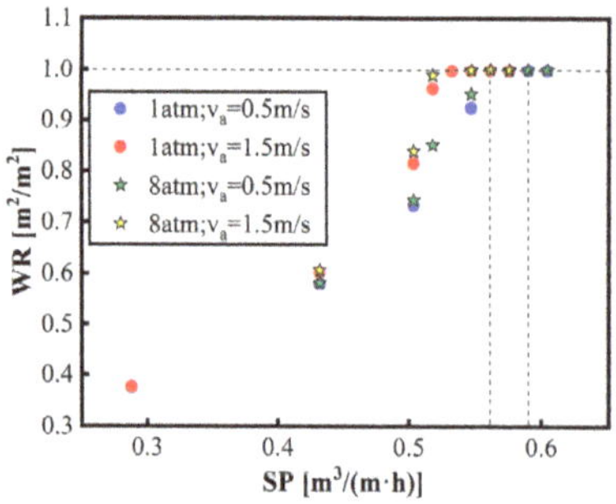


**Figure 8.** The effect of pressure on the critical amount of spray.

**Table 9.** The critical amount of spray for different air velocities under different pressures.

| Pressure (atm) | Air Velocity (m/s) | Critical Amount of Spray ($m^3/(m·h)$) |
|---|---|---|
| 1 | 0.5 | 0.59 |
| 1 | 1.5 | 0.56 |
| 8 | 0.5 | 0.59 |
| 8 | 1.5 | 0.56 |

As seen in Figure 8, the wettability under 8 atm is slightly better than that under 1 atm when a full-film flow is not reached. However, the stage where the *WR* is close to 1 but does not reach 1 is slightly longer under 8 atm than that under 1 atm; therefore, the critical amount of spray required for the liquid film to reach a full-film flow under 1 atm and 8 atm is the same. Thus, a higher pressure will cause a slight improvement in the wettability before the liquid film reaches a full-film flow, but it has no effect on the critical amount of spray.

We now compare our simulation results with Yin et al.'s experimental results regarding the effect of the amount of spray on the moisture removal rate during the dehumidification process, as seen in Figure 9 [7,8]. The moisture removal rate is proportional to the mass exchange area. For the flat falling-film liquid desiccant system, the mass exchange area is the area of the solution's film on the surface of the filler. Therefore, the moisture removal rate should be proportional to the *WR*. The circle points represent the experimental results of the $\Delta d$ changing with the *SP* under different pressures, and the star points represent the simulation results. As changes in pressure have little effect on wettability, the simulation results of the *WR* changing with the *SP* under 8 atm are taken as an example for comparative analysis.

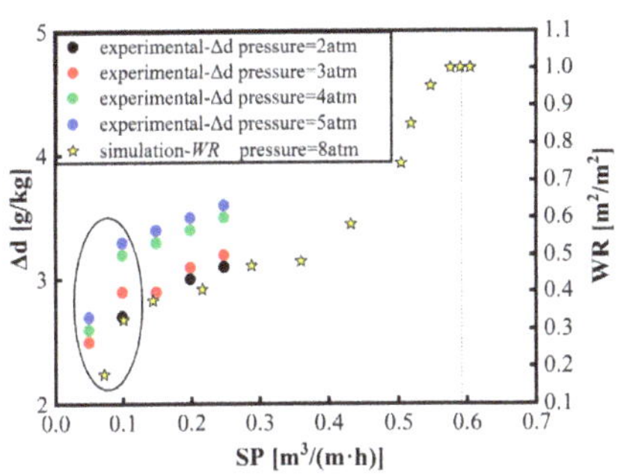


**Figure 9.** A comparison between the simulation results and the experimental results of references.

As seen in Figure 9, the $\Delta d$ increases with the increasing *SP* under all pressures. This is because the *SP* ranges of the experiment are smaller than the critical amounts of spray. Within the *SP* ranges of the experiment, the *WR* increases with the increasing *SP*, so that the mass exchange area increases. In addition, as indicated by the black circle in Figure 9, when the SP increases from 0.05 $m^3/(m·h)$ to 0.1 $m^3/(m·h)$, the growth rate of the $\Delta d$ is larger, and the WR also increases significantly. This is because of the changing flow pattern. Table 10 shows the phase diagrams at different *SP*s. When the *SP* increases from 0.072 $m^3/(m·h)$ to 0.101 $m^3/(m·h)$, the flow pattern of the liquid film changes from a drop flow to a multiple-streams flow, which results in a stepwise increase in the wettability and the amount of moisture removed. When the *SP* increases from 0.101 $m^3/(m·h)$ to 0.36 $m^3/(m·h)$, the flow state changes from a multiple-streams flow to single-stream flow. Thus, the wettability increases slowly in this range. When the *SP* > 0.36 $m^3/(m·h)$, the flow state is always a single stream until it reaches a full-film flow. When the *SP* is within the experimental range, the simulation results are reasonable for the compressed air drying system using liquid desiccants.

**Table 10.** The phase diagrams of the simulation results at different *SPs*.

| *SP* $m^3/(m{\cdot}h)$ | 0.072 | 0.101 | 0.36 | 0.59 |
|---|---|---|---|---|
| Phase diagram | 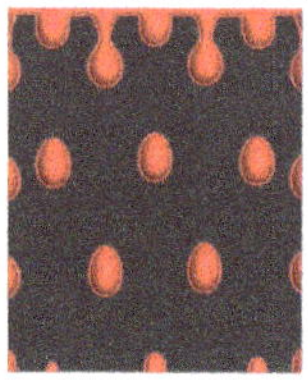 |  | 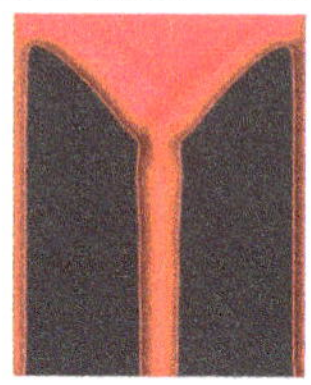 | 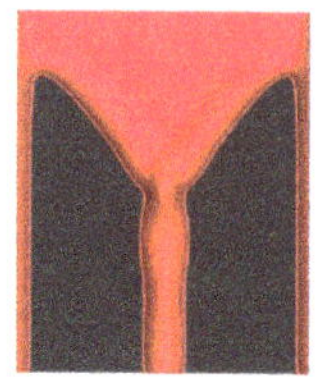 |

### 3.4. The Effects of Air Parameters on a Solution's Film Stability

We also simulate the solution's falling-film flow process along the flat plate under different air flow parameters. Based on the actual situation in the liquid desiccant air conditioning system, the study mainly simulates the falling-film flow process with an air flow rate of <2 m/s. The solution spray rate is 0.619 $m^3/(m{\cdot}h)$, which can form a full-film flow. And the solution condition is the most unfavorable dehumidification condition.

In order to study the solution's film stability, the gas-solution phase diagram on the cross-section of the falling-film plate at the center line in the x direction was studied. As shown in Figure 10, the yellow surfaces are the falling-film surface and the cross-section taken to research the phase diagram. Table 11 shows the gas–liquid phase diagrams of the falling-film flow process under a counter flow with air flow rates of 1 m/s and 2 m/s, and flow times of 0.1 s, 0.2 s, and 0.3 s. In the phase diagrams of Table 11, red represents the air phase and blue represents the solution phase. The liquid film with an air velocity of 2 m/s at 0.2 s is slightly disturbed at the front end of the liquid film compared to that under an air velocity of 1 m/s due to a significant air flow disturbance. However, due to the flat surface of the liquid film, the high surface tension of the solution, and the low air velocity under the operating conditions, the entire flow process is very stable. When the air flow velocity reaches its maximum (2 m/s), the solution's liquid film will not break due to the shear force of the air flow throughout the entire process. In addition, the simulation results show that the stability of the solution's liquid film is minimally affected under different pressures. The effect of air flow on the stability of the liquid film should be smaller in the counter flow than in the parallel flow. Thus, there will be no liquid film breakage within the given operating range. Therefore, for the flat falling-film flow, the shear force of the airflow is not the main cause of droplet entrainment during the liquid desiccant processes.

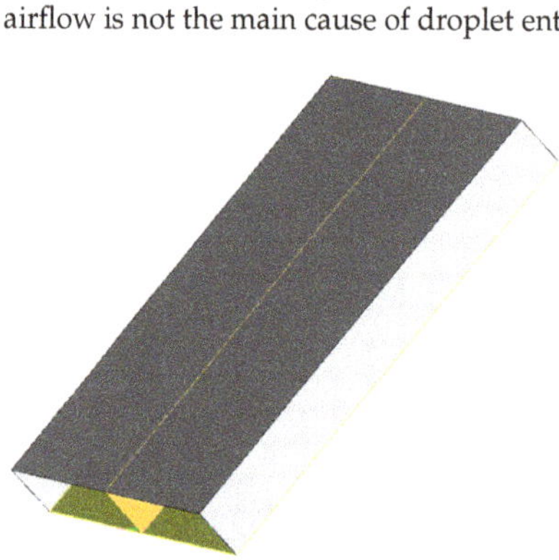

**Figure 10.** The cross-section taken to research the phase diagram.

**Table 11.** Phase diagrams of simulation results under different air flow rates.

| Time | 0.1 s | 0.2 s | 0.3 s |
|---|---|---|---|
| 1 m/s |  | 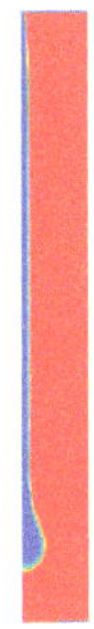 | 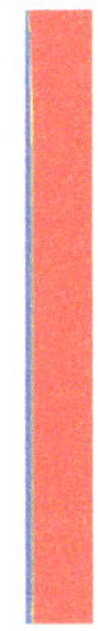 |
| 2 m/s |  |  | 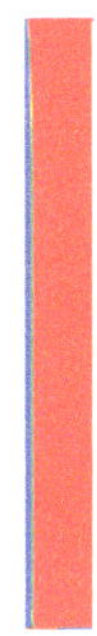 |

Considering whether the liquid film will undergo secondary fragmentation due to some accidental droplet impact on the liquid film, the flow evolution of a single droplet's impact on the flowing solution's film was simulated in this study. First, the main factors influencing the flow behavior when droplets hit the flowing solution's film in the operating conditions of the liquid desiccant air conditioning system were analyzed. Table 12 shows the main factors and their value ranges that affect droplet impact behavior. We then simulated each influencing factor separately.

**Table 12.** The main factors influencing droplet impact behavior and their value ranges.

| Main Factor | Value Range |
|---|---|
| Droplet size (mm) | 0.4–1.2 |
| Droplet impact velocity (m/s) | 0–2 |
| Droplet impact angle | 45–90° |
| Air velocity (m/s) | 0–1.5 (Counter flow/Parallel flow) |
| Solution velocity (m/s) | 0–0.6 |

Figure 11 compares the liquid film morphology at the moment when the liquid film deformation reaches its maximum when droplets of different particle sizes collide with a flowing liquid film. It can be seen that the larger the droplet size, the greater the liquid film's deformation. The liquid film's deformation reaches a maximum when the droplet size is 1.2 mm. At this point, the liquid film is most unstable, but it has not yet broken. The same method is used to study the effects of other factors separately. The results show that when the droplet impact velocity is 2 m/s, the gas–liquid two-phase flow behavior is that of a counter flow, the liquid film impact angle is 45°, the air velocity is 1.5 m/s, and the solution velocity is 0.6 m/s, in which case the flow pattern of the liquid film is most unstable when droplets collide with it. However, in separate studies of each factor, the liquid film does not break when the impact behavior occurs.

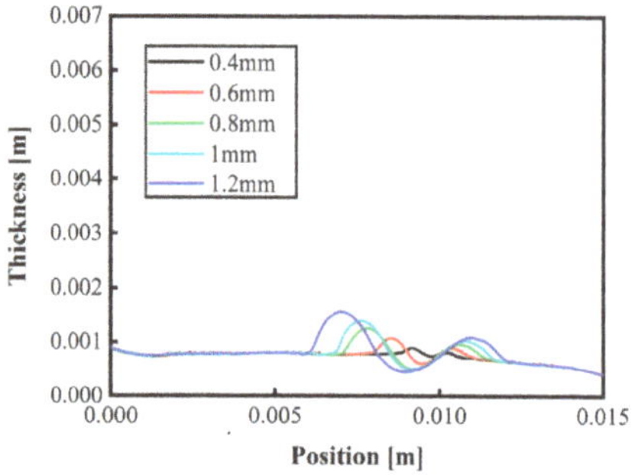


**Figure 11.** Comparison of liquid film morphology when droplets of different sizes collide with a flowing liquid film.

We further simulate the gas–liquid two-phase flow behavior when droplets collide with a flowing liquid film under the most extreme conditions, in which each influencing factor is taken as the value at which the liquid film flow is most unstable when the impact occurs. The simulation results are shown in Table 13. In the phase diagrams of Table 13, red represents the air phase and blue represents the solution phase. It can be seen that during the evolution of a gas–liquid two-phase flow, the difference in neck pressure generated during impact can cause the occurrence of a neck jet. However, due to the high surface tension of LiCl and the low solution flow rate, air flow rate, and droplet impact velocity, no liquid film fragmentation phenomenon occurs.

**Table 13.** The flow evolution of solution droplets impacting a liquid film under extreme conditions.

| | |
|---|---|
|  | 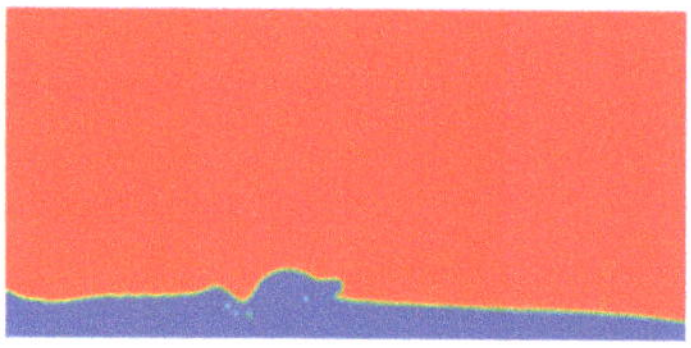 |
| t = 0.000 s | t = 0.003 s |
| 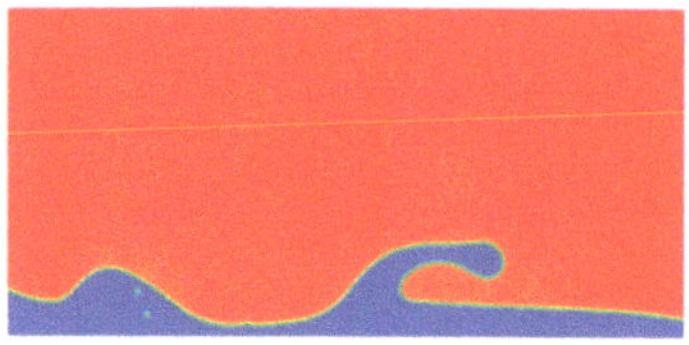 | 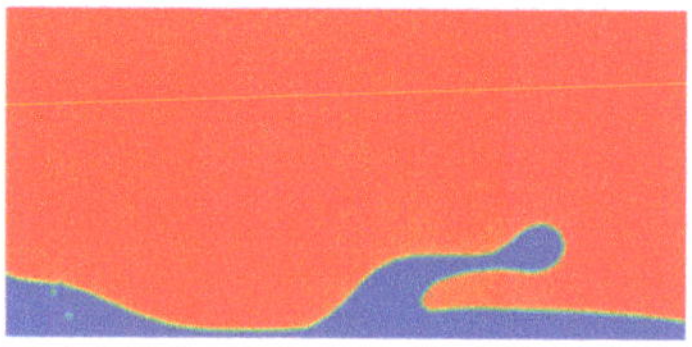 |
| t = 0.006 s | t = 0.009 s |
| 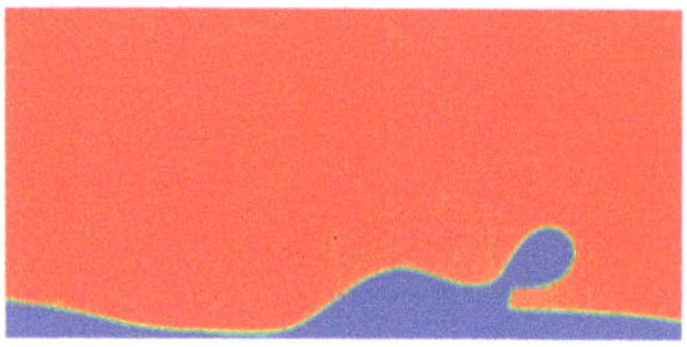 | 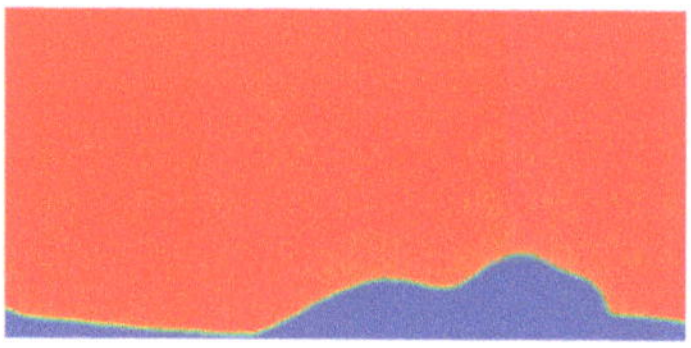 |
| t = 0.012 s | t = 0.015 s |

In addition to this, we also analyze the pressure within the watershed. Figure 12 shows the local phase diagram and pressure distribution when the droplet just hits the liquid film (0.003 s) under extreme conditions. It can be seen that due to the small impact velocity, the pressure difference generated is approximately 3000 Pa, which is much smaller than the 70,000 Pa found in the literature (the impact velocity in the literature is 10 m/s). In addition, the surface tension of the solution is higher than that of water; thus, there are no microdroplets generated due to the secondary fragmentation of the droplets.

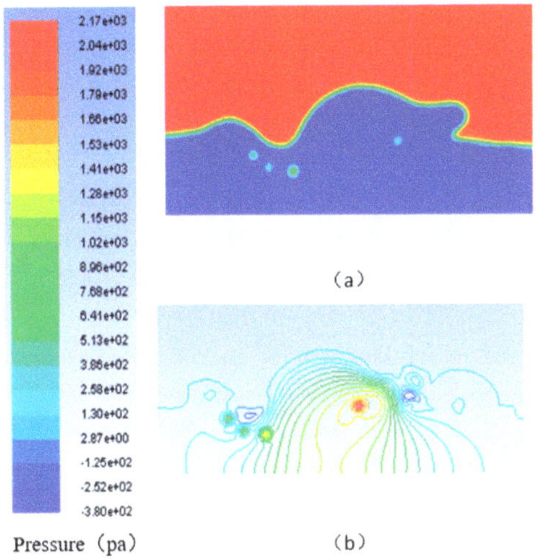


**Figure 12.** (**a**) A local phase diagram. (**b**) The local pressure distribution when a droplet hits the liquid film under extreme conditions.

To further validate the simulation results, we experimentally measured the mass concentration increment of the microparticles in the air before and after the dehumidification and regeneration processes at air volumes of 32.4 $m^3/h$, 64.8 $m^3/h$, 97.2 $m^3/h$, and 129.6 $m^3/h$. According to the previous research, the solution mainly absorbs particles with a diameter of 0–2.5 μm and releases particles with diameters of 2.5–10 μm [31]. Therefore, the size range of the microparticles analyzed in this part of the study is reduced to 2.5–10 μm. Define $\Delta C_{2.5\text{–}10\mu m}$ to represent the $\Delta C$ of the microparticles with a diameter of 2.5–10 μm. If the shear force of the air flow produces droplet entrainment, the $\Delta C_{2.5\text{–}10\mu m}$ will increase with the increasing air flow rate. Figure 13 compares the $\Delta C_{2.5\text{–}10\mu m}$ in the air at different air flow rates after the dehumidification and regeneration processes.

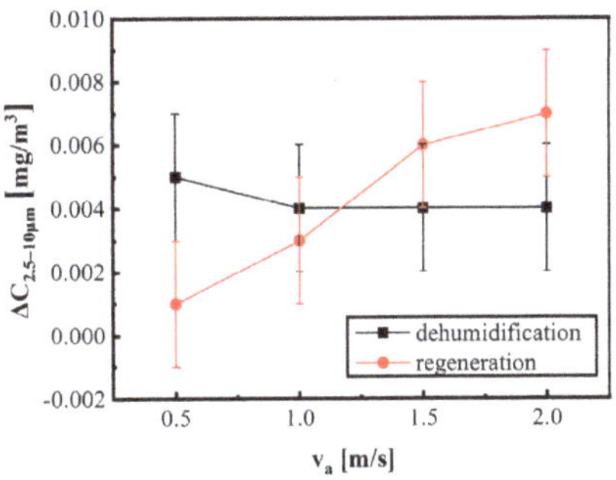


**Figure 13.** The effect of the airflow rate on the $\Delta C_{2.5\text{–}10\mu m}$ of microparticles.

As can be seen from Figure 13, for the dehumidification process, the $\Delta C_{2.5-10\mu m}$ is 0.005 mg/m$^3$ when the air flow rate is 0.5 m/s. When the air flow rate increases to 1 m/s, the $\Delta C_{2.5-10\mu m}$ decreases to 0.004 mg/m$^3$. When the air flow rate continues to increase, the $\Delta C_{2.5-10\mu m}$ does not change. Considering that the measuring instrument also exists at $\pm$0.002 mg/m$^3$ of zero drift, the dehumidification process has little effect on the microparticles. For the regeneration process, the $\Delta C_{2.5-10\mu m}$ increases with the increase in the air flow rate. Therefore, the experimental results indicated that the increase in microparticles is not directly related to changes in the air flow rate, but rather to the air treatment process. The higher the airflow rate, the greater the shear force of the airflow. This indicates that the shear force of the air flow is not the main reason for the release of microdroplets in the liquid desiccant of flat falling-film solutions, which is consistent with the simulation results.

## 4. Conclusions

To control droplet entrainment while improving the efficiency of the liquid desiccant air conditioning systems, this work mainly discusses the effects of air velocity, the air flow pattern, and pressure on the wettability and stability characteristics of the system during dehumidification and regeneration using LiCl-$H_2O$. The major conclusions of this work are as follows:

1. In terms of the film wettability, the main reason that the air velocity affects the wettability is that the shear stress generated by the direction of the air velocity disperses the direction of the surface tension and weakens its effect on the liquid film distribution. During dehumidification, the surface tension is the main factor that makes the wettability worse. Thus, the effect of air velocity on the improvement of wettability during dehumidification is apparent: the higher the air velocity, the better the wettability and the smaller the critical amount of spray. During regeneration, because the viscosity is the main factor that makes the wettability worse, the air velocity has little effect on the wettability. The gas-phase flow blocks the longitudinal spreading of the liquid film and destroys the stability of the liquid film in the counter flow. The adverse effect on stability is especially pronounced at the liquid outlet and the air inlet, which increases the critical amount of spray in the counter flow pattern. Therefore, from the viewpoint of wettability, parallel flow is better than counter flow. The pressure distribution is similar under different operating pressures when the flow is stable, and the differential pressures under different operating pressures are also very similar. Therefore, changing the pressure leads to a weak improvement in wettability but has no effect on the critical amount of spray. By comparing the simulation data under 8 atm with the experimental data from Yin et al. [7,8], it is found that the sudden increase in the amount of moisture removal when the *SP* changes from 0.05 m$^3$/(m·h) to 0.1 m$^3$/(m·h) in the experiment is caused by the change in the liquid film flow state.

2. In terms of the film stability, the simulation results show that the front end of the liquid film will be slightly disturbed by the shear force of the air flow during the falling-film process. In addition, the disturbance of the counter flow is greater than that of the parallel flow. However, due to the stable shape of the flat falling film, the larger surface tension of a salt solution and the smaller air flow rate, the shear force of the air will not break the liquid film. The experimental results also show that increasing the air flow rate during dehumidification does not increase the $\Delta C_{2.5-10\mu m}$. Therefore, for a flat falling-film flow, the shear force of the air flow is not the main cause of droplet entrainment during the dehumidification and regeneration processes. In addition, when a droplet collides with a flowing liquid film under the most extreme conditions, due to the high surface tension of LiCl and the low solution flow rate, air flow rate, and droplet impact velocity, the difference in neck pressure generated by the impact is far from the threshold for causing the solution's film to break. Thus, no liquid film secondary fragmentation phenomenon occurs during the liquid desiccant processes.

**Author Contributions:** Conceptualization, Y.L. and Y.Y.; data curation, Y.L.; investigation, Y.L.; resources, Y.L. and J.W.; software, Y.L.; validation, Y.L.; writing—original draft, Y.L.; writing—review and editing, Y.L. and J.W. All authors have read and agreed to the published version of the manuscript.

**Funding:** This research was funded by the Gansu Province Youth Science and Technology Fund Program, grant number 22JR5RA371.

**Data Availability Statement:** The data presented in this study are available on request from the corresponding author.

**Conflicts of Interest:** The authors declare no conflicts of interest.

## Nomenclature

$A_b$—the area of the mainly plate, [m$^2$]
$A_w$—the projection area of liquid film on the board, [m$^2$]
$\Delta d$—the difference of the air moisture content between the inlet and outlet, [g/kg]
$F$—the force, [N]
$Q$—volumetric flow rate of liquid, [m$^3$/s]
$SP$—the amount of spray, which is the volumetric flow rate per unit length per unit time, [m$^3$/(m·h)]
$v$—the velocity, [m/s]
$W$—the inlet width, [m]
$WR$—the wetting ratio, [m$^2$/m$^2$]
$\Delta C$—The mass concentration increment, [mg/m$^3$]
Greek symbol
$\alpha$—the volume fraction, [-]
$\kappa$—the curvature, [m$^{-1}$]
$\mu$—viscosity, [Pa·s]
$\rho$—density, [kg/m$^3$]
$\sigma$—surface tension, [mN/m]
Subscripts
$a$—air
$s$—solution
$vol$—volume

## References

1. Bazdar, E.; Sameti, M.; Nasiri, F.; Haghighat, F. Compressed air energy storage in integrated energy systems: A review. *Renew. Sustain. Energy Rev.* **2022**, *167*, 112701. [CrossRef]
2. Nehler, T. Linking energy efficiency measures in industrial compressed air systems with non-energy benefits—A review. *Renew. Sustain. Energy Rev.* **2018**, *89*, 72–87. [CrossRef]
3. Sureshkannan, V.; Vijayan, S.; Lenin, V.R. Design and performance analysis of compressed air adsorption dryer with heatless regeneration mode. *Heat Mass Transf.* **2022**, *58*, 631–641. [CrossRef]
4. Su, W.; Lu, Z.F.; She, X.H.; Zhou, J.M.; Wang, F.; Sun, B.; Zhang, X.S. Liquid desiccant regeneration for advanced air conditioning: A comprehensive review on desiccant materials, regenerators, systems and improvement technologies. *Appl. Energy* **2022**, *308*, 118394. [CrossRef]
5. Luo, J.L.; Yang, H.X. A state-of-the-art review on the liquid properties regarding energy and environmental performance in liquid desiccant air-conditioning systems. *Appl. Energy* **2022**, *325*, 119853. [CrossRef]
6. Oladosu, T.L.; Baheta, A.T.; Oumer, A.N. Desiccant solutions, membrane technologies, and regeneration techniques in liquid desiccant air conditioning system. *Int. J. Energy Res.* **2021**, *45*, 8420–8447. [CrossRef]
7. Yin, Y.G.; Zheng, B.J.; Yang, C.; Zhang, X.S. A proposed compressed air drying method using pressurized liquid desiccant and experimental verification. *Appl. Energy* **2015**, *141*, 80–89. [CrossRef]
8. Yin, Y.; Shao, B.; Zhang, X.S. Experimental investigation on compressed air drying performance using pressurized liquid desiccant. *Dry. Technol.* **2016**, *34*, 372–382. [CrossRef]
9. Peng, D.G.; Xu, S.H.; Yang, H.X. Heat and Mass Transfer Characteristics and Dehumidification Performance Improvement of an Evaporatively-Cooled Liquid Dehumidifier. *Appl. Therm. Eng.* **2020**, *178*, 115579. [CrossRef]
10. Liu, X.L.; Qu, M.; Liu, X.B.; Wang, L.S. Membrane-based liquid desiccant air dehumidification: A comprehensive review on materials, components, systems and performances. *Renew. Sustain. Energy Rev.* **2019**, *110*, 444–466. [CrossRef]
11. Mandow, W.; Micus, F.; Völker, L.; Fleig, D.; Jordan, U. Solar driven liquid desiccant dehumidification system: Measurements and annual system simulations. *Appl. Therm. Eng.* **2024**, *242*, 122485. [CrossRef]

12. Lee, J.H.; Cheon, S.Y.; Lee, S.J.; Cho, H.J.; Jeong, J.W. Applicability of a liquid-desiccant air-conditioning system in energy-efficient buildings with high latent load. *J. Build. Eng.* **2023**, *78*, 107608. [CrossRef]
13. Mortazavi, M.; Isfahani, R.N.; Bigham, S.; Moghaddam, S. Absorption characteristics of falling film LiBr (lithium bromide) solution over a finned structure. *Energy* **2015**, *87*, 270–278. [CrossRef]
14. Zhao, C.Y.; Zhang, P.; Guan, Q.; Qi, D.; Zhang, Y.; Jiang, J.M. Numerical study of falling film dehumidification performance on corrugated plates. *Int. J. Heat Mass Transf.* **2024**, *219*, 124843. [CrossRef]
15. Lu, X.B.; Liu, J.P.; Xu, X.W.; Chen, J.X. Change in wetting characteristic of heated refrigerant in falling film evaporation. *Int. J. Refrig.* **2020**, *117*, 198–208. [CrossRef]
16. Sakhnov, A.Y.; Volodin, O.A.; Pecherkin, N.I.; Pavlenko, A.N. Effect of contact angle on spreading of refrigerant mixture over the vertical cylinder. *Int. J. Heat Mass Transf.* **2023**, *215*, 124484. [CrossRef]
17. Kuznetsov, D.V.; Pavlenko, A.N.; Volodin, O.A. Effect of structuring by deformational cutting on heat transfer and dynamics of transient cooling processes with liquid film flowing onto a copper plate. *J. Eng. Thermophys.* **2020**, *29*, 531–541. [CrossRef]
18. Tan, L.Y.; Yuan, X.G.; Kalbassi, M.A. Computational fluid dynamics study of liquid distribution on structured packing surface. *Chem. Ind. Eng. Prog.* **2015**, *34*, 3221–3237.
19. Wen, T.; Luo, Y.M.; He, W.F.; Gang, W.J.; Sheng, L.Y. Development of a novel quasi-3D model to investigate the performance of a falling film dehumidifier with CFD technology. *Int. J. Heat Mass Transf.* **2019**, *132*, 431–442. [CrossRef]
20. Zhang, L.; Hihara, E.; Matsuoka, F.; Dang, C. Experimental analysis of mass transfer in adiabatic structured packing dehumidifier/regenerator with liquid desiccant. *Int. J. Heat Mass Transf.* **2010**, *53*, 2856–2863. [CrossRef]
21. Yu, D.; Cao, D.P.; Li, Z.Z.; Li, Q.S. Experimental and CFD studies on the effects of surface texture on liquid thickness, wetted area and mass transfer in wave-like structured packings. *Chem. Eng. Res. Des.* **2018**, *129*, 170–181. [CrossRef]
22. Wang, B.; Ke, B.Z.; Chen, B.W.; Li, R.; Tian, R. Study on the size of secondary droplets generated owing to rupture of liquid film on corrugated plate wall. *Int. J. Heat Mass Transf.* **2020**, *147*, 118904. [CrossRef]
23. Dai, J.F.; Fan, X.P.; Meng, B.; Liu, J.F. A coupled level-set and volume-of-fluid simulation for splashing of single droplet impact on an inclined liquid film. *Acta Phys. Sin.* **2015**, *64*, 094704.
24. LYu, Y.; Yin, Y.G.; Zhang, X.S.; Jin, X. Investigation of falling-film plate wettability characteristics under dehumidification and regeneration conditions using LiCl-$H_2O$. *Int. J. Refrig.* **2018**, *94*, 118–126. [CrossRef]
25. Van der Graaf, S.; Nisisako, T.; Schroën, C.G.P.H.; van der Sman, R.G.M.; Boom, R.M. Lattice Boltzmann simulations of droplet formation in a T-shaped microchannel. *Langmuir* **2006**, *22*, 4144–4152. [CrossRef]
26. Pozrikidis, C. Interfacial dynamics for Stokes flow. *J. Comput. Phys.* **2001**, *169*, 250–301. [CrossRef]
27. Taha, T.; Cui, Z.F. CFD modelling of slug flow in vertical tubes. *Chem. Eng. Sci.* **2006**, *61*, 676–687. [CrossRef]
28. Taha, T.; Cui, Z.F. CFD modelling of slug flow inside square capillaries. *Chem. Eng. Sci.* **2006**, *61*, 665–675. [CrossRef]
29. Brackbill, J.U.; Kothe, D.B.; Zemach, C. A continuum method for modeling surface tension. *J. Comput. Phys.* **1992**, *100*, 335–354. [CrossRef]
30. Huang, S.; Xu, J.; Liang, C.; Zhang, X. Size distribution measurement of packed tower drift based on hydrophobic materials. *Appl. Therm. Eng.* **2016**, *99*, 873–879. [CrossRef]
31. LYu, Y.; Yin, Y.G.; Zhao, X.W.; Zhang, C.B. Effects of liquid-desiccant air conditioning processes on the presence of inhalable particles in the air. *Build. Environ.* **2021**, *194*, 107662.


**Disclaimer/Publisher's Note:** The statements, opinions and data contained in all publications are solely those of the individual author(s) and contributor(s) and not of MDPI and/or the editor(s). MDPI and/or the editor(s) disclaim responsibility for any injury to people or property resulting from any ideas, methods, instructions or products referred to in the content.

buildings

MDPI

Article

# Experiment and Prediction of Pressure Drop in a Fiber–Powder Composite Material with Porous Structure for Energy Wheels and Air Cleaners

**Han Gao [1], Zhenhai Li [1,*], Xigang Zhou [2], Xiaolong Yin [2] and Mengmeng Shan [2]**

[1] School of Mechanical Engineering, Tongji University, Shanghai 200092, China
[2] Shandong Xuesheng Electric Appliance Co., Ltd., Yantai 261400, China
* Correspondence: lizhenhaioffice@163.com; Tel.: +86-13917357950

**Abstract:** Energy wheels and air cleaners play crucial roles in building air conditioning systems. The former is essential for conserving energy in air conditioning systems, while the latter is necessary for ensuring the quality of indoor air. Pressure drop is a crucial parameter for both energy wheels and air cleaners, and it is essential to conduct theoretical and experimental investigations to aid in their design. In this study, we focused on the study of pressure drop in a fiber–powder composite material which can be used for both total heat exchange and air purification. Experimental tests were initially conducted to examine the impact of different parameters on the pressure drop in the material. Subsequently, based on the special fiber–powder structure of the material, two pressure drop prediction methods with different prediction strategies were proposed. The two prediction strategies were compared by analyzing the prediction accuracy of the two methods. As tested by experimental data, for both methods, the absolute prediction error was less than ±6 Pa when the pressure drop was below 50 Pa, and the relative prediction error was less than ±8% for most data sets when the pressure drop was greater than 50 Pa. Moreover, the root mean square error ($RMSE$) and mean absolute percentage error ($MAPE$) values of prediction for both methods were less than 4 Pa and 7% respectively. The test results show that although the prediction strategies are different, both prediction methods can obtain acceptable prediction results, and both methods are practical. This study is intended to serve as a valuable reference for the design of energy wheels and air cleaners.

**Keywords:** air cleaner; energy wheel; porous materials; prediction model; pressure drop

## 1. Introduction

Energy wheels play a significant role in building energy conservation because they can reduce the fresh air load of air conditioning systems by recovering energy from indoor air. The results of previous experiments [1,2] have demonstrated the effectiveness of energy recovery wheels in reducing the fresh air load in both cold and hot climates. Figure 1 illustrates the working principle of energy wheels. Air cleaners are also widely used in building applications, where they play a key role in maintaining and improving the indoor air quality.

For energy wheels and air cleaners, the pressure drop is a critical parameter that significantly impacts the design of these devices. Therefore, it is meaningful to establish pressure drop prediction models to study the pressure drop of these devices. In [3–5], pressure drop models for energy wheels or rotary heat exchangers were developed, and they were utilized to investigate the thermal performance of the wheel. Moreover, in [6–8], pressure drop models for adsorption filters or air filters were developed for better design of the device. Dallaire et al. [4] examined the influence of a dimensionless pressure drop on the optimal values of two design variables of a rotary heat exchanger with a porous medium. Their findings indicated that the optimal length of the device is strongly affected by the dimensionless pressure drop. Harshe et al. [5] constructed pressure drop and heat and mass transfer models of desiccant wheels which can be utilized for energy recovery. In their study, heat and mass transfer coefficients were obtained by assuming that the Stanton number is proportional to the fractional coefficient. Zhang et al. [6] studied the pressure drop of honeycomb adsorption filters filled with granular adsorbents and built a prediction model of the filter. It was found that the pressure drop was mainly affected by the void ratio and the granular size and shape of the material.

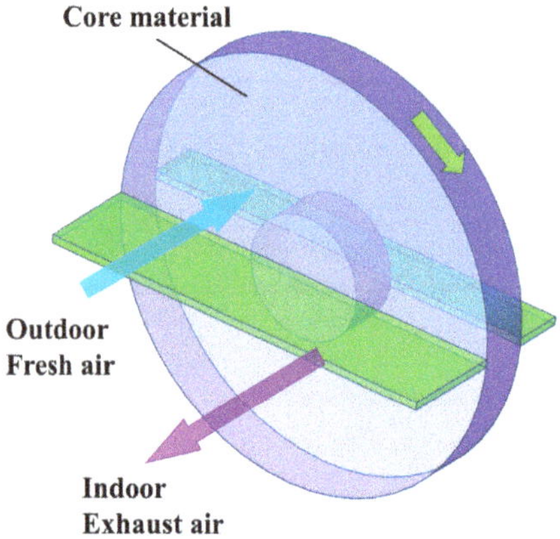


**Figure 1.** Schematic of energy wheels.

The material investigated in this study features a porous structure with a fibrous material serving as the substrate material, which is sprayed with adsorption material powder. Previous studies have shown that the material with this structure has good total heat exchange performance [9], which can be considered to be a potential high-performance energy wheel core material. In addition, the composite material can be used for air purification, as the sprayed adsorption material can absorb contaminants. As a result, the material is versatile and can be applied in a wide range of building applications, and it is meaningful to study its pressure resistance characteristics.

Studies have been conducted on the pressure drop of fibrous materials [10–17] and granular matrices [18–21], which contain experimental investigations and modeling of the materials' pressure drop. These studies can serve as a reference for the establishment of pressure drop models for porous materials. Liu et al. [10] developed a friction factor correlation of foam matrixes using experimental pressure drop data and found that the friction factor of granular matrixes is far more than that of foam matrixes. Watanabe et al. [11] studied pressure drop and heat transfer in a sintered fibrous porous media. In their study, the friction factor of the heat transfer tube was calculated, and a pressure drop model containing a quadratic function of velocity was built.

In the above research, the influence of various variables on the pressure resistance characteristics of porous materials was explored. For the prediction of pressure drop, it is important to know the relationship between various parameters and the materials' pressure drop. Wang et al. [17] simulated the pressure drop in a fibrous air filtration material and compared it with experimental results. An exponential relationship between the pressure drop and the fiber diameter and porosity of the material was found. Allen et al. [18] investigated the effect of different variables on packed bed pressure drop experimentally. Results showed that when building the pressure drop prediction model, the particle shape, surface roughness, and packing method of the material should be taken into consideration, because these factors have significant effects on pressure drop. Koekemoer and Luckos [19] studied the influence of particle size distribution and material type on the pressure drop of packed beds. In their study, the Ergun equation was modified to predict the pressure drop of packed beds filled with particles of multiple materials, and good prediction results were obtained.

However, most of the existing studies on pressure drop in porous materials have focused on porous material with a single medium, which means that the material does not possess a composite structure. Theoretical and experimental investigations on the pressure drop characteristics of fiber–powder composite materials are relatively rare. Due to both fiber and powder having impacts on the pressure drop of the material, the influence of both of them on the pressure drop needs to be considered when building a pressure drop prediction model.

The present study is aimed at studying the pressure drop in a fiber–powder composite material. Moreover, the study tries to find a high-accuracy pressure drop prediction method suitable for the studied material. Considering the specificity of the material structure, this paper tries to split the prediction of the pressure drop into several steps to accomplish it. As a result, the proposed prediction methods of the material are different from traditional prediction methods for porous material. The study began with an experimental investigation of the pressure drop of the composite material to study the impact of various parameters on the material's pressure drop. After that, a multi-step method for predicting the pressure drop of the material was proposed. The method divides the pressure drop of the material into two parts and predict them separately to increase prediction accuracy. After that, another prediction method with different prediction strategies was proposed; the method predicts the pressure drop of the material using one model but trained in two steps. The accuracy of the two methods was compared and their advantages were discussed. This research provides a reference for pressure drop prediction and the optimal design of energy wheels and air cleaners.

## 2. Methodology

In this section, the pressure drop test system and test method are introduced first. Subsequently, the pressure drop prediction models for the composite material are established, and finally, the prediction steps and methods using the built model are summarized.

### 2.1. Pressure Drop Experimental Setup

An experimental apparatus was utilized to test the air volume flowrate and pressure drop. Figures 2 and 3 show the schematic diagram and physical diagram of the experimental system, respectively. The system consists of an air flowrate measuring section and a pressure drop testing section, and the measured data can be analyzed using a data acquisition system. The details of the testing method of air flowrates and pressure drop are specified in GB/T 14295-2019 [22]. The pressure drop of the material was measured using a differential pressure gauge with a range of 0–1000 Pa. During the experiment, measurement points were established on the up- and down-wind side of the tested material to obtain the pressure drop, which can be calculated using the following equation:

$$\Delta p_t = p_{t,up} - p_{t,down} \tag{1}$$

During the experiment, the air volume flowrate was measured using nozzles. The total air flowrate through the tested material was obtained by summing the flowrates of the individual nozzles. The air velocity could be calculated using the measured air volume flowrate and cross-sectional area. The adjustable range of air velocity in the experiment was 0.2–1.5 m/s. The air volume flowrate of one nozzle and the total air volume flowrate can be calculated using the following equations, which are in accordance with the standard [23]:

$$Q_i = YC_iF_i\sqrt{2\Delta p_N u} \tag{2}$$

$$Q_N = \sum_{i=1}^{n} Q_i \tag{3}$$

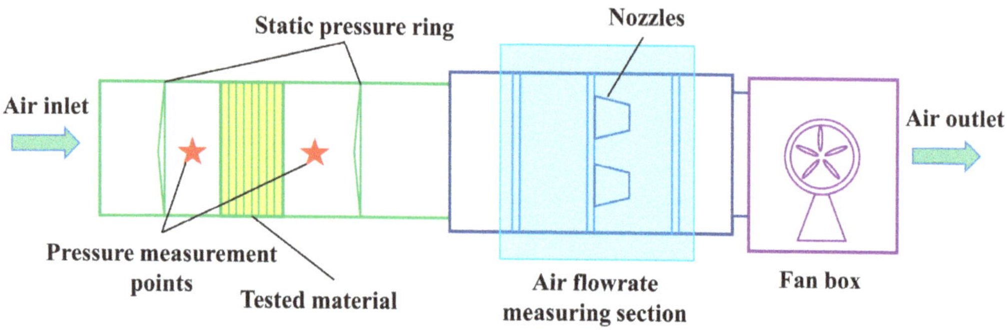


**Figure 2.** Schematic of the pressure drop test system.

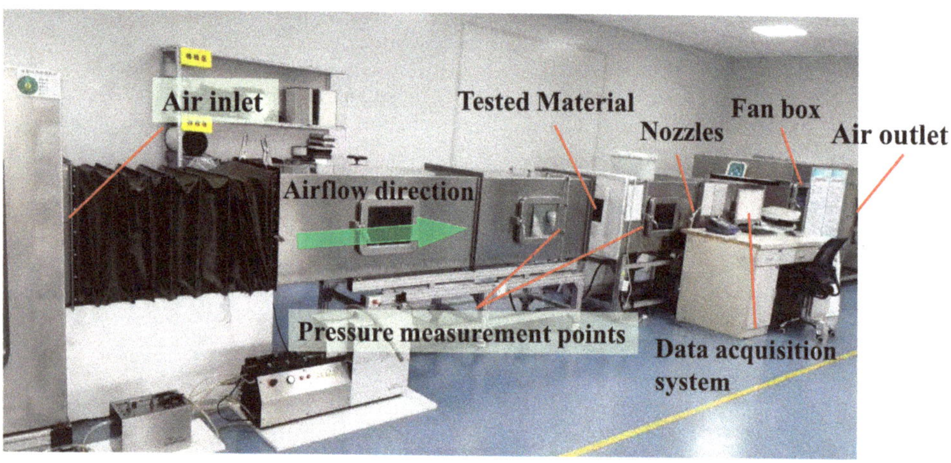


**Figure 3.** Physical map of the pressure drop test system.

Here, $Y$ is the expansion coefficient, where $Y = 0.452 + 0.548\left(1 - \frac{\Delta p_N \times 10^{-3}}{p_N}\right)$; $Q_i$ is the air volume flowrate through the ith nozzle; $Q_N$ is the total air volume flowrate; $F_i$ and $D_i$ are the cross-sectional area and diameter of the ith nozzle, respectively, where $F_i = \frac{\pi D_i^2}{4}$; $C_i$ is the flow coefficient of the $i$th nozzle; and $\Delta p_N$ and $p_N$ are the static pressure difference before and after the nozzle and the air pressure in front of the nozzle, respectively.

The tested material's substrate material is composed of polyester wadding, which has a fibrous structure. The powder material is uniformly sprayed onto the surface of the polyester wadding using a spraying process. After spraying and drying, the powder material adheres to the fiber filaments inside and on the surface of the substrate material. Activated carbon powder is used as the powder material. Considering its porous structure and adsorption characteristics, it can be used as a desiccant material for total heat exchange and an adsorption material for air purification. Figure 4 displays the physical image of the substrate material before and after the activated carbon is sprayed. More details of the tested material including the morphology and adsorption properties can be found in reference [9].

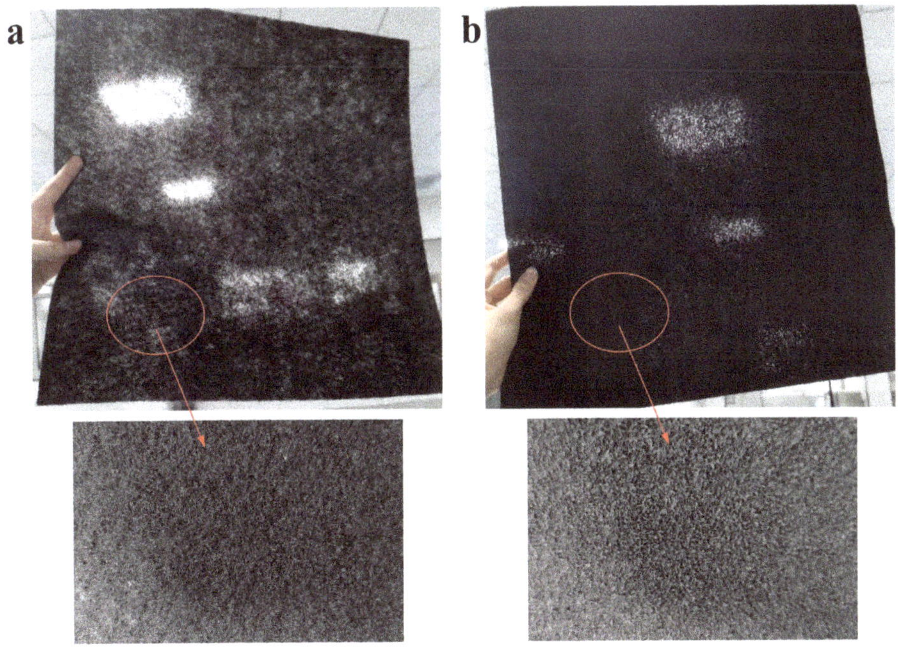


**Figure 4.** The tested material: (**a**) substrate material; (**b**) substrate material with activated carbon.

Four test materials with different amounts of carbon per unit area were prepared. All four test materials were based on the same substrate material. The thickness of the material was adjusted by changing the number of layers during the test, and the pressure drop of the multilayer materials was measured after stacking the multilayer materials together. In the experiment, the airflow passed vertically through the surface of each layer of material. The thickness and mass of a single-layer substrate material were about 3 mm and 0.027 kg, respectively.

*2.2. Pressure Drop Prediction Model for the Composite Material*

The main objective of this section is to establish a pressure drop prediction method capable of predicting material pressure drop under different adsorption material loading and operating conditions while using a fixed substrate. If the parameters of the substrate material are considered, more parameters will be introduced to the pressure drop prediction model. This will significantly increase the amount of experimental data needed to train the model, and the cost of experiment will be greatly increased. Moreover, the increase of parameter amounts will make the accurate prediction much more difficult. As a result, a fixed substrate is used for prediction. As the adsorption material loading increases, the pressure drop of the material increases based on the pressure drop of the substrate. This rise in pressure drop is attributed to the adhesion of adsorption materials. Given the wholly distinct shape and structure of the substrate material and adsorption material, the impact of various parameters on the substrate material pressure drop and the pressure drop increase may be different. As a result, the substrate material pressure drop and the pressure drop increase caused by adsorption materials are predicted separately, and the sum of them is calculated to obtain the total pressure drop of the material.

Figure 4 illustrates that the tested material possesses a fibrous and porous structure. Given the similarity in structure, the empirical equation of foam matrices' friction characteristics was utilized as a reference to establish the pressure drop model. The pressure drop and the friction factor can be calculated as follows [10]:

$$\Delta p = f_f \rho L u^2 \frac{1-\varepsilon}{D_p \varepsilon^3} \tag{4}$$

$$f_f = 22\frac{1-\varepsilon}{Re} + 0.22 \tag{5}$$

Here, *Re* is the Reynolds number, and the relationship between the Reynolds number and other parameters can be expressed as follows:

$$Re = (1-\varepsilon)\frac{D_p \rho u}{\mu} \tag{6}$$

Substituting Equations (5) and (6) into (4) yields

$$\Delta p = \left(22\frac{\mu}{D_p \rho u} + 0.22\right)\rho L u^2 \frac{1-\varepsilon}{D_p \varepsilon^3} \tag{7}$$

It can be seen from Equation (7) that when $\rho$ and $\mu$ are kept constant, $\Delta p$ is mainly related to $L$, $u$, $\varepsilon$, and $D_p$. The void ratio of single layer material can be calculated by the following equation:

$$\varepsilon = 1 - \frac{m_s}{\rho_s V_t} - \frac{m_d}{\rho_d V_t} \tag{8}$$

Here, $m_s$ and $m_d$ are the content of substrate material and adsorption material per unit area, and $V_t$ is the total volume of the material per unit area. Equation (8) shows that with a certain parameter of the substrate material and $V_t$, $\varepsilon$ is only influenced by $m_d$ when $\rho_d$ is considered as constant.

When the parameters of the substrate material are certain, $D_p$ is mainly affected by $m_d$. By increasing the amount of adsorption material, the porosity of the material being tested is reduced, and the size of the pores within the material is also impacted. Therefore, $u$, $L$, and $m_d$ can be considered as the most significant parameters affecting $\Delta p$. Since all three of these variables are easy to obtain, it is convenient to use them to train the model.

In the modeling of the pressure drop increase caused by adsorption materials, the structure of Equation (7) is referenced, and the following considerations are included:

1. In the $\frac{1-\varepsilon}{D_p \varepsilon^3}$ term in Equation (7), $D_p$ and $\varepsilon$ are mainly affected by $m_d$, and the term is positively correlated with $m_d$. To reduce computing expenses, the term is simplified to the form of an exponential function containing $m_d$.
2. The $D_p$ in the $22\frac{\mu}{D_p \rho u}$ term is simplified to the form of an exponential function containing $m_d$.
3. Referring to Equation (7), power functions are used to describe the relationship between $L$ and $\Delta p_d$ as well as the relationship between $u$ and $\Delta p_d$.

In conclusion, the pressure drop increase prediction model is formulated as

$$\Delta p_i = k_1\left(\frac{k_2 \exp(k_3 m_d)}{u} + k_4\right) L^{k_5} u^{k_6} (\exp(k_7 m_d) - 1) \tag{9}$$

Here, $k_i$s are undetermined coefficients which can be obtained through regression; $\mu$ and $\rho$ are considered as constants and merged into $k_i$s.

The prediction methods proposed in this work are based on the substrate material with constant porosity and material parameters. Therefore, the pressure drop of the substrate material is mainly influenced by the air velocity and material thickness.

The substrate material's pressure drop prediction model was developed based on the following equation [24]:

$$\frac{\Delta p}{L} = k_1 u + k_2 u^2 \tag{10}$$

The pressure drop prediction model of the substrate material is formulated as

$$\Delta p_s = (j_1 u + j_2 u^2) L^{j_3} u^{j_4} \tag{11}$$

Here, $j_i$s are undetermined coefficients which can be obtained through regression.

The total pressure drop of the material can be obtained by summing the pressure drop of the substrate material and the pressure drop increase caused by adsorption materials:

$$\Delta p_t = \Delta p_s + \Delta p_i \tag{12}$$

To simplify the modeling process and reduce the experimental data required for modeling, the pressure drop model of the substrate material can be trained first to obtain coefficients $j_i$s, and the coefficient $k_5$ in Equation (9) can be replaced by $j_3$. In this way, when training the prediction model for $\Delta p_i$, the experimental data needed in the modeling can be drastically reduced due to the reduction of one model parameter. Using this approach, the cost of modeling can be reduced significantly.

### 2.3. Pressure Drop Prediction Procedure

The total pressure drop of the tested material can be predicted following the steps below:

A. Obtain the tested data for prediction ($u, L, m_d, \Delta p_s, \Delta p_t$) and calculate $\Delta p_i$.
B. Train the substrate material pressure drop prediction model and obtain the coefficient $j_i$s.
C. Calculate $\Delta p_s$ with the trained model.
D. Train the adsorption material pressure drop prediction model and obtain the coefficients $k_i$s.
E. Calculate $\Delta p_i$ with the trained model and calculate the predicted $\Delta p_t$.

The above prediction process is depicted specifically in Figure 5. The input parameters required for training the model are specified in Table 1.

**Table 1.** Introduction of the model input parameter.

| Parameter | Acquisition Method | Use |
|---|---|---|
| Length of material ($L$) | Measured | Train the $\Delta p_s$ and $\Delta p_i$ prediction model |
| Air velocity ($u$) | Measured | Train the $\Delta p_s$ and $\Delta p_i$ prediction model |
| Desiccant material content ($m_d$) | Measured | Train the $\Delta p_i$ prediction model |
| Substrate material pressure drop ($\Delta p_s$) | Measured | Train the $\Delta p_s$ and $\Delta p_i$ prediction model |
| Material total pressure drop ($\Delta p_t$) | Measured | Train the $\Delta p_i$ prediction model |
| Material pressure drop increase ($\Delta p_i$) | Calculated | Train the $\Delta p_i$ prediction model |

Experimental data are used to train the $\Delta p_s$ and $\Delta p_i$ prediction model, and the parameters shown in Table 1 serve as the model inputs. With the help of the nonlinear regression method, the unknown coefficients $j_i$s and $k_i$s in the model can be obtained. Coefficients $j_i$s are the unknown coefficients of the substrate material pressure drop prediction model, and $k_i$s are the unknown coefficients of the pressure drop increase prediction model. In the prediction process, coefficients $j_i$s and input parameters $u$ and $L$ are used to calculate the predicted substrate material pressure drop, while coefficients $j_3$ and $k_i$s and input parameters $u$, $L$, and $m_d$ are used to calculate the predicted pressure drop increase. The predicted $\Delta p_s$ and $\Delta p_i$ are the outputs of the substrate material pressure drop prediction model and the pressure drop increase prediction model, respectively.

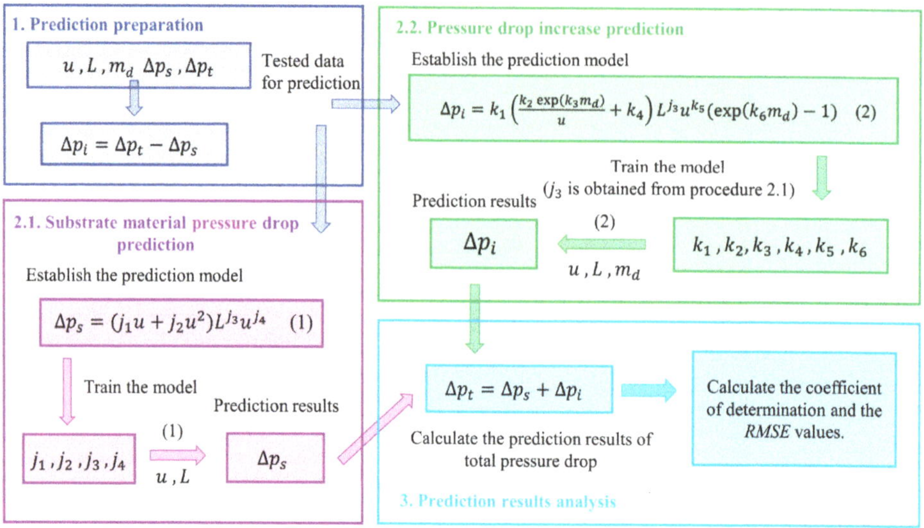


**Figure 5.** Schematic illustration of the pressure drop prediction method.

In the prediction process, the model is trained with the pressure drop data obtained from a series of experiments, which are called training data sets. The difference between the predicted and measured pressure drop results can be utilized to analyze the prediction accuracy of the model. In order to further validate the prediction accuracy of the model, the experimental data, which are called testing data sets, are obtained independently of the training sets. This part of the data can be used to independently validate the prediction accuracy of the model.

## 3. Results and Discussion

### 3.1. Experimental Results of Pressure Drop in Fibrous Core Materials

In this section, the pressure drop of the material was experimentally investigated. The impacts of varying the carbon content, the number of material layers, and the airflow velocity on material pressure drop were investigated. Figure 6 illustrates the variation of pressure drop with head-on air speed for different layers of substrate material.

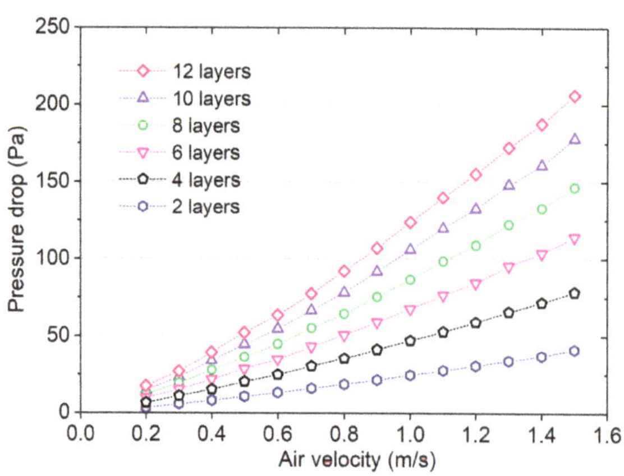


**Figure 6.** Pressure drop of substrate material with different numbers of layers.

Figure 6 shows that the pressure drop of the substrate material increased with the increase of the airflow velocity and the number of layers of the material, and the pressure drop did not vary linearly with the airflow velocity. For example, for the substrate material with 12 layers, the pressure drop increased from 40 Pa to 92 Pa when the airflow velocity increased from 0.4 m/s to 0.8 m/s, and it further increased to 156 Pa when the airflow velocity increased to 1.2 m/s. Figure 7 demonstrates the variation of pressure drop with airflow velocity of the tested material with different adsorption material amounts. The specific parameters of the four tested materials are shown in Table 2. The relative error of adsorption material amount of each layer material for the same material ID was measured to be within ±8% of the mean value. Test results showed that the adhesion amount of adsorption material has a large effect on the pressure drop. The higher the airflow velocity, the greater the influence of the adsorption material amount on the pressure drop. For example, the pressure drop of the material (ID:4) was about 2.6 times of that of the material (ID:1) when the airflow velocity was 1.4 m/s, while the pressure drop of the material (ID:4) was about 2.2 times of that of the material (ID:1) when the airflow velocity was 0.7 m/s. Test results show that when designing energy wheels and air cleaners with the studied material, the adsorption material amount of the material and the air flowrate should not be too high, as these can lead to excessive pressure drops of the devices and thus to high energy consumption of the fan.

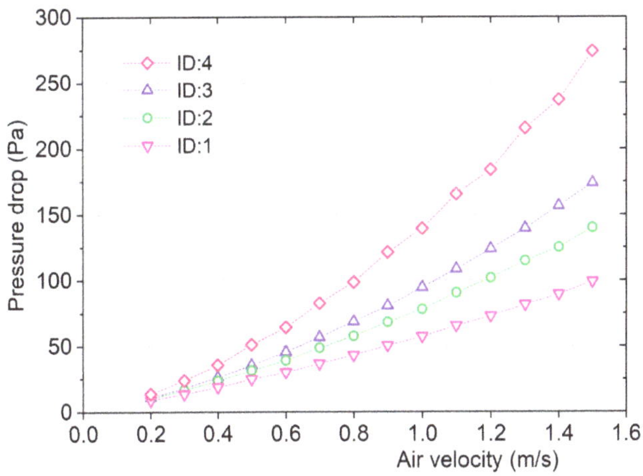


**Figure 7.** Pressure drop of the tested material with different adsorption material amounts.

**Table 2.** Test material parameters in Figure 7.

| Test Material ID (4 Layers) | 1 | 2 | 3 | 4 |
|---|---|---|---|---|
| Layer number of material | 4 | 4 | 4 | 4 |
| Material width W (cm) | 40 | 40 | 40 | 40 |
| Material height H (cm) | 40 | 40 | 40 | 40 |
| Substrate material content (kg) | 0.108 | 0.108 | 0.108 | 0.108 |
| Total content (kg) | 0.186 | 0.289 | 0.343 | 0.409 |

Figure 8 shows the pressure drop of the tested material with different numbers of layers. The specific parameters of each layer of the tested material are shown in Table 3. Figure 8 shows that at a certain airflow velocity, the pressure drop increases with increasing numbers of layers, which shows that the thickness of the material is also an important design parameter of the device. For example, the pressure drop of the single-layer material was 57 Pa when the airflow velocity was 1.4 m/s, while the pressure drop of the 7-layer material was increased to 398 Pa.

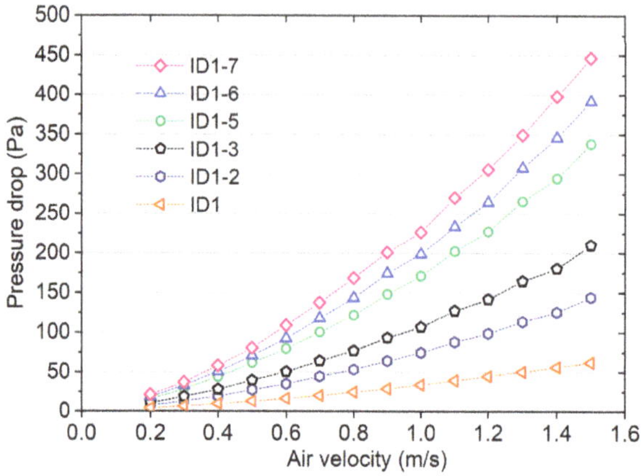


**Figure 8.** Pressure drop of the tested material with different numbers of layers.

**Table 3.** Test material parameters in Figure 8.

| Layer ID (Single Layer) | 1 | 2 | 3 | 4 | 5 | 6 | 7 |
|---|---|---|---|---|---|---|---|
| Material width $W$ (cm) | 40 | 40 | 40 | 40 | 40 | 40 | 40 |
| Material height $H$ (cm) | 40 | 40 | 40 | 40 | 40 | 40 | 40 |
| Substrate material content (kg) | 0.027 | 0.027 | 0.027 | 0.027 | 0.027 | 0.027 | 0.027 |
| Total content (kg) | 0.098 | 0.106 | 0.103 | 0.102 | 0.111 | 0.106 | 0.104 |

### *3.2. Analysis of the Prediction Results*

In this section, the total pressure drop of the material was predicted with the help of the method described in Figure 5, and the predicted results were analyzed. The pressure drop model of the substrate material was trained using the experimental data in Figure 6, while the pressure drop increase model was trained using the experimental data in Figure 7. Additional tests were conducted to verify the prediction accuracy of the model, and they were considered as testing data points. The test results are displayed in Table 4, and the parameters of the tested materials are listed in Table 5. Considering the limited difference in material thickness for different amounts of carbon on the material, the number of layers was used to express the material thickness for the convenience of modeling, and the influence of carbon amount on material thickness was merged into other terms. For multilayer materials, $m_d$ was taken as the average value of this parameter for each layer of material. When training the nonlinear model, the Levenberg-Marquardt method [25] was used. After the model was built, the prediction results were obtained for both the training data points and the testing data points with the help of the built model. Figures 9 and 10 show the comparison of the predicted and tested $\Delta p_i$ and $\Delta p_t$, respectively.

**Table 4.** Testing data sets.

| Data Point ID | $u$(m/s) | $\Delta p_i$(Pa) | $\Delta p_t$(Pa) | Data Point ID | $u$(m/s) | $\Delta p_i$(Pa) | $\Delta p_t$(Pa) |
|---|---|---|---|---|---|---|---|
| 1 | 0.2 | 8.1 | 11.6 | 15 | 0.2 | 6.1 | 18.5 |
| 2 | 0.3 | 11 | 17 | 16 | 0.3 | 10.7 | 30.3 |
| 3 | 0.4 | 14.9 | 23.2 | 17 | 0.4 | 16.3 | 44.4 |
| 4 | 0.5 | 20.4 | 31.3 | 18 | 0.5 | 22.4 | 59.2 |
| 5 | 0.6 | 25.5 | 38.9 | 19 | 0.6 | 30.1 | 75.2 |
| 6 | 0.7 | 32.4 | 48.6 | 20 | 0.7 | 37.6 | 92.9 |
| 7 | 0.8 | 38.8 | 57.6 | 21 | 0.8 | 45.3 | 110.1 |

**Table 4.** *Cont.*

| Data Point ID | $u$(m/s) | $\Delta p_i$(Pa) | $\Delta p_t$(Pa) | Data Point ID | $u$(m/s) | $\Delta p_i$(Pa) | $\Delta p_t$(Pa) |
|---|---|---|---|---|---|---|---|
| 8 | 0.9 | 47.1 | 68.8 | 22 | 0.9 | 55.4 | 131.2 |
| 9 | 1.0 | 54.2 | 79.1 | 23 | 1.0 | 65.7 | 152.6 |
| 10 | 1.1 | 64.7 | 92.6 | 24 | 1.1 | 75.4 | 174.2 |
| 11 | 1.2 | 72.8 | 103.8 | 25 | 1.2 | 88.7 | 197.9 |
| 12 | 1.3 | 84.5 | 118.7 | 26 | 1.3 | 100 | 222.6 |
| 13 | 1.4 | 93.6 | 130.9 | 27 | 1.4 | 115.5 | 248.7 |
| 14 | 1.5 | 108.5 | 150 | 28 | 1.5 | 128.3 | 275.2 |

**Table 5.** Test material parameters of testing data sets.

| Data Point ID | 1–14 | 15–28 |
|---|---|---|
| Layer number of material | 2 | 8 |
| Material width W (cm) | 40 | 40 |
| Material height H (cm) | 40 | 40 |
| Substrate material content (kg) | 0.054 | 0.216 |
| Total content (kg) | 0.204 | 0.622 |

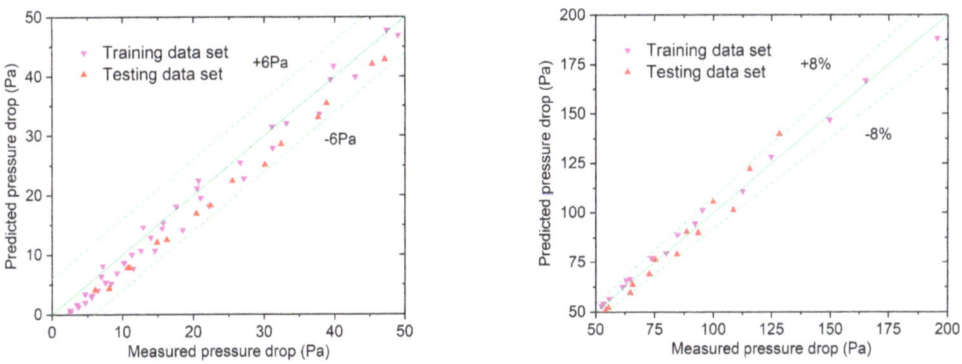


**Figure 9.** Comparison of the predicted and tested $\Delta p_i$ (**Left**: pressure drop range of 0–50 Pa; **Right**: pressure drop range of 50–200 Pa).

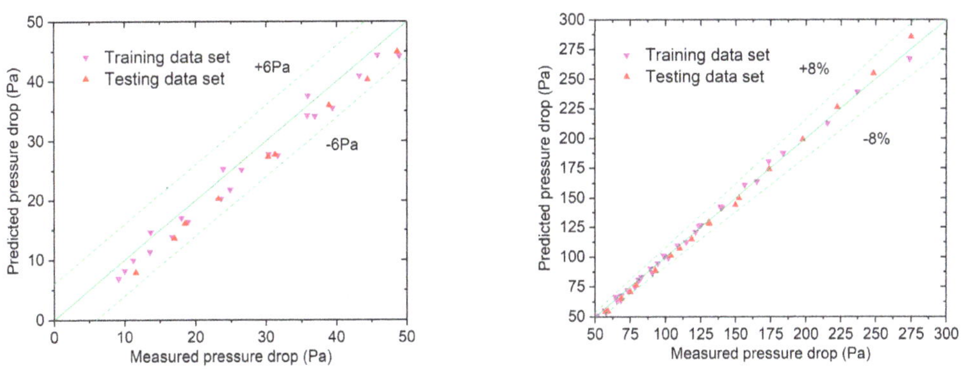


**Figure 10.** Comparison of the predicted and tested $\Delta p_t$ (**Left**: pressure drop range of 0–50 Pa; **Right**: pressure drop range of 50–300 Pa).

It can be seen from Figures 9 and 10 that when the pressure drop is less than 50 Pa, the absolute error of prediction can be controlled within ±6 Pa for both training sets and testing sets. When the pressure drop is greater than 50 Pa, the relative error of prediction is less than ±8% for most data sets. These data show that the prediction accuracy of the model is acceptable, and it is much higher than that of the previous study on porous materials [6].

The root mean square error (*RMSE*) and mean absolute percentage error (*MAPE*) were used to analyze the prediction accuracy of the model. These two indicators were used because *RMSE* can be used to reflect the absolute prediction error, and *MAPE* can be used to indicate the relative prediction error, and they can provide a comprehensive picture of prediction accuracy. They can be calculated using the following equations:

$$RMSE = \sqrt{\frac{\sum_{i=1}^{n}\left(x_{p,i} - x_{e,i}\right)^2}{n}} \tag{13}$$

$$MAPE = \frac{1}{n}\sum_{i=1}^{n}\left|\frac{x_{p,i} - x_{e,i}}{x_{e,i}}\right| \tag{14}$$

Here, $x$ is the pressure drop value, and subscripts p and e represent the predicted pressure drop and the tested pressure drop, respectively. Since training sets and testing sets are independent data sets, it is better to analyze the predictions results of them separately. As a result, the *RMSE* and *MAPE* values of training sets and testing sets were calculated separately. The calculation results are listed in Table 6.

**Table 6.** *RMSE* and *MAPE* values of the prediction model.

| Parameters | $\Delta p_t$ |
|---|---|
| *RMSE* training sets | 2.7 Pa |
| *RMSE* testing sets | 4.0 Pa |
| *MAPE* training sets | 5.2% |
| *MAPE* testing sets | 6.6% |

The absolute error of the predicted pressure drop is defined by

$$AE = x_p - x_e \tag{15}$$

Table 6 shows that the *RMSE* and *MAPE* values of training sets for $\Delta p_t$ are less than 3 Pa and 6%, and the *RMSE* values of testing sets are no more than 4 Pa and 7%.

### 3.3. Further Discussion of the Pressure Drop Prediction Method

The key idea of the method proposed in Section 2.3 is to divide the total pressure drop of the material into two separate parts for prediction. In an effort to simplify the prediction process, an alternative total pressure drop prediction method is proposed, and its prediction accuracy is compared with the accuracy of the method mentioned in Section 2.3. The method attempts to predict the total pressure drop directly on the basis of the substrate material pressure drop. First, the pressure drop prediction model of the substrate material is obtained using the method proposed in Section 2.3, after which the total pressure drop prediction model is directly established based on the pressure drop prediction model for the substrate material. In this way, the prediction process can be simplified.

Test results in Section 3.1 shows that the total pressure drop of the material increases with an increase in the amount of adsorption material, based on the substrate material pressure drop, and that the pressure drop rises with higher amounts of adsorption material. The total pressure drop model is simplified by multiplying the substrate material pressure drop model by a factor that contains the adsorption material amount. Consequently, the following total pressure drop prediction model is established:

$$\Delta p_t = (j_1 u + j_2 u^2) L^{j_3} u^{j_4} \left(1 + k_1 {m_d}^{k_2}\right)\left(1 + k_3 {m_d}^{k_4}\right) \tag{16}$$

Here, $j_i$s and $k_i$s are undetermined coefficients which can be obtained via regression.

To account for the potential difference in the impact of air velocity on the pressure drop characteristics of the substrate material and the material sprayed with adsorption materials, an additional term containing the air velocity is added in Equation (17). In this way, the relationship between airflow velocity and total pressure drop is determined through two regression processes. The modified prediction model of the total pressure drop can be expressed as follows:

$$\Delta p_t = (j_1 u + j_2 u^2) L^{j_3} u^{j_4} \left(1 + k_1 {m_d}^{k_2}\right)\left(1 + k_3 {m_d}^{k_4}\right) u^{k_5} \quad (17)$$

Using this method, the total pressure drop of the tested material can be predicted follow the steps below:

A. Obtain the tested data for prediction ($u, L, m_d, \Delta p_s, \Delta p_t$).
B. Train the substrate material pressure drop prediction model and obtain the coefficient $j_i$s.
C. Calculate $\Delta p_s$ with the trained model.
D. Train the total pressure drop prediction model and obtain the coefficient $k_i$s.
E. Calculate the predicted $\Delta p_t$ with the trained model.

The above prediction process is depicted specifically in Figure 11.

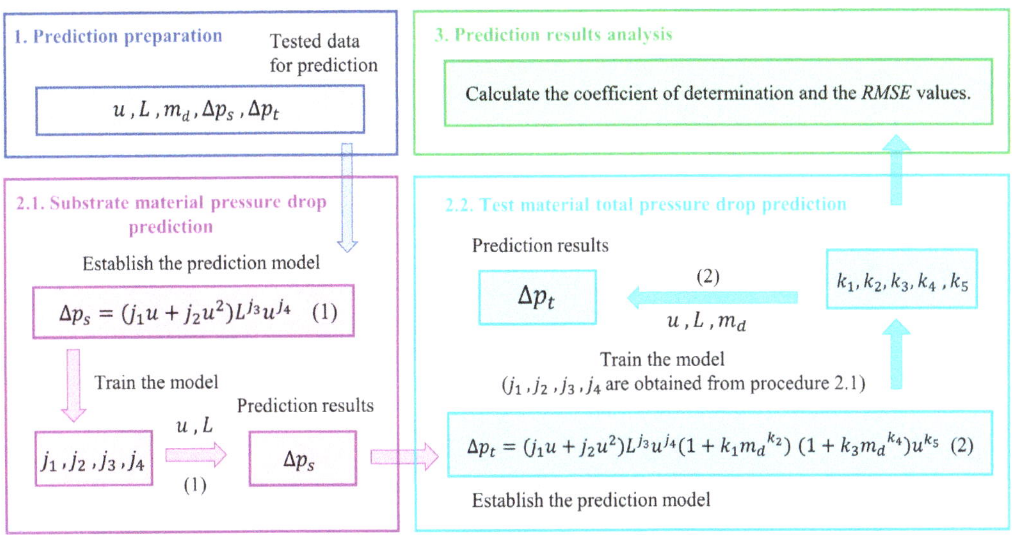


**Figure 11.** Schematic illustration of the prediction method proposed in Section 3.3.

The total pressure drop of the material was predicted using the method described in this Section. The training data points and the testing data points are the same as those used in Section 3.2. To study the effect of model modification on the improvement of prediction accuracy, the total pressure drop was predicted using Equations (16) and (17), respectively. The prediction results of $\Delta p_t$ using Equations (16) and (17), are shown in Figures 12 and 13, respectively, and Table 7 shows the calculated results of the accuracy indices of the method proposed in this section using Equations (16) and (17).

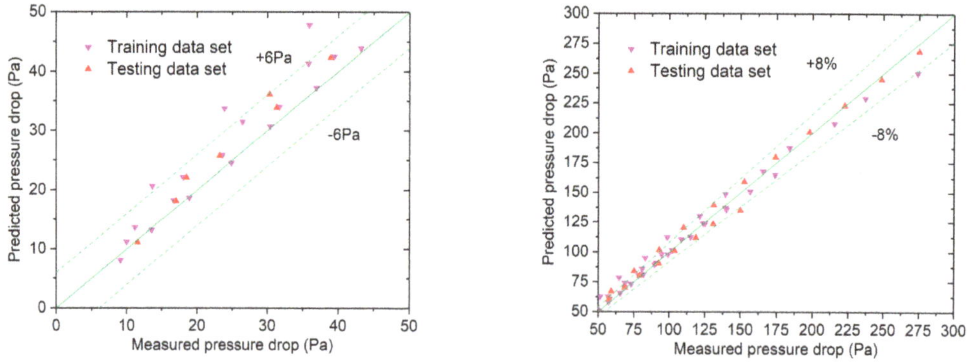


**Figure 12.** Comparison of the predicted and tested $\Delta p_t$ using the method proposed in Section 3.3 with Equation (16) (**Left**: pressure drop range of 0–50 Pa; **Right**: pressure drop range of 50–300 Pa).

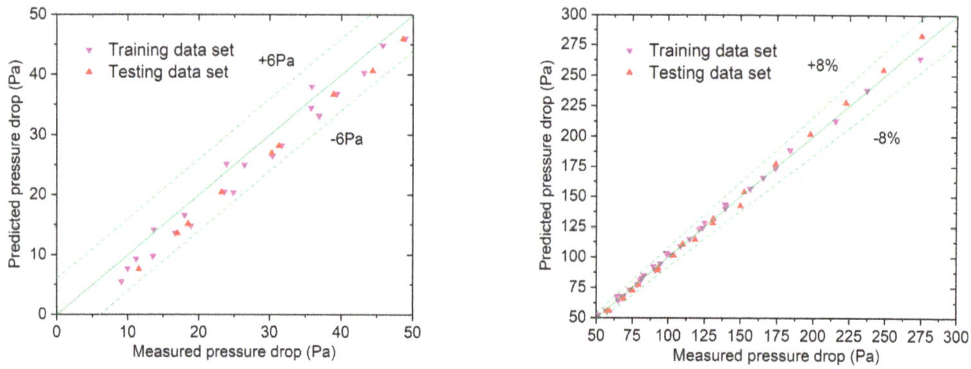


**Figure 13.** Comparison of the predicted and tested $\Delta p_t$ using the method proposed in Section 3.3 with Equation (17) (**Left**: pressure drop range of 0–50 Pa; **Right**: pressure drop range of 50–300 Pa).

**Table 7.** *RMSE* and *MAPE* values of the method proposed in Section 3.3.

| Parameters | $\Delta p_t$ | $\Delta p_t$ (After Modification) |
|---|---|---|
| *RMSE* training sets | 6.3 Pa | 2.8 Pa |
| *RMSE* testing sets | 6.0 Pa | 3.6 Pa |
| *MAPE* training sets | 8.2% | 5.9% |
| *MAPE* testing sets | 7.1% | 6.1% |

It can be seen from Figures 12 and 13 that the prediction accuracy of the method proposed in this section using Equation (17) was obviously higher than that using Equation (16). Moreover, Table 7 shows that after integrating the modified model, as expressed by Equation (17), the *RMSE* and *MAPE* values of both training sets and testing sets decreased significantly. For example, the *MAPE* value of training sets decreased from 8.2% to 5.9%, and the *MAPE* value of testing sets decreased from 7.1% to 6.1%. This indicates that determining the relationship between airflow velocity and total pressure drop through two regression processes helps to mitigate prediction errors effectively. Figures 10 and 13 demonstrate that the proposed method, utilizing the modified model, exhibits a similar level of prediction accuracy as that in Section 2.3. For example, when the pressure drop is below 50 Pa, the absolute prediction error can be controlled within ±6 Pa, and when the pressure drop is greater than 50 Pa, the relative prediction error is within ±8% for most data sets.

Figure 14 depicts the absolute prediction error of two distinct prediction methods. Figure 14 clearly illustrates that the range of absolute prediction error using Equation (17) is noticeably smaller than that using Equation (16). The absolute prediction error of testing sets using Equation (17) ranges from −7.6 Pa to 7.6 Pa, while that using Equation (16) ranges from −14.8 Pa to 10.3 Pa. The primary advantage of this method is its reduced number of prediction steps compared to that in Section 2.3. However, it should be noted that the predicted total pressure drop in this method is based on the predicted results of the substrate material pressure drop. Consequently, this approach is more reliant on the predicted data of the substrate material pressure drop.

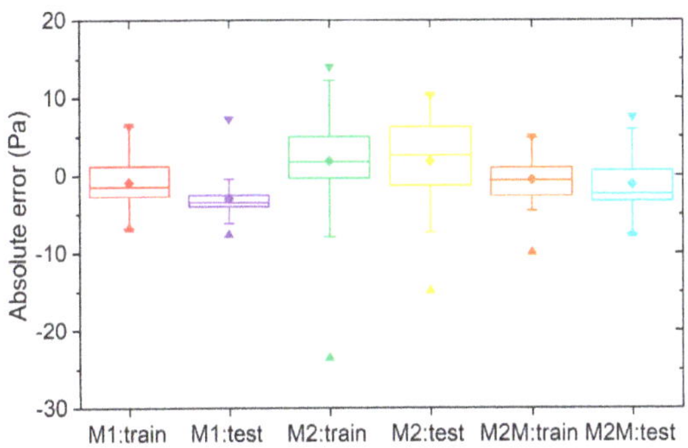


**Figure 14.** Comparison of the absolute prediction error of two different prediction methods (M1: the method proposed in Section 2.3; M2: the method proposed in Section 3.3 using Equation (16); M2M: the method proposed in Section 3.3 using Equation (17)).

## 4. Conclusions and Future Work

This study investigated the pressure drop in a fiber–powder composite material. Referring to the fiber–powder structure of the material, two pressure drop prediction methods with different prediction strategies were proposed, and their prediction accuracies were analyzed. Both methods take into account the differences in pressure drop characteristics between the substrate and composite materials. As verified by experimental data, both proposed methods demonstrated acceptable prediction results. Specifically, the absolute prediction error of them was within ±6 Pa when the pressure was below 50 Pa, and the relative prediction error was within ±8% for the majority of data points when the pressure exceeded 50 Pa. Each method has its advantages; the first proposed method is less reliant on the predicted results of the substrate material pressure drop, while the second proposed method involves fewer prediction steps.

The proposed prediction methods were validated within certain ranges. For example, the airflow velocity ranged from 0.2 m/s to 1.5 m/s, and the number of material layers ranged from 2 layers to 8 layers. In the future, more test data will be used to further verify the model and to explore whether its applicability can be expanded. Prediction results in Section 3.3 show that determining the relationship between airflow velocity and total pressure drop through two steps instead of one step helps to reduce prediction errors, and as a result, more studies on the optimization of the prediction process can be conducted in the future to further improve the prediction method.

**Author Contributions:** Conceptualization, H.G., Z.L. and X.Z.; Methodology, H.G., Z.L. and X.Z.; Supervision, Z.L and X.Z.; Formal analysis, H.G.; Investigation, X.Y. and M.S.; Data curation, H.G., X.Y. and M.S.; Writing—original draft, H.G. All authors have read and agreed to the published version of the manuscript.

**Funding:** This research received no external funding.

**Data Availability Statement:** The data presented in this study are available on reasonable request from the corresponding author.

**Conflicts of Interest:** The authors declare no conflict of interest.

## Nomenclature

| | |
|---|---|
| $C$ | Nozzle flow coefficient |
| $D$ | Nozzle diameter (m) |
| $D_p$ | Equivalent spherical diameter of porous media (m) |
| $f_f$ | Friction factor |
| $F$ | Cross-sectional area of nozzle (m$^2$) |
| $L$ | Length of material (m) |
| $m$ | Material content (kg/m$^2$) |
| $p$ | Air pressure (Pa) |
| $Q$ | Air volume flowrate (m$^3$/s) |
| $Re$ | Reynolds number |
| $u$ | Airflow velocity (m/s) |
| $V$ | Material volume (m$^3$) |
| $Y$ | Expansion coefficient |
| *Greek symbols* | |
| $\varepsilon$ | Porosity |
| $\mu$ | Viscosity (Pa.s) |
| $\rho$ | Fluid density (kg/m$^3$) |
| *Subscripts* | |
| $d$ | Adsorption material |
| $e$ | Tested data |
| $p$ | Predicted data |
| $s$ | Substrate material |
| $t$ | Tested material |

## References


1. Antonellis, S.D.; Intini, M.C.; Joppolo, M.; Pedranzini, F. Experimental analysis and practical effectiveness correlations of enthalpy wheels. *Energy Build.* **2014**, *84*, 316–323. [CrossRef]
2. Men, Y.; Liu, X.; Zhang, T. Frost prevention research of the enthalpy wheel in air conditioning systems and industrial heat recovery systems. *Build. Environ.* **2022**, *222*, 109428. [CrossRef]
3. Mioralli, P.C.; Ganzarolli, M.M. Thermal analysis of a rotary regenerator with fixed pressure drop or fixed pumping power. *Appl. Therm. Eng.* **2013**, *52*, 187–197. [CrossRef]
4. Dallaire, J.; Gosselin, L.; Silva, A. Conceptual optimization of a rotary heat exchanger with a porous core. *Int. J. Therm. Sci.* **2010**, *49*, 454–462. [CrossRef]
5. Harshe, Y.M.; Utikar, R.P.; Ranade, V.V.; Pahwa, D. Modeling of rotary desiccant wheels. *Chem. Eng. Technol.* **2005**, *28*, 1473–1479. [CrossRef]
6. Zhang, R.; Li, Z.; Zeng, L.; Wang, F. Pressure drop in honeycomb adsorption filters filled with granular activated carbon. *Powder Technol.* **2021**, *393*, 550–558. [CrossRef]
7. Zhang, R.; Li, Z.; Zeng, L.; Wang, F. Prediction of pressure drop in adsorption filter using friction factor correlations for packed bed. In *Proceedings of the International Symposium on Heating, Ventilation and Air Conditioning*; Springer: Singapore, 2019.
8. Liu, M.; Claridge, D.E.; Deng, S. An air filter pressure loss model for fan energy calculation in air handling units. *Int. J. Energy Res.* **2003**, *27*, 589–600. [CrossRef]
9. Gao, H.; Li, Z.; Zhou, X.; Yin, X. Experimental study of a LiCl-modified fibrous core material for energy wheels. *Energy Build.* **2022**, *275*, 112449. [CrossRef]
10. Liu, J.F.; Wu, W.T.; Chiu, W.C.; Hsieh, W.H. Measurement and correlation of friction characteristic of flow through foam matrixes. *Exp. Therm. Fluid Sci.* **2006**, *30*, 329–336. [CrossRef]
11. Watanabe, R.; Kobayashi, T.; Otomo, Y.; Akisawa, A.; Ueda, Y.; Enoki, K. Experimental investigation into the heat transfer and pressure drop performance of sintered high porosity media. *Appl. Therm. Eng.* **2021**, *196*, 117284. [CrossRef]
12. Mancin, S.; Zilio, C.; Cavallini, A.; Rossetto, L. Pressure drop during air flow in aluminum foams. *Int. J. Heat Mass Transfer* **2010**, *53*, 3121–3130. [CrossRef]
13. Fourie, J.G.; Du Plessis, J.P. Pressure drop modelling in cellular metallic foams. *Chem. Eng. Sci.* **2002**, *57*, 2781–2789. [CrossRef]

14. Dukhan, N. Correlations for the pressure drop for flow through metal foam. *Exp. Fluids* **2006**, *41*, 665–672. [CrossRef]
15. Du Plessis, J.P.; Montillet, A.; Comiti, J.; Legrand, L. Pressure predictions for flow through high porosity metallic foams. *Chem. Eng. Sci.* **1994**, *49*, 3545–3553. [CrossRef]
16. Wang, H.; Guo, L. Experimental investigation on pressure drop and heat transfer in metal foam filled tubes under convective boundary condition. *Chem. Eng. Sci.* **2016**, *155*, 438–448. [CrossRef]
17. Wang, X.; Kim, K.; Lee, C.; Kim, J. Prediction of air filter efficiency and pressure drop in air filtration media using a stochastic simulation. *Fibers Polym.* **2008**, *9*, 34–38. [CrossRef]
18. Allen, K.G.; Von Backström, T.W.; Kröger, D.G. Packed bed pressure drop dependence on particle shape, size distribution, packing arrangement and roughness. *Powder Technol.* **2013**, *246*, 590–600. [CrossRef]
19. Koekemoer, A.; Luckos, A. Effect of material type and particle size distribution on pressure drop in packed beds of large particles: Extending the ergun equation. *Fuel* **2015**, *158*, 232–238. [CrossRef]
20. Oschmann, K.T.; Wirtz, S.; Kruggel-Emden, H. Pressure drop investigations in packings of arbitrary shaped particles. *Powder Technol.* **2015**, *271*, 109–124.
21. Cerantola, D.J.; Lane, C.D. The existence of universal pressure loss and heat transfer correlations for packed beds. *Appl. Therm. Eng.* **2022**, *212*, 118468. [CrossRef]
22. *GB/T 14295-2019*; Air Filter. Standardization Administrastion of the People's Republic of China: Bejing, China, 2019.
23. *GB/T 37292-2019*; Air Enthalpy Testing Facility For Test—General Technical Requirements. Standardization Administrastion of the People's Republic of China: Bejing, China, 2019.
24. Forchheimer, P. Wasserbewegung durch boden. *Zeit. Ver. Deutsch. Ing.* **1901**, *45*, 1781–1788.
25. Sauer, T. *Numerical Analysis*, 2nd ed.; China Machine Press: Beijing, China, 2014.


**Disclaimer/Publisher's Note:** The statements, opinions and data contained in all publications are solely those of the individual author(s) and contributor(s) and not of MDPI and/or the editor(s). MDPI and/or the editor(s) disclaim responsibility for any injury to people or property resulting from any ideas, methods, instructions or products referred to in the content.

*buildings* MDPI

*Article*

# Theoretical and Experimental Study of Positive-Pressure Condensation Heat and Mass Transfer Processes in Bent-Tube Heat Exchangers

**Jiaming Xing and Qing Cheng ***

School of Energy Science and Engineering, Nanjing Tech University, Nanjing 211816, China; xingjiaming@njtech.edu.cn
* Correspondence: chengqingny@njtech.edu.cn; Tel.: +86-137-7061-8138

**Abstract:** Condensation dehumidification is currently the mainstream means of dehumidification, and the idea is to precipitate moisture by cooling the air below the dew point temperature; however, this process requires the use of a chiller to provide a low-temperature cooling source, which triggers reheat losses. By positive-pressure condensation, the dew point temperature can be increased, thereby increasing the cooling source temperature. In this paper, the dehumidification process in the bent-tube heat exchanger is investigated theoretically and experimentally. The bent-tube heat exchanger efficiently removes moisture from the air and increases the dehumidification efficiency through positive-pressure condensation. Experiments on positive-pressure condensation and dehumidification were conducted at varying pressures, with the results demonstrating that the model's accuracy is within ±17%. As the fluid flow rate and pipe diameter rise, so do the dehumidification capacity and heat transfer coefficient. Furthermore, the findings show that the air humidity after dehumidification drops from 16.2 g/kg to 12.9 g/kg, meaning it is just over half of the value at atmospheric pressure, within the pressure that ranges from 100 kPa to 800 kPa. Increasing pressure enhances the heat transfer coefficient, while increasing humidity exacerbates this effect. With a 20% increase in wet air humidity, the heat transfer coefficient varies between 18% and 37%.

**Keywords:** condensation and dehumidification; heat and mass transfer; positive pressure; bent-tube heat exchanger

## 1. Introduction

With the rapid development of society, carbon emissions have increased. The increase in carbon emissions has led to global warming, causing major changes in the Earth's climate system, sea-level rise, and ecological damage. Reducing the carbon footprint is key to achieving global sustainable development goals and combating global warming, extreme weather, and ecosystem collapse [1].

Energy consumption and carbon emissions are increasing dramatically in industrial production and everyday life. Projections show that by 2040, global primary energy consumption will be five times higher than in 1970 and $CO_2$ emissions will be 1.5 times higher than in 2017 (2200% increase from 1900) [2]. As an essential aspect of air conditioning, dehumidification plays a crucial role in creating a healthy and comfortable living environment, as well as ensuring effective industrial humidity control [3,4]. In industries such as lithium battery manufacturing, paint shops, injection molding shops, semiconductor manufacturing, and pharmaceuticals, where maintaining low- and ultra-low-humidity environments is critical to the process, reducing the energy consumption of dehumidification is even more critical [5].

The current common dehumidification methods are solution dehumidification, membrane dehumidification, and condensation dehumidification. Solution dehumidification [6] involves mostly corrosive liquid that needs to be regularly replaced; there is a high cost of equipment maintenance, after the dehumidification solution of the air may be attached to the liquid. In the membrane dehumidification [7], in the relative cost of dehumidification membrane, put into use, the operation process of low energy consumption, and dehumidification membrane in the long term after the work, there will be a reduction in permeability flux, leading to a decline in dehumidification efficiency. Condensation dehumidification [8] is not suitable for deep dehumidification; the equipment has a higher cost and energy consumption, making it unsuitable for places where the temperature is too high or too low. In addition, maintenance is more difficult.

Nowadays, condensation dehumidification is a commonly used method [9]. The atmospheric condensation dehumidification process relies on a chiller to generate a low-temperature cooling source, followed by a reheating step [10]. As a result, energy losses are prevalent in the atmospheric pressure condensation dehumidification process, which in turn leads to high energy consumption [11]. During deep dehumidification operations, the dew point temperature drops significantly, sometimes below 0 °C, a phenomenon that can lead to equipment frosting and a consequent reduction in energy efficiency.

The efficiency of condensation and dehumidification of humid air is significantly influenced by the heat exchanger's surface structure [12]. Different surface wettability characteristics—such as hydrophilic, hydrophobic, and heterogeneous surfaces—exhibit unique behaviors, which, in turn, significantly impact the effectiveness of the condensation process [13].

The hydrophilicity and hydrophobicity of the heat exchanger surface directly determine the behavior and efficiency of water vapor condensation and the discharge of condensate, thus affecting condensation and dehumidification performance. Both hydrophilic and hydrophobic surfaces affect the dehumidification efficiency of a heat exchanger. Vivekh et al. [14] coated a finned tube heat exchanger with a polymer desiccant to test its dehumidification and thermal efficiency and assessed energy savings when combined with a chiller under different climatic conditions. Indrawan et al. [15] explored the changes in the thermophysical properties of ionic liquids in desiccant air conditioning systems, particularly the changes in humidity ratio during dehumidification and regeneration. Liu et al. [16] investigated heat exchangers with finned tubes coated with desiccant and discovered that lower cooling water temperatures decreased the humidity ratio of the output air, while higher-quality desiccant enhanced dehumidification. Yang et al. [17] created a desiccant-coated heat pump system for humidification and dehumidification, testing it under summer and winter conditions and analyzing performance at various compressor frequencies and cycles. Myoung et al. [18] studied condensation in vertical channels with hydrophilic–hydrophobic patterns, analyzed the effects of pattern parameters on thermal performance and pressure loss, and derived a relationship between flow rate and droplet motion angle. Wang et al. [19] explored vapor condensation on nanoscale hydrophilic dot substrates through molecular dynamics simulations. Larger spots accelerate nucleation, while higher spot density enhances nucleation but slows cluster growth. Mostafa et al. [20] explored the effects of hydrophilic, hydrophobic, and mixed surfaces on vapor condensation, aiming to enhance heat transfer by combining both properties on a single surface. Wan et al. [21] studied heat transfer in finned tubes with hydrophilic/hydrophobic surfaces, finding that external hydrophilic surfaces reduce nucleation barriers, while internal hydrophobic surfaces enhance droplet condensation.

Furthermore, the heat exchanger structure plays a crucial part in the condensation and dehumidification process. Ahmadi et al. [22] explored the role of ambient temperature and fin spacing in the dehumidification process of finned-tube heat exchangers. They observed that higher ambient temperatures promoted a faster dehumidification process, while increasing the fin spacing somehow optimized the heat transfer efficiency. They found that higher ambient temperatures accelerate dehumidification, while increased fin spacing improves heat transfer up to a certain limit. Li et al. [23] investigated the heat and mass transfer mechanisms of a five-row tube-and-fin heat exchanger in a dehumidification scenario, in particular analyzing the effect of water vapor mass fraction and inlet velocity on this process. Wang et al. [24] studied the corrugated finned-tube heat exchangers with large-diameter tubes under dehumidification conditions, and the results showed that their heat transfer performance was better than that of dry conditions, especially when the tube rows were larger or the fin spacing was smaller. Zhuang et al. [25,26] developed a numerical model to predict condensate droplet behavior on heat exchanger fins under dehumidification, including droplet formation, growth, and movement. The model validated by experiments accurately predicts droplet behavior and matches experimental heat and mass transfer coefficients. Koca [27] proposed and experimentally studied a hybrid refrigeration system combining a condensing panel and a radiant refrigeration system. The dehumidification performance was analyzed by measuring radiation, convection, and condensation heat transfer coefficients. Zong et al. [28] used theoretical analyses and practical experiments to evaluate the operating performance of a heat exchanger for dehumidification and heating in a greenhouse.

By raising the pressure, the dew point temperature of humidity can be increased, thus increasing condensation efficiency, simplifying the system, and reducing energy consumption. Bent-tube heat exchangers can improve heat transfer efficiency and space-saving, improve fluid distribution uniformity, enhance seismic, and have the ability to withstand high-temperature and high-pressure conditions. Hence, bent-tube heat exchangers in many complex conditions have irreplaceable advantages, especially suitable for high-performance industrial heat transfer dehumidification.

This study focuses on an in-depth study of the process of condensation and dehumidification under positive pressure. The study mainly focuses on the process of condensation and dehumidification under positive pressure to carry out in-depth research. Using experimental and theoretical analyses, the positive-pressure condensation process of wet air in the bent-tube heat exchanger under different operating conditions and tube diameter conditions is investigated, with the aim of providing a theoretical basis for the improvement of the heat transfer performance of wet air.

## 2. Methods

### 2.1. Modeling of Positive-Pressure Condensation

#### 2.1.1. Physical Modeling

It is expected that the velocity and pressure inside the enclosure will not change. As shown in Figure 1, a model of wet air flow is depicted. With an overall length of L = 300 mm, the aluminum casing wall creates a heat exchanger with an internal tube diameter of D1 = 10 mm and an external tube diameter of D2 = 20 mm. The calculation region refers to the region where humid air flows in the inner tube wall versus the region where low-temperature cooling water flows in the outer tube wall. In Figure 1, a model of the bent-tube heat exchanger is displayed.

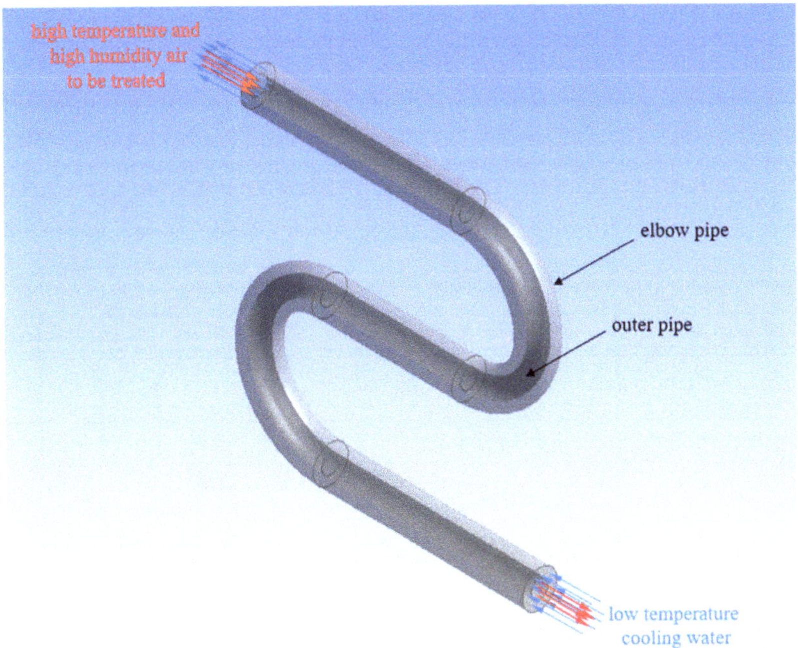


**Figure 1.** Bent-tube heat exchanger model.

Pressure selection for positive-pressure condensing and dehumidifying heat exchangers: Positive-pressure condensing and dehumidifying heat exchangers are often used as part of industrial processes as well as in domestic living, where too high a pressure may lead to equipment damage or safety risks. Positive-pressure condensing and dehumidifying heat exchangers need to be at a certain pressure in order to perform heat transfer and dehumidification operations effectively, and the material and manufacturing costs of the equipment may increase as the operating pressure increases. Therefore, material availability and cost considerations need to be taken into account when selecting the operating pressure range. Controlling the pressure range within an appropriate range will result in more stable and reliable operation of the equipment, too high a pressure will result in unstable operation or degradation of performance. The pressure range was ultimately controlled to be 100–1500 kPa, which is more suitable for condensation dehumidification scenarios.

In our practical experiments, the gas–liquid separator helps to extract the liquid from the air efficiently and ensures maximum separation of moisture. The positive-pressure condensation dehumidification system is shown in Figure 2, which consists of a compressor, an expander, an air–air heat exchanger, a dehumidifier, a cooling unit, and an electric motor, which are connected to each other by a common shaft between the compressor and the expander. The process starts when air enters the compressor from the user end to increase its pressure and temperature. The pressurized air is then initially cooled in an air-to-air heat exchanger. The air is then further cooled by passing through a cooling unit, causing condensed moisture to accumulate and be released into a dehumidifier. The air then flows through an expander, resulting in a reduction in pressure and temperature. The energy released by the expander cancels out the energy required for compression via coaxial transmission. Eventually, the cooler air leaving the expander enters the hotter portion of the air-to-air heat exchanger, where it is heated to the desired temperature. This produces air with adjustable humidity and temperature, which is subsequently delivered to the user. Ensuring air quality is a key factor when working with compressed air systems, especially in sensitive environments. To address this, we would like to emphasize the importance of incorporating appropriate air treatment components such as oil filters, air dryers, and particulate filters into dehumidification systems. These effectively remove oil and other contaminants from compressed air before it is released into the environment.

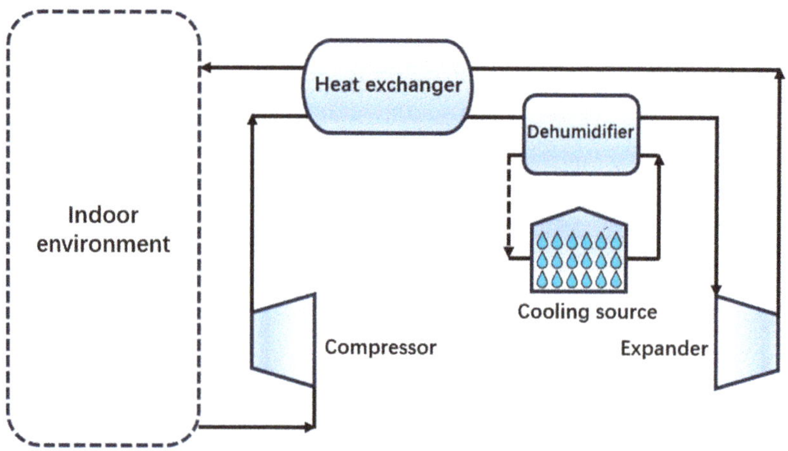


**Figure 2.** Positive-pressure condensation dehumidification system.

The damp air and the cold wall surface undergo a simultaneous phase shift process that interacts with one another. The significant change in the physical properties of water vapor during the transformation of gaseous water into liquid water complicates the modeling of the process, and this challenge must be met by simplifying the actual phenomenon. During condensation, water vapor cools to its dew point and turns into liquid water, releasing latent heat. This heat is absorbed by the surrounding air but causes only a small temperature change. Instead of significantly warming the air, the latent heat primarily affects the humidity, with minimal impact on the air's temperature. The hypotheses of this research are as follows:

(1) Air and water vapor are combined to form wet air; a single-phase, multi-component fluid that is considered an indestructible gas combination.
(2) The physical parameters of the wet air may be considered to remain constant during the condensation procedure in the theoretical analysis because the temperature of the wet air varies very little throughout the condensation procedure and because the proportion of the volume of water vapor in the wet air is extremely low [29–31].

2.1.2. Heat and Mass Transfer Model

The wet air phase change heat transfer process studied in this paper needs to be realized by the component transport model, which includes the combined transport model, premixed combustion, multiple component transport, and so on. The premixed combustion model is suitable for the process of combustion reaction; it is obvious that this paper is not suitable for this model. The combination of transport models is suitable for the process of ignition, melting, etc., and is not suitable for this paper's model. The transport of various components is suitable for the process of phase change, suitable for this paper's model.

The multicomponent transport model allows for a more accurate representation of systems, where multiple species are present and interact, such as in our study. It accounts for the different diffusivities, interaction effects, and concentration gradients of the components, which are critical for modeling complex transport phenomena (heat and mass transfer in humid air or mixed gas flows). This model enables a more realistic prediction of the behavior of the system, especially when dealing with mixtures of gases or vapors.

All heat and mass transfer processes must comply with three physical laws, namely, conservation of energy, conservation of mass, and conservation of momentum. For the processes of wet air flow in the heat exchanger and condensation and dehumidification in the tube in this paper, a Cartesian coordinate system is established, and the different physical quantities in the fluid are described by the controlling equations with respect to time and space.

Mass conservation equations:

$$\frac{\partial \rho}{\partial t} + \frac{\partial(\rho u)}{\partial x} + \frac{\partial(\rho v)}{\partial y} + \frac{\partial(\rho w)}{\partial z} = 0 \tag{1}$$

where $\rho$ signifies the fluid density, $t$ means time; $u, v, w$ correspond to the directional elements of the velocity vector in the $x$, $y$, $z$ directions, respectively.

Conservation of momentum equation:

$$\frac{\partial(\rho u)}{\partial t} + div(\rho u u) = div(\mu grad u) + S_u - \frac{\partial P}{\partial x} \tag{2}$$

$$\frac{\partial(\rho v)}{\partial t} + div(\rho v u) = div(\mu grad v) + S_v - \frac{\partial P}{\partial y} \tag{3}$$

$$\frac{\partial(\rho w)}{\partial t} + div(\rho w u) = div(\mu grad w) + S_w - \frac{\partial P}{\partial z} \tag{4}$$

where $\mu$ means dynamic viscosity, $S_u, S_v, S_w$ means source terms for the momentum equation, and $P$ is pressure.

The energy conservation equation:

$$\frac{\partial(\rho T)}{\partial t} + div(\rho u T) = div(\frac{k}{c_p} grad T) + S_T \tag{5}$$

where $T$ implies the fluid temperature, $k$ denotes the fluid's thermal conductivity, $c_p$ means the fluid's specific heat capacity, and $S_T$ is a viscous dissipative term.

The continuity equation:

$$\frac{\partial \rho}{\partial t} + \frac{\partial(\rho u_i)}{\partial x_i} = S_m \tag{6}$$

where $\rho$ is density, $\mu$ is velocity, and $S_m$ is a quality source item.

In addition to the continuity equation, the equations for energy, mass, and momentum conservation, the component conservation equations must be added because the condensation and transfer of heat of humid air in the tube of air in the model are accompanied by a phase change. When the component conservation equations are used in calculations, the double precision method is required.

Component conservation equations:

$$\frac{\partial}{\partial t}(\rho \omega_i) + \nabla \cdot (\rho \overline{v} \omega_i) = -\nabla \overline{J}_i + R_i + S_i \tag{7}$$

where $S_i$ is a component source phase of $i$ substance; $R_i$ is the net production rate of substance $i$.

The wet air is approximated as an incompressible ideal gas, and the liquid droplets and liquid film condensed on the wall surface will flow away with the flow of wet air, so the condensation velocity formulas for the combined effect of convection and diffusion of wet air in the pipeline should be considered as follows.

$$m = \frac{\rho D}{1 - \omega_N} \frac{\partial \omega}{\partial n}\bigg|_W \tag{8}$$

where the diffusion coefficient is denoted by $D$.

$$D = \frac{2.24 \times 10^{-5}}{P_0}\left[\frac{T}{273.15}\right]^{1.8} \tag{9}$$

where $P_0$ is standard atmospheric pressure.

Quality source items:

$$S_m = m \tag{10}$$

Energy source items:

$$S_h = m \cdot r \tag{11}$$

where $r$ indicates the water vapor's latent heat in wet air.

There is the presentation of equations other than conservation of energy, conservation of mass, conservation of components, and conservation of momentum.

The formula for water vapor concentration in saturated humid air is as follows:

$$\omega_w = \frac{P_s(t_w)}{\rho R_v T_w} \tag{12}$$

where $P_s$ stands for the water vapor saturation pressure in wet air, $T_w$ is the wall temperature, and $R_v$ stands for the water vapor gas constant.

The equation for the saturation pressure of water vapor in wet air is as follows:

$$P_s = 610.78 \exp(17.2694 \frac{t}{t + 238.3}) \tag{13}$$

The formula for the Reynolds number $Re_{Dc}$ is as follows:

$$Re_{Dc} = \frac{D_c v_{\max} \rho}{\mu} \tag{14}$$

where $D_c$ is the pipe diameter of the elbow, $\mu$ is the wet air viscosity, and $v_{\max}$ stands for maximum flow rate.

The following is the formula for the total transfer of heat coefficient:

$$h_0 = \frac{q_0}{\overline{T}_{air} - \overline{T}_c} \tag{15}$$

where $q_0$ is the total heat flow, and the denominator represents the distinction between the mean wet air temperature and the mean wall temperature.

The formulas for the latent heat transfer coefficient and the sensible heat transfer coefficient are as follows:

$$h_s = \frac{q_s}{\overline{T}_{air} - \overline{T}_c} \tag{16}$$

$$h_f = \frac{q_t}{\overline{T}_{air} - \overline{T}_c} \tag{17}$$

where $q_t$ indicates the latent heat flux, $q_s$ represents the sensible heat flux, the total heat flux, and $q_0$ is equal to the sum of the latent heat flux and the sensible heat flux:

$$q_0 = q_s + q_t \tag{18}$$

The formula for the mass transfer coefficient is as follows:

$$h_m = \frac{m}{\overline{\omega}_{air} - \overline{\omega}_c} \tag{19}$$

where $m$ is the surface condensation rate, and the denominator is expressed as the distinction between the average water vapor mass fraction of the wet air inlet and outlet and the water vapor mass fraction of the saturated wet air for the average temperature.

The following are the mass transfer factor equation and the heat transfer factor equation:

$$j_h = \frac{h_s}{G_{max} c_p} Pr^{\frac{2}{3}} = St \cdot Pr^{\frac{2}{3}} \tag{20}$$

$$j_m = \frac{h_m}{G_{max}} Sc^{\frac{2}{3}} = St_m \cdot Sc^{\frac{2}{3}} \tag{21}$$

where $G_{max}$ represents the maximum mass flow rate, $Pr$ is the number of Prandtl, and $Sc$ is the number of Schmidt, and $G_{max}$ and $Sc$ are calculated as follows:

$$G_{max} = v_{max} \cdot \rho \tag{22}$$

$$Sc = \frac{\mu}{\rho D} \tag{23}$$

The convective mass flux is expressed in Equation (24):

$$m_{av} = \rho \omega v_n \tag{24}$$

Adding the source term to the condensation process, Equation (25) is used to calculate the condensation mass flow rate per unit volume:

$$S_m = m_{av} \frac{A_{face}}{V_{cell}} \tag{25}$$

where $A_{face}$ represents the area of neighboring cells on the condensing wall surface, $V_{cell}$ represents the neighboring cell volume, and $S_m$ represents the condensed mass flow rate of vapor.

The equation for the vapor source term is as follows:

$$S_v = \omega S_m \tag{26}$$

The source terms of the energy equation are formulated as follows:

$$S_e = \gamma S_m \tag{27}$$

where $\gamma$ represents the latent heat of vapor.

In the thermal convection problem, Richardson numbers indicate the importance of natural convection relative to forced convection:

$$R_i = \frac{Gr}{Re^2} \tag{28}$$

$$\mathrm{Re} = \frac{\rho VD}{\mu} = \frac{VD}{\nu} \tag{29}$$

$$G_r = \frac{g\alpha_v \Delta t L^3}{v^2} \tag{30}$$

where $V$ is the flow rate, $D$ is pipe length, and $v$ is the kinematic viscosity.

#### 2.1.3. Boundary Conditions

Performing a simulation solution is a computational fluid dynamics method based on the finite element method, which works by dividing the solution domain into small grids and applying the governing equations in order to compute the solution results. Throughout the solution process, it is crucial to set appropriate boundary conditions, especially for non-stationary problems, and to define the initial conditions explicitly. In this process, a variety of common boundary conditions are provided, each with their applicable scenarios. For different types of flow fields and problems, a reasonable selection and setting of boundary conditions will help the convergence of the simulation and provide a deep understanding of the actual physical processes. Collectively, the boundary conditions selected are shown in Table 1. The inlet boundary condition is set according to the experimental data, and the inlet is selected as velocity inlet; based on the assumption that the system reaches a steady state or equilibrium at the outlet, and assuming that the outlet pressure is fixed, the selection of the pressure outlet boundary condition is able to reduce the unsteady or reflux phenomenon in the numerical simulation by controlling the outlet pressure; the heat transfer occurs between the wall and the fluid, which directly affects the temperature distribution, turbulence characteristics, and flow pattern of the fluid. The wall boundary condition is selected as the thermal coupling boundary condition.

**Table 1.** Boundary condition type.

| District | Boundary Condition | Clarification |
|---|---|---|
| Inlet | Velocity-inlet | Ensures the flow rate and pressure conditions of the inlet humid air. |
| Outlet | Pressure-outlet | Guaranteeing the magnitude of the pressure and solving the problem of computational convergence. |
| Condensing wall surface | Thermally coupled wall | The heat exchange process on the condensing wall is accompanied by the generation of condensate. |

This section describes the conditions under which the bent-tube heat exchanger [32] operates and explores the condensation phenomena on both the cooling water end and the air end by means of numerical simulation. In addition, the role of inlet pressure and relative humidity on the condensation characteristics is investigated. Through the in-depth discussion of the principle, the targeting method, and the optimization strategy, under the set simulation conditions, the pressure variation range of the simulation is defined from

101.325 kPa to 1500 kPa [33,34], the moisture content range is from 60% to 100%, and the gravitational acceleration is constant at 9.81 $m/s^2$. In the study, the inner diameter of the bent-tube heat exchanger is set to 10 mm, while the inner diameter of the outer tube is set to 20 mm. The water temperature is 285 K, and the wet air temperature is 310 K. Inside the casing, the velocity of the water was specified to be 4 m/s, while the flow rate of the humid air was similarly set to 4 m/s. Figure 3 illustrates the schematic micro-model of the flow of the humid air through the bent-tube heat exchanger.

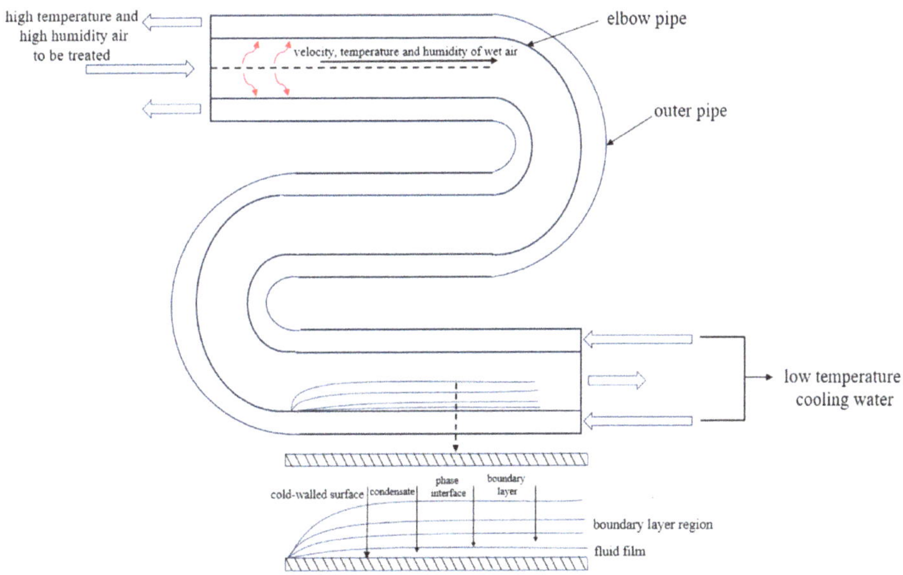


**Figure 3.** Schematic diagram of a microscopic model of bent-tube heat exchanger.

In this paper, user-defined functions are used, which are compiled into an external program connected to the system within the solver. The advantages and disadvantages of compiled and interpreted types are considered, and based on specific needs and performance considerations, the interpreted type is ultimately chosen. This choice is made to improve performance while maintaining as much compatibility as possible with different environments. In the source term of the condensation model, the entire calculation process is clearly shown in Figure 4.

In this paper, the validation of mesh independence was achieved by modeling the heat flux and condensation with different first-layer mesh thicknesses at the wall. Based on the results of calculating the total heat flux and total condensation, the appropriate first-layer grid thickness has been selected. Figure 5 illustrates the mesh model, and Figure 6 shows the verification of the irrelevance for different mesh thicknesses. When mesh independence is verified, models with first-layer mesh sizes from 0.0001 mm to 0.05 mm have been calculated. It can be seen from the graph that the heat flux and condensation amount increase as the grid thickness decreases, while the grid thickness levels off when it reaches 0.0001 mm and does not change when it reaches 0.0001 mm. So, 0.0001 mm is set as the first mesh thickness of the model, and the mesh under this thickness saves computational resources and is accurate. Under the condition that the quality of the mesh remains unchanged, as the mesh size decreases, the calculation results will be about accurate; however, it will also increase the calculation time accordingly. Therefore, it is necessary to choose the appropriate mesh size for the actual calculation.

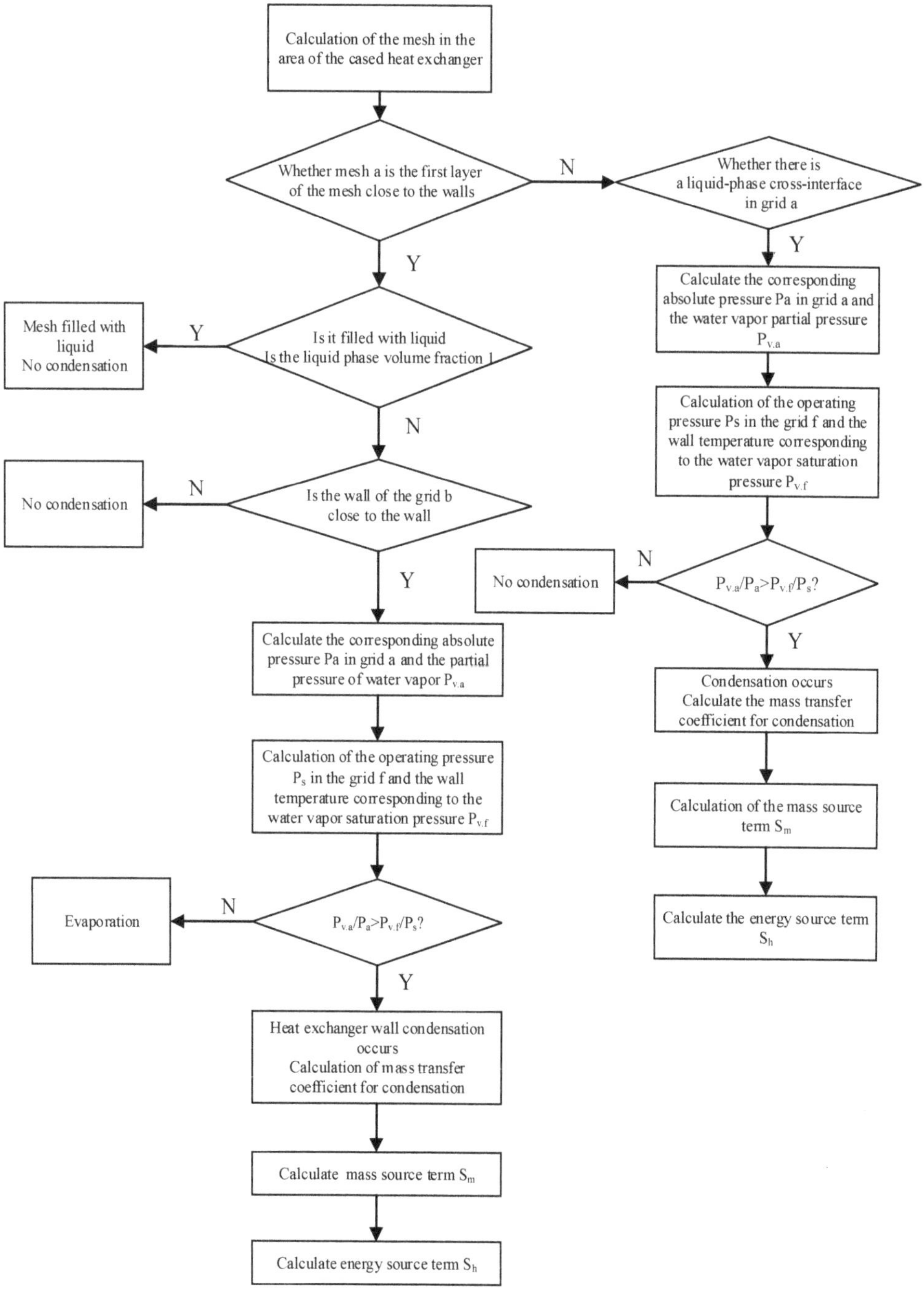


**Figure 4.** Calculation process.

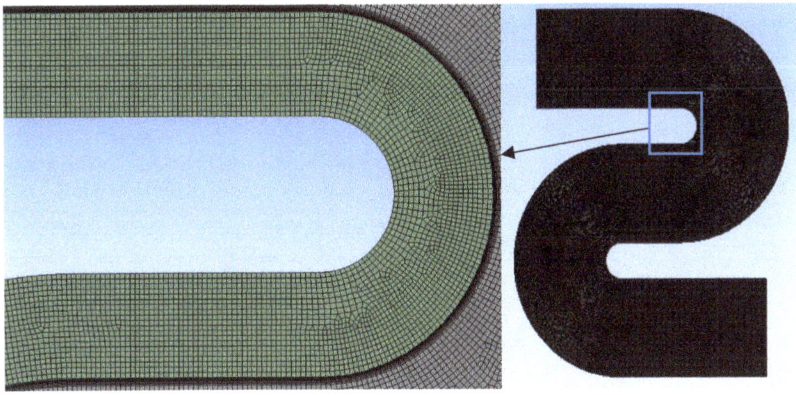

Figure 5. Grid model.

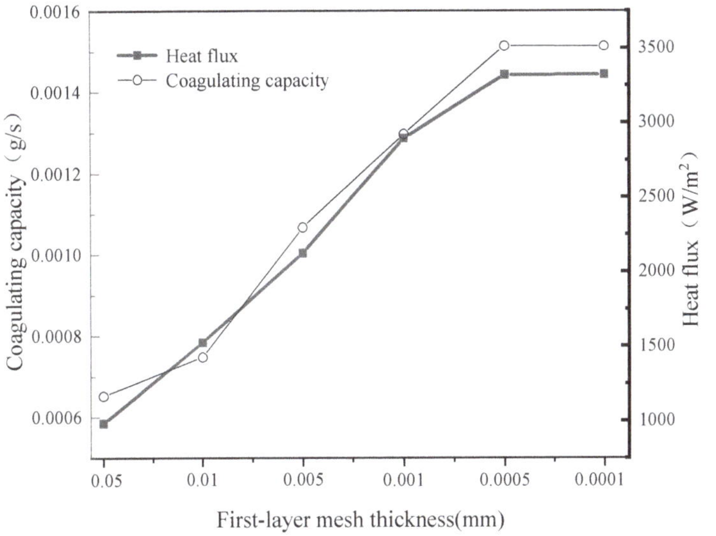


Figure 6. Grid independence verification.

## 2.2. Experimental Study

### 2.2.1. Experimental Method

Wet air pretreatment and positive-pressure condensation are the two main modules that make up the demonstration system. As shown in Figure 7, the layout of the test bench is schematic. The system consists of a condenser, humidification box, submersible pump, air chamber, and air-routing hot air cannon. In order to humidify the air, a submersible pump is positioned at the bottom of the humidification box and creates a fine mist of water. The condenser saturates the air with water, while the humidifier maintains a constant temperature and provides a uniform air flow. After being pressured by the compressor, the humidified air from the pre-treatment apparatus passes through the heat exchanger for condensation and dehumidification. Lastly, to effectively separate the two, a gas–liquid separator is employed.

**Figure 7.** Experiment diagram.

Specifications of key components are listed in Table 2. Table 3 details the measurement tools used during the experiments. Specialized sensors were used to detect temperature, relative humidity, and pressure directly. Float flowmeters were used specifically to measure the flow rate of liquids, in contrast to differential pressure sensors, which were suitable for detecting the flow rate of gases. After the data have stabilized, a plan is put in place to take five readings overall, recording data at all measurement locations every 10 min under all operational situations. The final measurement is obtained by averaging the values.

**Table 2.** Components specifications.

| Name | Brief Statement |
|---|---|
| Compressor | The centrifugal variable speed compressor is equipped with an inlet and an outlet, and the flow rate of the compressor can be smoothly adjusted. |
| Condenser | Plate heat exchanger with the cold side connected to water, which acts as a cooling source. |
| Cooler | A 2-m-long cased heat exchanger with an outer tube diameter of 20 mm and an inner tube diameter of 12 mm has a heat transfer area of 0.08 $m^2$. |
| Dehumidifier | Air–water separators that store moisture from wet air. |

**Table 3.** Measurement tools and parameters.

| Parameter | Meter | Range | Accuracy |
|---|---|---|---|
| Temperature | Pt100 resistance thermometer | −200~420 °C | ±0.1 °C |
| Relative humidity | Electronic humidity sensor | 0~100% | ±3% RH |
| Pressure | Piezometer | 0~250 kPa | 0.2% FS |
| Water flow rate | Float flowmeter | 40~440 L/h | 0.2% FS |
| Air flow rate | Pressure transmitter | 0~50 Pa | ±0.25% FS |

2.2.2. Experimental Conditions

The test circumstances, which include six operational scenarios in total, are shown in Table 4. The range of the pressure is 146.09 kPa to 293.6 kPa. About 37 °C is the intake air temperature and about 26 °C is the water at the entry temperature of the heat exchanger. In the experimental setup, the heat exchanger has an inner diameter of 0.01 m, a tube wall thickness of 0.001 m, a tube length of 2 m, a mass flow rate of 0.026 kg/s for air, and a volumetric flow rate of 400 L/h for water.

**Table 4.** Test conditions.

| Test | Pressure (kPa) | Heat Exchanger Outlet Temperature (°C) | Humidity (g/kg) |
|---|---|---|---|
| 1 | 146.09 | 27.34 | 15.89 |
| 2 | 184 | 27.56 | 12.71 |
| 3 | 205.63 | 26.85 | 10.88 |
| 4 | 229.95 | 27.47 | 10.08 |
| 5 | 248 | 26.7 | 8.92 |
| 6 | 293.6 | 27.33 | 7.8 |

## 3. Results and Discussion

The gravitational acceleration is 9.81 $m/s^2$, and a bent-tube heat exchanger diameter of 10 mm is considered, as well as a water temperature of 285 K and an inlet temperature of 310 K for accurate calculations and analyses. In a detailed analysis of the outcomes of numerical simulations of air-side condensation in a condensing heat exchanger under wet conditions, we have looked closely at how inlet wet air pressure and moisture content affect the properties of air-side condensation, demonstrating how these factors greatly influence the condensation process' efficiency and features. In considering the pressure variation interval of the model, the range is set from 101.325 to 1500 kPa, while the relative humidity range is selected from 60% to 100%, and the flow speed of the humid air is determined to be 3–5 m/s.

### *3.1. Validation with Experiment*

In order to more accurately simulate the procedure of wet air condensation and dehumidification in the heat exchanger, the two-dimensional model established in this paper was further verified through a comparative analysis between experiment and simulation. Comparison with the experimental data was carried out to verify the accuracy of the model by comparing and analyzing the wet air outlet temperature and outlet humidity content obtained from the experimental and simulation results.

Figures 8 and 9 compare the simulation and experimental results [35] for outlet humidity and temperature under six different operating conditions, respectively. The errors between the simulated and experimental values of outlet moisture content and temperature are minimal, both within 17%. The export temperature and the experimental values have a smaller error, and the relative error is within 8%, indicating that the export temperature is well predicted in the model. The relative error of humidity is within 17%. The deviation in individual experimental values from the simulated calculated values is large, which may be due to the limitation of experimental conditions and the small amount of data on experimental values. The model established in this paper meets the requirements for use and verifies the reasonableness of the model.

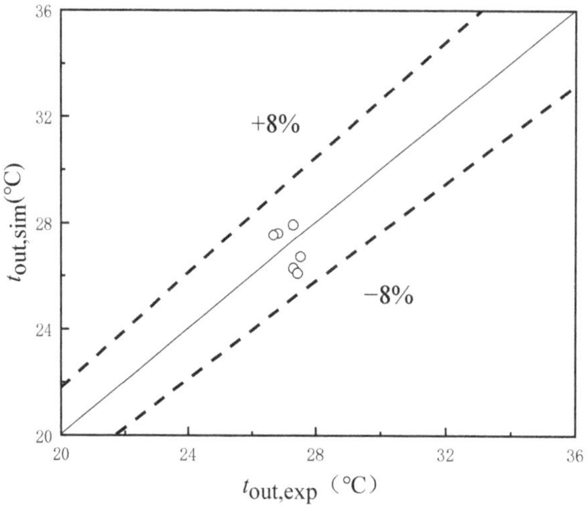


**Figure 8.** Comparison of the simulated and experimental results of the outlet temperature.

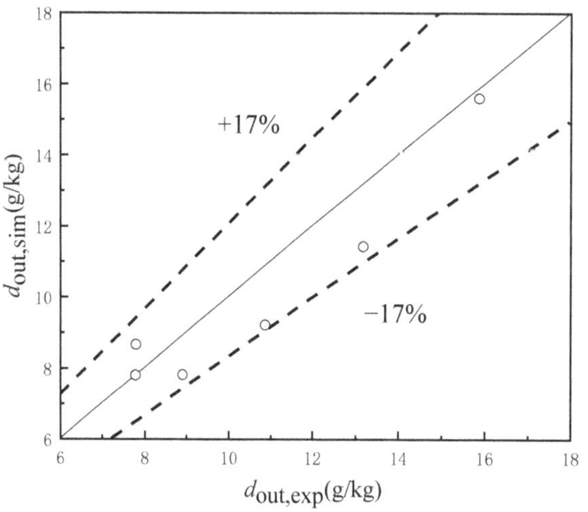


**Figure 9.** Comparison of the simulated and experimental results of the outlet moisture content.

### 3.2. Effect of Pipe Diameter

In this section, when dehumidifying condensed wet air with positive pressure, the impact of a bent-tube heat exchanger's pipe diameter variation on the heat and mass transfer process is examined. This study analyzes the variation of humidity near the exit of the bent-tube heat exchanger and its effect on the heat transfer coefficient for pipe diameters of 6–10 mm, respectively. In a particular environment where the initial air temperature is set at 37 °C, the wet air velocity is 3 m/s, the relative humidity is 27.2 g/kg, and the water's temperature is 12 °C. Figures 10 and 11 show how pressure variation affects the dynamic relationship between the heat transfer coefficient and the moisture content of the wet air during the condensation process. In the presence of positive pressure, the condensation heat and mass transfer rates of the wet air show a significant increase in the range of 2.5 to 3.4 times compared to the atmospheric environment. The air at the exit of the bent-tube heat exchanger ranges from 12.9 g/kg to 16.2 g/kg, which is significantly lower than half the air's humidity at the same atmospheric pressure. Figure 8 shows the pressure dependence of the heat transfer coefficient of condensed moist air at positive pressure and the correlation between the tube diameter and the moisture content at the exit of the moist air: As the diameter of the tube decreases, the humidity at the exit of the moist air decreases, a phenomenon which is explained by the fact that the enlargement of the diameter of the tube increases the area of the heat transfer, which in turn improves the efficiency of the heat transfer. At high pressures (up to 800 kPa and above), the humidity of the wet air outlet does not change much; in contrast, the heat transfer coefficient shows a tendency to increase with increasing pressure.

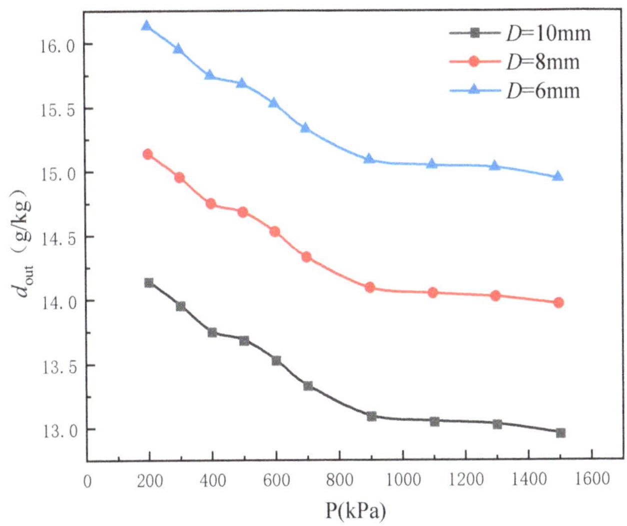


**Figure 10.** Effect of pressure and pipe diameter on the humidity content of wet air.

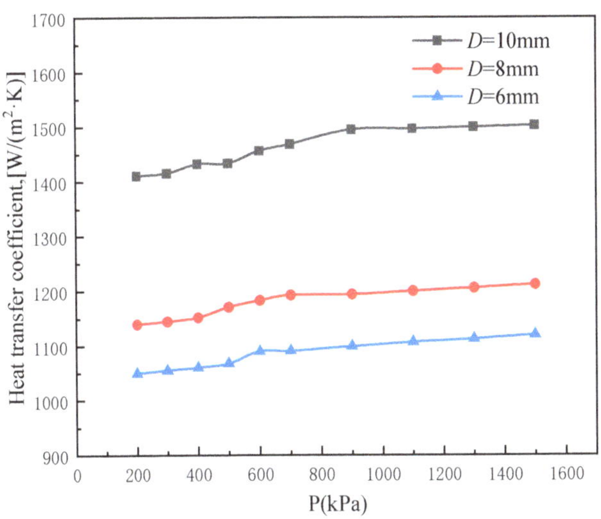


**Figure 11.** Effect of pressure and pipe diameter on heat transfer coefficient.

### 3.3. *Effect of Humid Air Flow Velocity*

Figure 12 reveals the mass fraction distribution of wet air at flow rates of 3–5 m/s, respectively. Through the flow of wet air through the bent-tube heat exchanger, the increase in inlet velocity leads to an increase in the mass fraction of water vapor in the wet air at the outlet due to the condensation of water vapor in the front half of the bent-tube heat exchanger and the formation of a region of low mass fraction at the tail end of the bent-tube heat exchanger accompanied by a high mass fraction of air, which attenuates the condensation effect.

**Figure 12.** Mass fraction of change in wet air flow rate.

Figures 13 and 14 reveal the effect of variation in wet air velocity on the moisture content and heat transfer coefficient at the heat exchanger outlet during positive-pressure condensation dehumidification. When the pressure ranges from 100 to 800 kPa, it is observed that the moisture content at the exit of the wet air shows a significant change, while the heat exchanger heat transfer efficiency is optimized. This phenomenon is attributed to the fact that the humid air has enough time to fully exchange heat with the heat exchanger. Positive-pressure condensation dehumidification was most effective at a flow rate of 3 m/s. The moisture content near the heat exchanger's outflow increased with increasing velocity. However, as the velocity increases, the moisture content of the wet air outlet increases. Nonetheless, the overall dehumidification effect is more desirable in larger dehumidification environments due to the increased rate of mass flow rate change. The increased pressure improves the conduction, convection, and diffusion characteristics of the moist air, allowing it to exchange heat more efficiently with the dehumidification equipment, thus speeding up the dehumidification process.

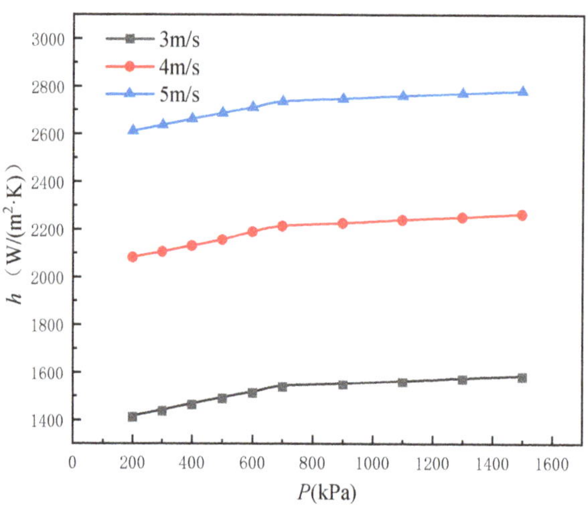


**Figure 13.** Effect of pressure and flow rate on humidity content of wet air.

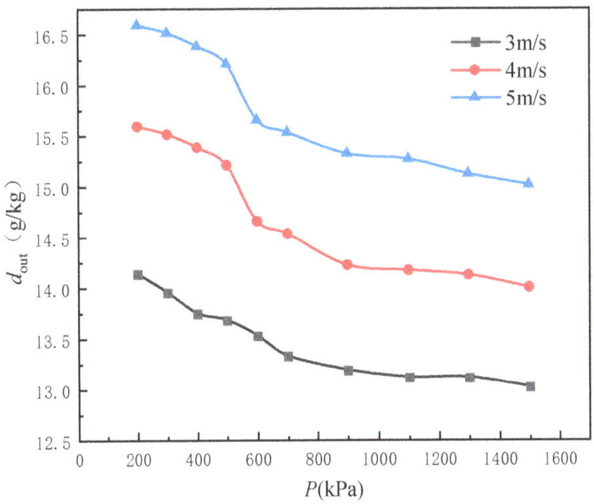


**Figure 14.** Effect of pressure and flow rate on heat transfer coefficient.

Figure 15 shows the change in cold source water temperature during the wet air positive-pressure condensation dehumidification process. The change in cold source water temperature due to variations in wet air flow rate is very small, because the cold source water has a larger specific heat capacity than the wet air. As a result, the effects of flow rate and pressure changes on the cold source water temperature are minimal.

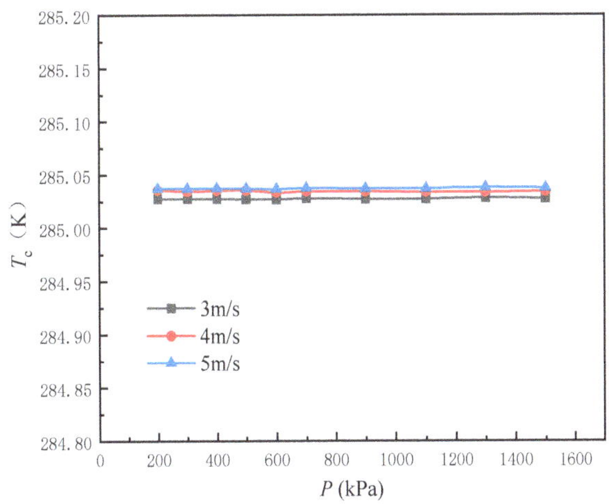


**Figure 15.** Effect of pressure and flow rate on cold source outlet temperature.

### 3.4. Effect of Humid Air Humidity

Figure 16 presents the variation of vapor mass fraction in the bent-tube heat exchanger under different relative humidity conditions. A uniform decreasing trend in the humidity of the wet air was observed when the wet air flowed through the bent-tube heat exchanger with the pressure set at 300 kPa, the temperature maintained at 310 K, the inlet velocity up to 4 m/s, and the relative humidity at 60%, 80%, and 100%, respectively.

**Figure 16.** Humid air quality scores at 60%, 80%, and 100% relative humidity.

In a constant relative humidity environment, the humidity content at the outlet shows a decreasing trend as the pressure increases; at the same time, the increase in relative humidity at the inlet significantly improves the dehumidification effect. The heat flux and heat transfer rate exhibit an increasing trend as the humidity level of the air rises. This is because an increase in relative humidity also affects the air's dew point temperature, which causes the water molecules in the air to accumulate and become more saturated. This leads to an increase in the temperature differential between the air's dew point temperature and the surface layer of the droplets in the same pressure environment. This evolution clearly shows how variations in ambient humidity can impact wet air condensation. Figures 17 and 18 reveal the effect of pressure and relative humidity on the heat transfer coefficient and moisture content at the outlet of the bent-tube heat exchanger, which shows the most significant dehumidification effect and the most significant change in the heat transfer coefficient in the pressure variety of 200–800 kPa. The heat transfer coefficient shows a gradual increase until the pressure reaches 800 kPa, and the heat transfer coefficient tends to stabilize. The heat transfer coefficient changes significantly when the relative humidity is increased from 60% to 80% while the pressure remains constant.

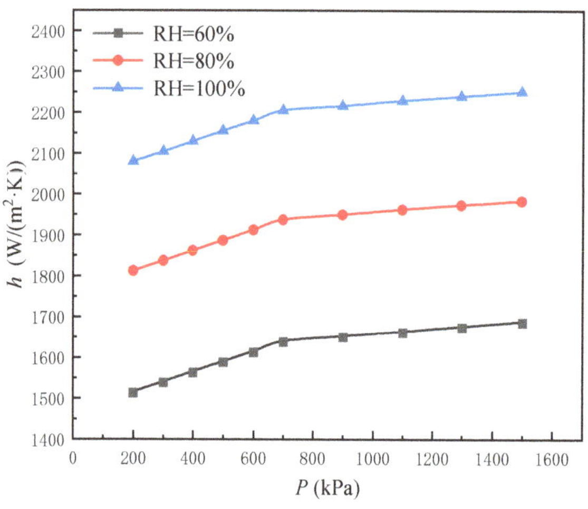


**Figure 17.** Effect of pressure and relative humidity on heat transfer coefficients.

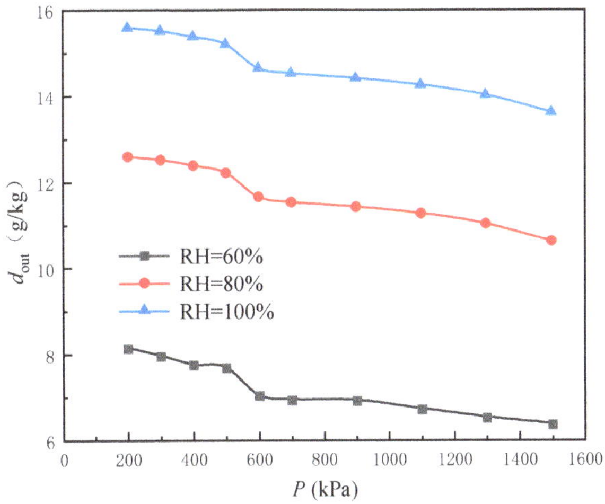


**Figure 18.** Effect of pressure and relative humidity on moisture content at the outlet.

### 3.5. Temperature Distribution in Humid Air

Temperature distribution is a key factor in the heat exchange process. Analyzing the temperature distribution ensures uniformity, preventing local overheating and maintaining heat transfer efficiency. Under the conditions of a pressure setting of 300 kPa and inlet velocity of 4 m/s, the differences in temperature distribution at different humidity levels were observed and presented by varying the inlet relative humidity to 60%, 80%, or 100%, as shown in Figure 19. The bent-tube heat exchanger's moist air temperature dispersion is more consistent, and changes in moisture content have less of an impact on it. By observing the distribution condition of the cloud diagram, the flow rate of the wet air is low near the wall, and turbulence is generated when it flows through the bends of the heat exchanger, slows down, and tends to be stable near the exit of the heat exchanger.

**Figure 19.** 60%, 80%, and 100% wet air temperature distribution.

Figure 20 reveals the temperature distribution of the wet air inside the duct of the bent-tube heat exchanger as the inlet velocity reaches 3 m/s, 4 m/s, and 5 m/s, respectively, at a pressure of 300 kPa and a relative humidity of 100%. The heat exchanger's moist air temperature distribution is essentially the same, and the temperature field in the center of the tail increases slightly with the increase in velocity, while in other regions, the temperature changes observed from the cloud diagrams do not vary much with the wet air flow velocity.

**Figure 20.** Temperature distribution for wet air velocity of 3 m/s, 4 m/s, and 5 m/s.

### 3.6. Threaded Bend-Tube Heat Exchanger

The threaded bent-tube heat exchanger is modeled in Figure 21, and it can be observed that due to the condensation of steam, the mass fraction of the steam varies greatly in the first half of the tubes, and there is a tail flow region at the back of each tube. The tail flow region has a low mass fraction of steam and a high mass fraction of air, and hence, the condensation is weak. The vapor mass fraction of the bent-tube heat exchanger at different relative humidities when the pressure is 300 kPa, inlet velocity is 4 m/s, and relative humidity is 60%, 80%, and 100% is shown below. In the process of wet air flowing through the bent-tube heat exchanger, as the pressure increases, the humidity content of the wet air outlet decreases, and the dehumidification effect is better and is more suitable for deep dehumidification; with the increase in humidity, the humidity content of the wet air outlet decreases significantly. Compared to an ordinary bent-pipe heat exchanger of the same size, the threaded bent-pipe heat exchanger has a larger heat transfer area, and the threaded bent-pipe heat exchanger's dehumidification effect is better, offering significant advantages. Although the threaded elbow heat exchanger has a better heat transfer performance than the ordinary elbow heat exchanger, in the case of positive-pressure condensation, the ordinary elbow heat exchanger is already sufficient to achieve the dehumidification performance of daily life.

**Figure 21.** Threaded bend-tube heat exchanger and ordinary bend-tube heat exchanger.

### 3.7. Discussion

Based on the simulation analysis, raising the pressure level significantly enhances the condensation and dehumidification efficiency of wet air. As the pressure increases, the temperature differential between the air and the cold source decreases, but the dehumidification effect remains constant; the dehumidification pressure, heat exchanger tube size, humid air flow velocity, and relative humidity all have a significant impact on the heat and mass transfer coefficients during humid air condensation and dehumidification. Wet air in the flow through the bend-tube heat exchanger close to the wall will have turbulence; the centrifugal force generated in the bend compared to the straight tube significantly improves the performance of heat and mass transfer, and in the bend, the turbulence in the flow leads to significant changes in the turbulence, enhancing the rate of heat and mass transfer. Based on the discussion in this paper, in a positive-pressure setting, the heat and mass transfer coefficients for condensing and dehumidifying wet air rise with increasing pressure, thereby lowering the energy required for these processes.

## 4. Conclusions

The performance of positive-pressure condensation heat transfer of humid air in a bent-tube heat exchanger, particularly under varying pressure and relative humidity circumstances, was examined using a bent-tube heat exchanger model, and its validity was confirmed experimentally. The analyses show how the bent-tube heat exchanger's surface heat transfer coefficient and dehumidification capacity changed when pressure, relative humidity, flow rate, and tube diameter changed:

(1) It was found that raising the pressure level was effective in reducing the moisture contained in the dehumidified air. The deviation between the simulated and experimental results was kept within ±17%, which confirms the validity of the framework for the positive-pressure condensation dehumidification process.

(2) The bent-tube heat exchanger's influence on the heat and mass transfer performance during the condensation and dehumidification of humid air against the backdrop of pressure increase is greatly improved by increasing the tube diameter. The relative changes in heat and mass fluxes exhibited a minimum as the pressure climbed to 900 to 1500 kPa, although the absolute changes in heat and mass fluxes were found to peak between 200 and 900 kPa within the heat exchanger's operational range.

(3) The heat transfer coefficient rose by 44% when the moist air velocity increased from 3 m/s to 5 m/s, which is a 33% increase in velocity. When the flow rate was reduced to 3 m/s, the humidity near the exit of the bent-tube heat exchanger was reduced by 15% compared to the case where the flow rate was 5 m/s.

(4) For every 20% increase in humidity, the heat transfer coefficient in humid air rises by around 18% to 37%. Changes in relative humidity at the inlet of a bent-tube heat exchanger have a negligible effect on the internal temperature distribution.

(5) When the temperature and humidity in the bent-tube heat exchanger are certain, positive-pressure condensation leads to faster condensation and higher heat transfer efficiency; 800 kPa dehumidification is about 1.1 times that of 200 kPa, and the heat transfer coefficient is about 1.1 times that of 200 kPa.

**Author Contributions:** Conceptualization, Q.C.; methodology, Q.C.; software, J.X.; validation, J.X.; formal analysis, J.X.; investigation, J.X.; writing—original draft preparation, J.X.; writing—review and editing, Q.C.; supervision, Q.C.; project administration, Q.C. All authors have read and agreed to the published version of the manuscript.

**Funding:** This research received no external funding.

**Data Availability Statement:** The original contributions presented in the study are included in the article; further inquiries can be directed to the corresponding author.

**Acknowledgments:** Thanks to all those who contributed to the articles.

**Conflicts of Interest:** The authors declare that they have no known competing financial interests or personal relationships that could have appeared to influence the work reported in this paper.

## References


1. Yu, F.; Yuan, Q.; Sheng, X.; Liu, M.; Chen, L.; Yuan, X.; Zhang, D.; Dai, S.; Hou, Z.; Wang, Q.; et al. Understanding Carbon Footprint: An Evaluation Criterion for Achieving Sustainable Development. *Chin. J. Popul. Resour. Environ.* **2024**, *22*, 367–375. [CrossRef]
2. Tian, S.; Su, X.; Geng, Y. Review on Heat Pump Coupled Desiccant Wheel Dehumidification and Air Conditioning Systems in Buildings. *J. Build. Eng.* **2022**, *54*, 104655. [CrossRef]
3. Shaikh, J.S.; Ismail, S. A Review on Recent Technological Advancements in Humidification Dehumidification (HDH) Desalination. *J. Environ. Chem. Eng.* **2022**, *10*, 108890. [CrossRef]
4. Zhang, Q.; Li, Y.; Zhang, Q.; Ma, F.; Lü, X. Application of Deep Dehumidification Technology in Low-Humidity Industry: A Review. *Renew. Sustain. Energy Rev.* **2024**, *193*, 114278. [CrossRef]
5. Fix, A.J.; Gupta, S.; Braun, J.E.; Warsinger, D.M. Demonstrating Non-Isothermal Vacuum Membrane Air Dehumidification for Efficient next-Generation Air Conditioning. *Energy Convers. Manag.* **2023**, *276*, 116491. [CrossRef]
6. Chao, Y.; Wang, C.; Zhou, Z.; Yang, X.; Du, Y.; Liang, Y.; Deng, N.; Liu, J. Recent Advance of Adsorbent Materials for Indoor Dehumidification. *Sep. Purif. Technol.* **2025**, *357*, 130096. [CrossRef]
7. Labban, O.; Chen, T.; Ghoniem, A.F.; Lienhard, J.H.; Norford, L.K. Next-Generation HVAC: Prospects for and Limitations of Desiccant and Membrane-Based Dehumidification and Cooling. *Appl. Energy* **2017**, *200*, 330–346. [CrossRef]
8. Abdullah, S.; Zubir, M.N.B.M.; Muhamad, M.R.B.; Newaz, K.M.S.; Öztop, H.F.; Alam, M.S.; Shaikh, K. Technological Development of Evaporative Cooling Systems and Its Integration with Air Dehumidification Processes: A Review. *Energy Build.* **2023**, *283*, 112805. [CrossRef]
9. Tian, S.; Geng, Y.; Huang, L.; Li, S.; Wang, Q.; Su, X. Fundamental and Perspectives of Thermo-Responsive Materials for Dehumidification and Water Harvesting. *J. Clean. Prod.* **2023**, *412*, 137446. [CrossRef]
10. Chun, L.; Liao, Z.; Wang, G.; Xiao, Y.; Huo, J.; Liu, D.; Jiang, B. Operational Characteristics and Controlling Strategies of a Novel Dual-Return-Air Dehumidification Evaporative Cooling System (DDEC). *Energy* **2024**, *309*, 133173. [CrossRef]
11. Huang, Z.; Xiao, Y.; You, H.; Chen, D.; Hu, B.; Li, G.; Han, J.; Lysyakov, A. Performance Analysis and Multi-Objective Optimization of a Novel Solid Oxide Fuel Cell-Based Poly-Generation and Condensation Dehumidification System. *Energy Convers. Manag.* **2024**, *319*, 118935. [CrossRef]
12. Zhang, Y.; Wang, W.; Zheng, X.; Cai, J. Recent Progress on Composite Desiccants for Adsorption-Based Dehumidification. *Energy* **2024**, *302*, 131824. [CrossRef]
13. Akhtar, M.U.S.; Fadlallah, S.O.; Khan, M.I.; Asfand, F.; Al-Ghamdi, S.G.; Mishra, R. Sustainable Humidity Control in the Built Environment: Recent Research and Technological Advancements in Thermal Driven Dehumidification Systems. *Energy Build.* **2024**, *304*, 113846. [CrossRef]
14. Vivekh, P.; Islam, M.R.; Chua, K.J. Experimental Performance Evaluation of a Composite Superabsorbent Polymer Coated Heat Exchanger Based Air Dehumidification System. *Appl. Energy* **2020**, *260*, 114256. [CrossRef]
15. Indrawan, W.; Lubis, A.; Nasruddin; Alhamid, M.I. The Experimental Study of Dehumidification and Regeneration Processes in a Fin and Tube Liquid Desiccant System. *Case Stud. Therm. Eng.* **2022**, *39*, 102440. [CrossRef]
16. Liu, M.; Prabakaran, V.; Bui, T.; Cheng, G.G.; Pang, W. Three-Dimensional Numerical Analysis of Fin-Tube Desiccant-Coated Heat Exchanger for Air Dehumidification in Tropics. *Appl. Energy* **2023**, *331*, 120346. [CrossRef]
17. Yang, K.-S.; Lin, K.-H.; Tu, C.-W.; He, Y.-Z.; Wang, C.-C. Experimental Investigation of Moist Air Condensation on Hydrophilic, Hydrophobic, Superhydrophilic, and Hybrid Hydrophobic-Hydrophilic Surfaces. *Int. J. Heat Mass Transf.* **2017**, *115*, 1032–1041. [CrossRef]
18. Han, M.H.; Yang, J.S.; Lee, J.S.; Min, J.K. Numerical Study on Condensation Heat Transfer in Vertical Channel with Hydrophilic-Hydrophobic Patterned Surfaces. *Int. Commun. Heat Mass Transf.* **2023**, *144*, 106807. [CrossRef]
19. Wang, Z.-J.; Wang, S.-Y.; Wang, D.-Q.; Yang, Y.-R.; Wang, X.-D.; Lee, D.-J. Water Vapor Condensation on Substrates with Nanoscale Hydrophilic Spots: A Molecular Dynamics Study. *Int. J. Heat Mass Transf.* **2023**, *205*, 123929. [CrossRef]
20. Mohamed, M.A.; Ahmed, S.A.; Emeara, M.S.; Mesalhy, O.; Saleh, M.A. Experimental Study for Enhancing Condensation on Large-Scale Surface Using Hybrid Hydrophilic-Hydrophobic Patterns. *Case Stud. Therm. Eng.* **2023**, *45*, 102984. [CrossRef]
21. Wan, Z.; Hu, X.; Wang, X.; He, Z. Experimental Study on the Boiling/Condensation Heat Transfer Performance of a Finned Tube with a Hydrophilic/Hydrophobic Surface. *Appl. Therm. Eng.* **2023**, *229*, 120494. [CrossRef]

22. Ahmadi, B.; Arasteh, H.; Yaghoubi, M.; Rahbari, A. Experimental Visualization and Analysis of Transient Heat and Mass Transfer during Natural Dehumidification of Air around Square Finned Tubes under Various Operating Conditions. *Energy Convers. Manag.* **2022**, *255*, 115295. [CrossRef]
23. Li, M.; Zhou, W.; Wei, J.; Tao, W. 3D Numerical Simulation of Heat and Mass Transfer of Fin-and-Tube Heat Exchanger under Dehumidifying Conditions. *Int. J. Heat Mass Transf.* **2018**, *127*, 597–610. [CrossRef]
24. Wang, C.-C.; Liaw, J.-S. Air-Side Performance of Herringbone Wavy Fin-and-Tube Heat Exchangers under Dehumidifying Condition—Data with Larger Diameter Tube. *Int. J. Heat Mass Transf.* **2012**, *55*, 3054–3060. [CrossRef]
25. Zhuang, D.; Ding, G.; Hu, H.; Fujino, H.; Inoue, S. Condensing Droplet Behaviors on Fin Surface under Dehumidifying Condition. *Appl. Therm. Eng.* **2016**, *105*, 336–344. [CrossRef]
26. Zhuang, D.; Ding, G.; Hu, H.; Fujino, H.; Inoue, S. Condensing Droplet Behaviors on Fin Surface under Dehumidifying Condition. Part II: Experimental Validation. *Appl. Therm. Eng.* **2016**, *105*, 345–352. [CrossRef]
27. Koca, A.; Atayılmaz, Ş.Ö. Experimental Investigation of Heat Transfer and Dehumidifying Performance of Novel Condensing Panel. *Energy Build.* **2016**, *129*, 120–137. [CrossRef]
28. Zong, C.; Li, H.; Song, W.; Zhao, S.; Lu, J.; Ma, Q. Study on the Double-Coil Heat Exchanger for Dehumidification and Heating in Greenhouses: Modeling, Analysis, and Optimization. *Comput. Electron. Agric.* **2024**, *227*, 109569. [CrossRef]
29. Lei, X.; Lin, Z.; Wang, Y.; Li, X.; Liu, Y. Numerical Investigation of Condensation Characteristics of a Moist Air-to-Air Plate-Fin Heat Exchanger and Experimental Validation. *Int. J. Heat Mass Transf.* **2022**, *191*, 122846. [CrossRef]
30. Feurhuber, M.; Neuschwander, R.; Taupitz, T.; Schwarz, V.; Frank, C.; Hochenauer, C. A Computational Fluid Dynamics (CFD) Model to Simulate the Inactivation of Geobacillus Stearothermophilus Spores in Different Moist Heat Sterilization Environments. *Phys. Med.* **2021**, *12*, 100039. [CrossRef]
31. Talukdar, P.; Iskra, C.R.; Simonson, C.J. Combined Heat and Mass Transfer for Laminar Flow of Moist Air in a 3D Rectangular Duct: CFD Simulation and Validation with Experimental Data. *Int. J. Heat Mass Transf.* **2008**, *51*, 3091–3102. [CrossRef]
32. Dalkılıç, A.S.; Mercan, H.; Özçelik, G.; Wongwises, S. Optimization of the Finned Double-Pipe Heat Exchanger Using Nanofluids as Working Fluids. *J. Therm. Anal. Calorim.* **2021**, *143*, 859–878. [CrossRef]
33. Yang, B.; Yuan, W.; Kong, X.; Zheng, T.; Li, F. Mass Transfer Study on High-Pressure Membrane Dehumidification Applied to Aircraft Environmental Control System. *Int. J. Heat Mass Transf.* **2023**, *202*, 123680. [CrossRef]
34. Yin, S.; Zhu, M.; Liu, Q.; Huang, X.; Wang, H. Release Behaviour of a High-Pressure Vapor Vessel with Condensation: Test and Modeling Study. *Appl. Therm. Eng.* **2022**, *200*, 117647. [CrossRef]
35. Zhang, S.; Wei, X.; Cheng, Q. Theoretical and Experimental Research on Heat and Mass Transfer Process of Positive Pressure Condensation Dehumidification for Humid Air. *Int. J. Refrig.* **2024**, *159*, 74–87. [CrossRef]


**Disclaimer/Publisher's Note:** The statements, opinions and data contained in all publications are solely those of the individual author(s) and contributor(s) and not of MDPI and/or the editor(s). MDPI and/or the editor(s) disclaim responsibility for any injury to people or property resulting from any ideas, methods, instructions or products referred to in the content.

MDPI AG
Grosspeteranlage 5
4052 Basel
Switzerland
Tel.: +41 61 683 77 34

*Buildings* Editorial Office
E-mail: buildings@mdpi.com
www.mdpi.com/journal/buildings

Disclaimer/Publisher's Note: The title and front matter of this reprint are at the discretion of the Guest Editors. The publisher is not responsible for their content or any associated concerns. The statements, opinions and data contained in all individual articles are solely those of the individual Editors and contributors and not of MDPI. MDPI disclaims responsibility for any injury to people or property resulting from any ideas, methods, instructions or products referred to in the content.

www.ingramcontent.com/pod-product-compliance
Lightning Source LLC
Chambersburg PA
CBHW040317270926
51850CB00007BC/1508
*9783725870202*